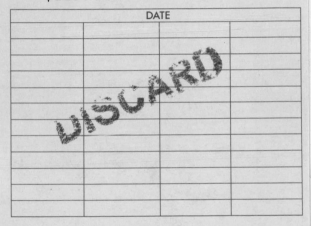

Biohazards
Management Handbook

OCCUPATIONAL SAFETY AND HEALTH

A Series of Reference Books and Textbooks
on Occupational Hazards ● Safety● Health ●
Fire Protection ● Security ● and Industrial Hygiene

Series Editor
ALAN L. KLING
Loss Prevention Consultant
Jamesburg, New Jersey

Other Volumes in Preparation

Biohazards Management Handbook

edited by

Daniel F. Liberman
Environmental Medical Service
Massachusetts Institute of Technology
Cambridge, Massachusetts

Judith G. Gordon
Gordon Resources Consultants, Inc.
Reston, Virginia

MARCEL DEKKER, INC.   New York and Basel

Library of Congress Cataloging-in-Publication Data

Biohazards management handbook / edited by Daniel F. Liberman, Judith
  G. Gordon.
        p. cm.--   (Occupational safety and health ; 17)
   ISBN 0-8247-7897-9 (alk. paper)
   1. Environmental health--Handbooks, manuals, etc. 2. Industrial
hygiene--Handbooks, manuals, etc. I. Liberman Daniel F.
   II. Gordon, Judith G.   III. Series: Occupational safety and
health (Marcel Dekker, Inc.) 17.
RA566.22B56
604.7--dc19   89-1048
                                                   CIP

MARCEL DEKKER, INC.
270 Madison Avenue, New York, New York 10016

Current printing (last digit):
10   9   8   7   6   5   4   3   2   1

PRINTED IN THE UNITED STATES OF AMERICA

Preface

Dramatic developments in biology over the past several years have
sparked unprecedented interest in basic applied biology. Following
the first successful directed insertion of foreign nucleic acid into microorganisms, researchers around the world recognized the potential of
these novel genetic techniques. Concern about the hazards—real,
potential, or imaginary—was such that federal agencies, universities,
hospitals, and industry responded by developing safety programs designed to manage "biohazards." Biological safety has matured from its
beginning as a topic of concern and now ranks alongside industrial
hygiene and health physics as a safety discipline. This is only natural
since a particle is a particle is a particle. Whether it is chemical, biological or radioactive characteristics that are being investigated, particles still behave the same and the control strategies applied are similar if not identical.

The objective in developing this monograph is to provide guidance,
explanation, and technical information to those who share in the responsibility for the development and maintenance of biosafety programs.
The monograph is timely because the time for compliance with OSHA
laboratory standards, hazard communication standards, NIH policies,
and recommendations from the National Academy of Science is fast
approaching.

The value of this book goes beyond biosafety: the principles are
applicable to research and developmental activity in general.

Daniel F. Liberman
Judith G. Gordon

iii

Contents

Contents

Contributors

Gary D. Alpert Department of Environmental Health and Safety, Harvard University, Cambridge, Massachusetts

Stanley F. Bielicki National Hazards Control Institute, Easton, Pennsylvania

James L. Boyland Biotech Services, Inc., Clayton, Missouri

Alan M. Ducatman Environmental Medical Service, Massachusetts Institute of Technology, Cambridge, Massachusetts

Melvin W. First Department of Environmental Science and Physiology, Harvard School of Public Health, Boston, Massachusetts

Diane O. Fleming* Office of Safety and Environmental Health, The Johns Hopkins University School of Hygiene and Public Health, Baltimore, Maryland

Robyn M. Gershon** Department of Biosafety, Yale University, New Haven, Connecticut

---

Current affiliations:
*Sterling Drug, Great Valley, Pennsylvania
**School of Hygiene and Public Health, The Johns Hopkins University, Baltimore, Maryland

Lawrence M. Gibbs*   Department of Chemical Safety, Yale University, New Haven, Connecticut

Rose H. Goldman   Occupational and Environmental Health, Cambridge Hospital, Cambridge, Massachusetts

Judith Gordon   Gordon Resources Consultants, Inc., Reston, Virginia

Everett Hanel, Jr.   Environmental Control and Research Program, NCI-Frederick Cancer Research Facility, Program Resources, Inc., Frederick, Maryland

Lynn Harding   Department of Environmental Health and Safety, Harvard University, Cambridge, Massachusetts

Ralph W. Kuehne   U.S. Army Medical Research Institute of Infectious Diseases, Fort Detrick, Maryland

Martin A. Levitt**   Department of Environmental Health and Safety, Boston College, Chestnut Hill, Massachusetts

Daniel F. Liberman   Environmental Medical Service, Massachusetts Institute of Technology, Cambridge, Massachusetts

Neil S. Lipman   Department of Comparative Medicine, Massachusetts Institute of Technology, Cambridge, Massachusetts

David Lupo   B & V Testing, Inc., Waltham, Massachusetts

Christian E. Newcomer   Department of Comparative Medicine, School of Veterinary Medicine, Tufts University, Boston, Massachusetts

Franklin Pearce   Department of Fire and Safety, Memorial Sloan-Kettering Cancer Center, New York, New York

Robert A. Spurgin   BFI Medical Waste Systems, Santa Ana, California

---

Current affiliations:
*Department of Environmental Health and Safety, The University of Connecticut, Storrs, Connecticut
**Director of Industrial Hygiene and Training, NAACO, North Andover, Massachusetts

Harlee S. Strauss* H. Strauss Associates, Inc., Natick, Massachusetts, and Center for Technology, Policy and Industrial Development, Massachusetts Institute of Technology, Cambridge, Massachusetts

Joseph Van Houten Department of Safety and Industrial Hygiene, Schering Corporation, Bloomfield, New Jersey

David Vlahov The Johns Hopkins University, Baltimore, Maryland

Maury I. Wolfe Dimarinisi & Wolfe, Boston, Massachusetts

_____

Current affiliation: H. Strauss Associates, Inc., Natick, Massachusetts

I. Facility Considerations

1

Design of the Facility

MAURY I. WOLFE *Dimarinisi & Wolfe, Boston, Massachusetts*

I. INTRODUCTION

A. Basic Considerations

The design of a laboratory facility might, without reflection, appear
to be a relatively straightforward matter. With reflection and a quick
scanning of this chapter, the reader should recognize that a single approach to laboratory design does not exist. The purpose of this section is to point out some very basic divisions among the types of laboratory projects in order to establish both a context and an approach
for the discussion of laboratory design.

B. Facility Considerations

General Comments

The institutional setting for a biological laboratory must be understood
before the laboratory design is begun. An academic laboratory ordinarily is designed to accommodate a mixture of teaching and research
functions; a hospital laboratory may be designed for research with a
teaching component or it may be a support service to patient care; a
commercial/institutional laboratory may be intended for research, testing, demonstration, production, or a combination thereof.

The first task then is to establish the purposes the laboratory is
to serve. All decisions about the laboratory must be understood and
approved by an appropriate authority within the institution. It is
good practice to set up a building committee with full authority to ap-

prove planning and design decisions. It is the responsibility of the
committee members to communicate the institution's requirements to the
designer and to review and approve the designer's work. A conscien-
tious committee, working in concert with a responsible designer, will
limit misunderstandings as well as expedite the design process.

Research Facilities versus Production Facilities

The design of a research facility must be sensitive to the research tasks
to be undertaken as well as to the character of the researchers, their
habits and research styles. Such laboratories frequently are planned
and laid out in accordance with the idiosyncracies of the researcher.
This custom design feature can meet the preferences of only one or a
few scientists.

In contrast to this customized design, a production facility must
meet the requirements of the production process(es) and its equipment.
Unlike the idiosyncracies of people, those of the process and equipment are usually objectively determinable and fixed.

In a broad sense both testing laboratories and those producing a
commercial product can be considered production laboratories in that
there are specific processes for which the laboratory is designed. This
is in contrast to the research laboratory, which is designed to house
researchers who determine, as they see fit, the activities to be undertaken in the laboratory.

New Construction versus Renovation

The constraints placed upon laboratory design are dependent upon
whether the laboratory is to be built new or is a renovation. Ordinarily, renovation within an existing building will be more constrained
than new construction. A very thorough preliminary investigation
should be made of any existing building considered for renovation into
laboratories before any significant effort is put into the design. While
there are few wonders that unlimited funds cannot accomplish, unlimited funds are themselves a wonder. All project planners should
understand that renovations can cost more than equivalent new construction—in addition to the new work to be built, the existing construction may require demolition and removal, adjacent spaces may require
protection and existing services may require maintenance, augmenta-
tion, or, more commonly, may simply have to be worked around.

Free-standing Facility versus Facility
Contained in Another Complex

A free-standing laboratory facility is generally less constrained than
one that is contained within another complex. Every design must conform to the existing conditions in which it is placed, and the plumbing,

HVAC, and other service and environmental requirements for a research laboratory are usually in excess of those demanded by ordinary building funciton. Building a laboratory within another complex, unless the entire complex is designed for laboratories, affects the laboratory and the rest of the complex in both floor plan and cost.

C. Summary of Basic Considerations

It is essential to a successful design process that the basic constraints upon the project be established and understood before design is begun. The laboratory's owner (university, hospital, government agency, company, individual, etc.) must make explicit to the designer the full nature of the institutional context in which the laboratory will function, the purposes the laboratory will serve, the physical context in which it will be built, and the proposed budget for the laboratory's construction.

It is the designer's responsibility to question the owner thoroughly enough that he or she is satisfied that the project's basic character and constraints are understood. It is the owner's responsibility to be fully forthcoming about the project, its intent, scope, and budget. The failure of either party is almost certain to cause complications, delays, and cost escalations in the project.

II. THE DESIGN AND CONSTRUCTION PROCESS

A. Overview

For the purposes of understanding the typical process by which a laboratory facility will be designed and built, it is not necessary to delve into all the possible variations on the ways in which the work necessary for the implementation of the project can be done. It is necessary, however, to understand what the basic tasks in the implementation are and how they should be gone about.

The following discussion is based upon usual American practice, where the owner hires a designer. Typically, this is an architect, who is responsible for both the facility design and the development of construction documents, which are used by contractors to base their proposed price to build the facility. The actual construction is done by the general contractor and his subcontractors; usually, the architect administers the contract for construction acting as the owner's agent.

It should be noted that there are other procedures for design and construction of a facility; for example, in a design-build process the contractor and the designer work together, often with the designer as the contractor's consultant instead of the owner's.

B. Hiring an Architect

Architect Selection

In order to select an architect, the owner should get a list of reliable
architectural firms with experience in laboratory design. The list can
be compiled from the recommendations of others who have had a laboratory designed recently, from research in professional journals, from
reliable acquaintances or professional contracts of the owner, or any
other credible source. The owner should then interview the architects,
see examples of their work, and check their references.

The principal criteria for final selection of the architect should be
the owner's assessment of the architect's abilities to carry out the project and the owner's sense of how well he and the architect can communicate and work together. These criteria should be met in the selection of the architect, since the laboratory's design is sure to be compromised if the architect's capacities are lacking or the owner and architect do not work very well together.

Architect's Fees

The proposed fees for services should not be a principal determinant
in the architect's selection. There may be some variation among architects in the fees they charge, but, as a rule, competent professionals
will charge the same order of fee for the same order of services. The
architect is most likely to propose either a fixed fee for the design
services or a fee based upon a percentage of the construction cost of
the project. The actual dollar amount of the fee in either case should
be approximately the same for a specific project.

The fee for an architect's basic servcies will pay for the architect
and his principal consultants, that is, civil, structural, mechanical,
electrical engineers, landscape architects, and cost estimators.

As a percentage of construction cost, the architect's fee will range
form approximately 7.5% for very large laboratory projects to 15.0% for
fairly small laboratory design projects.

It should be noted that the first step in a project's implementation,
programming, is not part of the basic services of the architect and is
not included in the base fee. If the architect does the programming
of the laboratory, then he should receive an additional fee for this
service. The architect's proposal for this service is most likely to be
either a fixed fee or an hourly rate with an upper limit. Fees for programming services can range very widely, depending upon the extent
of the project being programmed, the thoroughness of the program to
be developed, and the owner's review and presentation requirements
for the completed program document. As a very rough rule of thumb,
the oner should expect to pay from 1 to 2% of the project's construction
cost for the development of a program.

C. Developing a Facility Program

The Nature of a Program

Programming is the systematic establishment of the requirements for the project. The development of an acceptable program should be the first step in the design of a laboratory.

A completed program should include clear statements about all of the determinable features of the project. This requires the cooperation of the owner's staff—building committee, department heads, and all others responsible for determining the project's features. It is the programmer's responsibility to question these individuals about the project, determine all of the features of the project, codify them in a systematic way, and produce a completed program document. This program document is then reviewed by the owner's staff to make certain that it truly represents the project's needs. This program, once approved, will serve as the basis for the architect's design of the project.

The programmer does not have to be the architect. In fact, specialists in programming can be hired by the owner before the architect is engaged. Such programming consultants do not produce a design but rather produce the program for review by the owner and for presentation to the architect as the basis for the design. On the other hand, the program can be prepared by the architect as a preliminary service prior to the beginning of the basic design services.

The advantage of having the design architect do the programming is that he will be very familiar with the project's requirements and there should not be any misunderstandings of the programmatic document. There are two possible disadvantages of having the design architect do the programming. These are that (1) not all good designers are equally adept at programming, and (2) designers are prone to be biased, even unconsciously, toward making the project as large as possible. The decision whether the designer should also be the programmer has to be made on a case-by-case basis and is dependent upon the good judgment of the owner and his consultants and on the forthrightness of the architect.

The completed facility program should contain the following sections:

Program Summary—a tabulation of the major components of the facility, each space/room within the component, and a net square footage for each.
Relationship Diagrams—diagrammatic representation of the relationships among all of the major components and diagrammatic representations of the relationships among the spaces/rooms within each component.
Detailed Program for Each Space/Room—a program for each space within the facility giving a description of the space's use; required

relationships to other spaces; furnishing and/or equipment in the space; services required for the space and its equipment; specific dimensional requirements for the space (if any); special lighting, acoustic, HVAC, electrical, plumbing requirements (if any); floor, wall, and ceiling finish requirements; net square footage needed for the space; and any other specific requirements which would have to be met in the space's design.

Staffing Chart—a summary of the staff for the facility, including simplified job descriptions (job titles, if explanatory, will do), numbers of staff in each position, shift breakdowns, male/female staff ratios, heirarchical organizational staffing diagram; an estimate should also be made of the numbers of other nonstaff people who will be coming to the facility, including deliveries, visitors, inspectors, auditors, stockholders, etc.

Code/Regulation/Ordinance Summary—a very concise listing of the codes, regulations, ordinances, etc. that affect the facility design. Where such codes apply specifically to a space within the facility, the applicable code section should be cited in the detailed program description of the space (e.g., plumbing codes specify the number of toilet fixtures that must be supplied for specific numbers of employees; the applicable seciton should be cited in the program for the staff toilets).

Exterior Programmatic Requirements—a summary of the facility's requirements outside the building, including parking, loading/delivery, refuse storage/disposal, pedestrian access, landscaping, building image, and any other specific feature or requirement that emerges from the programming interviews and investigations.

Understanding the Program

The process and results are greatly facilitated if the program is developed in a standardized format. Figure 1-1 is an example of a standardized format sheet for the programming of specific spaces within a facility. Each programmer usually develops his own format; the owner's only concern should be that the format yields clear and complete programming information. Similarly, any diagrams produced in the course of programming should be clear and informative. Each programmer will diagram in his own style or manner; therefore, the owner must be certain that the diagrams are an accurate and understandable representation of the proposed facility.

Square Footage

If the programming is done correctly, the program summary will yield a total figure for the net square footage required by the facility. Net square footage usually excludes the area in the building required for structure, exterior walls, interior partitions, corridors, stairs, eleva-

ROOM / SPACE NAME: _____

USE / FUNCTION: _____

DIMENSIONS / AREA: _____

RELATIONSHIPS: _____ Adjacencies _____

Separations _____

MATERIALS / FINISHES: ____ Floor_____

Base _____

Ceiling _____

Walls _____

SERVICES: _____ Lighting _____

Power_____

HVAC _____

Plumbing _____

Fire Protection _____

Other_____

FURNITURE: _____

EQUIPMENT: _____

REMARKS: _____

FIGURE 1-1 Standard format for programming.

tors, mechanical and electrical support spaces. Toilets are, according
to the preference of the programmer, sometimes included in net areas,
sometimes not. The gross square footage that the facility will occupy
is usually based on the programmer's informed opinion of what the
efficiency ratio for the facility should be obviously, the actual gross
square footage of the facility will be dependent upon the actual designed and constructed layout. The various formulas the programmer
uses are:

$$\text{Efficiency Ratio} = \frac{\text{Net Square Footage}}{\text{Gross Square Footage}}$$

$$\text{Grossing Factor} = \frac{\text{Gross Square Footage}}{\text{Net Square Footage}}$$

$$\text{Grossing Factor} = \frac{1}{\text{Efficiency Ratio}}$$

Efficiency ratios vary greatly according to nature of the facility;
for example, a laboratory constructed within an existing building that
already contains most of the laboratory support and staff spaces elsewhere in the building will have a high efficiency ratio. This is in contrast to a laboratory facility constructed as a new multistory building,
which will have a lower efficiency ratio because of the amount of space
in the facility that will have to be dedicated to stairs, elevators, plumbing and HVAC shafts, structure, etc. As a very rough rule of thumb,
the efficiency ratio of a laboratory facility should be between 65 and 85%.

Flexibility of Program

An important issue not to be overlooked in programming is the provision for future expansion of the facility. In some cases the owner
has very clear and specific intentions for future expansion–knowing,
for example, that certain rooms or spaces will be added to or dupli-
cated at some specific future time. In other cases, the owner may
want the facility design to be such that it does not preclude or render
difficult possible future expansion. Whatever is known or expected
with regard to the facility's expansion must be recorded explicitly in
the program so that the designer can allow for this in the layout of
the facility.

Budget

From the program the owner should be able to develop a realistic budget for the facility's construction since the program will have provided
a gross area for the project. A knowledgeable cost estimator should
be able to provide a reasonably accurate order-of-magnitude cost estimate for the laboratory facility if he knows its location, size, and

character. If it's known that the program calls for a 20,000 square foot new laboratory building, that the site available for the building does not appear to present any peculiarly difficult construction problems, and that new laboratory building construction is averaging about $150 per square foot without furnishings and equipment, then the owner should be prepared to spend on the order of $3,000,000 for the construction.

If the program yields a facility that exceeds the owner's funds for construction, then the owner has the opportunity either to seek additional funds or to reduce the facility before proceeding into design. This confirmation of the fit between the project budget and the owner's desires for its scope is one of the most important results of a thorough programming phase in the project's implementation; it avoids innumerable difficulties later on in the project.

D. Design

Schematic Design

Once a satisfactory program has been developed, the architect can begin the design process. If the architect has not been the programmer, his first step will be to review the program document and the project budget. The architect must ascertain for himself that the program is sufficiently clear to proceed and that the owner's funds are sufficient to allow the project, as programmed, to be built.

Upon confirmation of the program and the availability of sufficient funds, the architect will prepare for the owner's review of the schematic design. The schematic design iwll consist of drawings that show the size and relationships of the facility's spaces. The extent to which the schematic design will look like a building is dependent upon the design style and practices of the architect. Some designers propose alternate schematic designs to allow the owner to choose among various basic options for the facility's organizing scheme; for example, is it to be a long, linear scheme, a courtyard scheme, a tower scheme, etc. Once such basic schematic matters are settled, another level of schematic design can be addressed. At this level, sometimes called preliminary design, the architect will produce drawings that will represent a building, not just a schematic organization.

The owner will be required to review all of the schematic designs and to inform the architect of his preferences and any difficulties he sees. The schematic design can serve as a test of the program since it should yield an approximate gross area and volume for the project. If the programmer has made an error in estimating the grossing factor for the facility, it should be apparent from the schematic design. Any discrepancy between the program and the project budget must be addressed before the architect proceeds to the next design phase.

Design Development

Upon approval of a set of schematic design documents by the owner,
the architect will prepare design development documents. These documents show the size and character of all parts of the facility. The
architectural, structural, mechanical, and electrical systems for the
facility, its materials of construction, its floor plans and elevations
will be shown sufficiently enough for the owner to appraise realistically
what the completed project will look like and how it will function.

The owner must again review the design and communicate to the
designer his concerns about any part of it so that changes can be made
before the next phase of the design process begins. A more refined
estimate of the project's construction cost can be made at this stage.

Construction Documents

The next step in the design process is the production of the construction documents. Construction documents can be divided into two categories: drawings and specifications. They are the documents on which
contractors will base their proposed price to build the facility and from
which their workmen will actually construct the facility. As a result,
the construction documents must be as complete and accurate as possible. It is the designer's responsibility to see that these documents
are true to the design intent and that they give the contractor unambiguous instructions as to what is to be built.

The owner should review the construction documents as they proceed to completion; a check at 25, 50, 75, and 100% completion is good
policy for a project of the complexity of a laboratory. A full set of
construciton document drawings for a new building will include civil
engineering drawings, landscape drawings, architectural drawings,
structural drawings, HVAC (heating, ventilation, and air-conditioning)
drawings, plumbing drawings, fire-protection (sprinklers) drawings,
and electrical drawings. There may also be specialized drawings for
equipment, telecommunications, data processing, and security systems.
The owner should not expect that he will be able to review the documents in very great detail—that responsibility lies with the designer.
However, the owner should check them to see that they are being faith-
ful to the approved design development documents. For his part, the
architect should keep the owner informed of any changes in the design
or problems that arise as the construction documents are prepared.

It is very prudent for the owner to have an independent check of
critical features of the project. Therefore, when there are very special requirements that must be met, such as the performance of the
HVAC system to maintain proper biological containment standards,
the owner should engage his own independent consultants to review
the construction documents in greater detail.

   At the completion of the construction documents, the owner will
receive a final estimate of the construction cost of the project.
   When the construction documents are completed and approved,
the owner, in conjunction with the architect, will put together a package of materials in order to solicit contractor's proposals for doing the
actual construction work. The soliciting of such proposals is the next
phase of the designer's services.

E. Construction

Bidding or Negotiation

The owner may choose to have contractors bid on the project and to
award the contract to whichever bidder the owner finds most responsive to his needs. Owners should note that the lowest bid may not
necessarily be the most responsive; a very low bid, one well below
the estimated construciton cost, may be evidence that the contractor
has not understood the project's requirements or that he intends to
secure the job by bidding low and then to make a profit by overcharg-
ing for any changes.
   In cases where the circumstances allow, owners can avoid the potential for such problems by choosing to negotiate with one or more
contractors whose experience and reputation show them to be reliable,
competent, and capable of building the facility as it is designed.
   In either case, bidding or negotiation, the drawings and specifications will be put together with other documents, which will set the
conditions under which the contract for construction will be awarded,
present what the contract itself will contain, set the conditions of its
administration, and make clear any other requirements or conditions
the owner wishes to place upon the contracotr. The owner, his law-
yers, insurance agents, and other such parties will generally be as-
sisted by the architect in putting together these documents, sometimes
called the "bid package."
   The intended result of the bidding or negotiations is the execution
of a contract between a contractor and the owner for the construction
of the facility. Once this has been done, the final phase of the architect's services will commence.

Construction Process

The architect and his engineers will usually be responsible for the
administration of the contract while construction is underway; this
requires the architect to act as the owner's representative in interactions with the contractor. In that capacity the architect will visit the
project site at regular intervals to determine in general the work's
progress and to answer questions from the contractor. The architect
does not supervise the construction work; that is the responsibility
of the contractor.

During construction the contractor will submit to the architect for approval shop drawings, product information, and samples of the materials he proposes to use in the construction. The architect and his engineers will review this information in order to make certain that the work the contractor proposes to do will conform to the construction documents and thereby to the design of the facility that the owner has approved.

If for some reason the work of construction has to be changed from that shown in the construction documents, the architect will prepare a change order for the contractor. Such changes must be reviewed and approved by the owner and the contractor since they often have effects on both the length of time for the construction and the amount of money the contractor receives for his work. Change orders can arise from errors or omissions in the contract documents, from changes in the owner's requirements for the project during the construction, from the discovery of unforeseen conditions, or from suggestions by the contractor. If the owner has expressed his needs for the project carefully and clearly to the architect and the architect and his engineers have done a thorough job in the preparation of the construction documents, then the owner should expect the value of the change orders to be relatively small, several percent at most of the total cost of the project.

As the owner's representative the architect will review the contractor's requisitions for payment and certify the amounts which should be paid. The architect will also review the project at its completion in order to make certain that the work has been properly concluded.

On larger projects the owner may wish to have the work of the contractor observed more closely than the periodic visits of the architect would allow. For example, the architect and his engineers would typically bisit a project weekly. For more complete oversight, the owner may wish to engage a "clerk-of-the-works" for the construction period. The clerk may be hired by the architect or by the owner. In either case, the owner will pay an extra amount for this service.

After the construction is completed, the owner can occupy and use the new facility. It is normal for the owner to receive a warranty of one year on the work by the general contractor. This warranty is different from that which may be provided by the supplier's of materials and equipment used in the construction. The contractor's warranty means that the owner has the right to call him back to correct any defects in the work that show themselves during the first year of the project's new life.

F. Cost Estimates

Throughout the design and construction document preparation process, the owner will receive cost estimates for the project. It is important to emphasize that these estimates are just that, estimations of the project's cost. The actual cost of the project is determined in the contract for construction agreed to by the owner and the general contractor selected to do the construction work.

It is generally good practice for the owner to insist that the architect engage an independent cost estimator to prepare the cost estimates. This independent estimator should be familiar with the project type so that both the architect and the owner can rely upon his estimates. If the estimator is competent and there are no peculiar circumstances that affect the bidding or negotiations with the contractors, then the owner should expect the contract price for the project to be ±5% of the final estimate made from the completed construction documents.

In budgeting for the project, the owner should allow a contingency for any change orders that may arise during the construction process. For new construction, 5% contingency should be sufficient; for renovations, where the likelihood of unforeseen complications arising is greater, a contingency of 10% is more appropriate. The contingency should be in addition to the contract sum agreed to with the contractor. If the contractor has fairly figured his proposal and the construction process proceeds smoothly, then the owner should not have to expend the full amount of the contingency. Under such happy circumstances the remaining contingency funds can be used in furnishing and equipping the facility or returned to the owner's general funding source.

III. SITING OF THE FACILITY

A. General Comments

After the determination of the project's programmatic needs and budget, the most important considerations affecting the development of a laboratory facility are those directly concerned with the facility's location. Whether the laboratory is free-standing or contained within an existing facility, the demands it will make upon its surroundings and the capacity of those surroundings to respond to the facility's requirements can be the difference between a satisfactory completed project and one that was infeasible from its inception.

B. Zoning

Though there are a few exceptions, most communities in which a laboratory is to be built will have a set of regulations governing what sort of uses particular areas may be put to. Zoning regulations are a community's means for controlling its development and are a legitimate exercise of the community's right to protect the safety and welfare of its inhabitants.

Zoning regulations typically include the establishment of districts within the community in which specific categories of land use are allowed or disallowed; for example, a residential district may allow single family homes, apartment buildings, and small professional offices but prohibit retail stores, industrial buildings, etc. The zoning regulations usually also include dimensional regulations which specify limitations on how a prticular parcel can be developed; for example, zoning regulations will give minimum lot sizes, setbacks, maximum building heights, percentages of open space, and maximum building area (expressed as a maximum Floor Area Ratio, or F.A.R.). Zoning regulations will also usually set the number of parking spaces and loading docks required for a specific building use and size; for example, one parking space will be required for each 300 square feet of an office building and one loading dock truck space for each 10,000 square feet of building space.

It is essential to any proposed development that the zoning requirements for the project site are completely understood. The regulatory environment must also be understood. The degree to which any set of regulations are adhered to varies considerably from community to community. Some communities require all projects to meet the zoning regulations in all respects. Other communities can be more lenient and will grant variances to the regulations if a reasonable case can be made for the exception. It is always prudent to discuss the proposed project and the zoning regulations affecting it with the appropriate community officials—those with the responsibility and authority to interpret and enforce the zoning regulations. These official's titles vary from community to community, they can be zoning officer, building inspector, planning board, development director, etc.

Zoning issues can be especially problematic for biological laboratories. Very often, particularly in small communities, regulations may not have a specific land-use category into which the laboratory fits; for example, is a research laboratory to be classified as a medical laboratory, a light industry, or an office use? The practical difference between being placed in one category rather than another can be the difference between being allowed to build the project and being barred from using the land for a laboratory.

The problem of zoning can become even more complex if the community is sensitive to the fact that the laboratory work will be dealing with biologically hazardous materials. Communities may have adopted special regulations affecting the development of a project that will house recombinant DNA research work. These regulations will be additional constraints on the siting of the project beyond those contained in the ordinary zoning regulations. Other portions of this monograph address such regulatory issues in greater detail; nonetheless, it is essential that the laboratory owner and designer be fully aware of these regulations, their potential effects on the project's siting and layout, and the efforts that will have to be made to comply with them in the design.

As part of the preliminary planning process for the facility, the architect should provide the owner with a zoning analysis of any particular site on which the laboratory is proposed to be built. This analysis should list all of the zoning regulations and their effect on the project. For example, if a zoning regulation requires one off-street parking space for each 300 square feet of building, then a 30,000 square foot building will require 100 off-street parking spaces. The parking of this many vehicles requires either a substantial amount of land area on the site or the contruction of a multilevel parking structure. The strict enforcement of such regulations can make a project either financially or physically impossible.

The only proper way to deal with zoning issues is to talk directly with the responsible local officials. Delaing with these officials from the very start of a project can save considerable amounts of time and money. When a dispute arises with a local official, for example, over the interpretation of some part of the zoning code, it always pays to have already established a cordial working relationship with that official. Local officials often have a great deal of latitude and discretionary powers in the interpretation of local codes and ordinances, including zoning regulations and building codes. This means that, whatever the words in the regulations seem to mean, the local official stands as the penultimate interpreter of that meaning; if the official says that he interprets a laboratory to be a business and that business uses are not allowed in the zone in which the laboratory is to be built, then the laboratory owner will not be allowed to build the facility unless he can win an appeal to a higher authority. If that higher authority is an elected or appointed board, then its decisions will be made in a political context and are likely to be supportive of the local official. The owner's only recourse then would be to the ultimate authority in such matters, the courts. As most people know, it is generally advisable to avoid the courts. Recourse to them is almost invariably expensive and time-consuming; in addition, the outcome of zoning cases is not always certain, no matter how apparently self-evident the meaning of the regulations may seem. The best course of action for any project is to be certain from its earliest stages that it will conform to the local regulations as they are understood, interpreted, and enforced by the local officials.

C. Access

A good site for a laboratory facility has good access for all of the types of people and services the facility will need. In planning, it is important to list all of the types of access that will be needed and to assess a proposed site in light of these requirements. These requirements may vary from facility to facility; for example, a large production facility may have an essential need for easy access for frequent truck deliveries and pick-ups, while a research laboratory might need easy access to local universities and hospitals.

Part of the basic planning should include consideration of the facility's dependencies upon local and regional transportation networks. Is the fcility going to need good access to public transportation to attract staff? Is the facility site capable of providing public access and service access at the same time without uneasy conflicts? Is the local road network sufficient to handle the types and volumes of traffic the facility will generate?

In the case of a laboratory facility located within another complex, the issues of access may be more localized but are equally critical. Can materials be delivered to the laboratory in a safe and efficient manner? Can laboratory waste be removed safely and cleanly? Will access to the laboratory by staff and visitors be simple and easy or will it be circuitous, confusing, or otherwise constrained?

All of the needs for access should be established in the programming phase of the project's design. Once these requirements are known, then the owner and the designer can reasonably assess the suitability of any particular proposed facility location. As in most situations, the owner should be prepared to make some compromises with regard to access as long as the essential operational requirements of the laboratory facility can be met by the means of pedestrian and vehicular access available.

D. Abutters and Neighbors

The abutters to a proposed facility location should always be carefully assessed. Both the adjacent and nearby neighbors can be either an asset to a laboratory lcoation or a liability. For example, it can be a significant asset to a laboratory to be near a university: the laboratory staff can be recruited form there, the university library can be used as a facility resource, and research ideas can be stimulated by conversations among the laboratory and university populations. Similarly, locations of a recombinant DNA laboratory in an area where there are other such laboratories may make its placement in the community much easier than if it is the first such facility to be located there.

On the other hand, if a laboratory owner wishes the facility to establish itself with a clearly separate image, then it might best be located away form other laboratories of similar nature.

Some abutting or nearby uses may be incompatible with a laboratory. In communities that are sensitive to the potential hazards of biological research, it would be ill-advised to consider the location of a laboratory adjacent to a school. Nearby uses that are noisy, produce noxious fumes, vibrations, or other such conditions can make the building of

a laboratory more expensive than ordinarily expected in order to miti-
gate these environmental problems. Such problems can also make the
facility a less attractive location for recruiting and holding staff.

In the case of a facility located within a larger complex, the issues
associated with abutting or nearby uses can be especially critical. The
services to an adjacent part of the building may not be suited to location of the laboratory, or deliveries to the laboratory would be in conflict with already established patterns of use in the existing complex,
or the laboratory might become accessible to the general public in ways
that would be detrimental to its operation or to the larger complex's
public image.

Again, the point is that the nature of the abutters and other neighbors to a laboratory must be understood and carefully considered in
the determination of a particular location's suitability for the laboratory's
development.

E. Utilities

Laboratories can place special demands upon the utilities that serve
them. The service requirements for the laboratory must be established
reasonably accurately before the laboratory site is selected. If the
electric power demands of the laboratory equipment exceed that available at the site, then the laboratory's development will not be feasible
unless the power available is increased or the laboratory's power demand is reduced. In some cases these options are precluded either by
operational, cost, or regulatory factors.

Similar constraints will apply to the other utilities that normally
serve a site; that is, water, sewer, and gas services. The laboratory
requirements for all these services must be assessed by the appropriate
engineering specialists, the site services available must be checked
with the local utility companies, the local regulations affecting the provision of these services to a biological laboratory must be reviewed
and the costs must be estimated, and then the feasibility of the laboratory's development can be reasonably considered.

Telephone and other telecommunication services to a site are generally not a problem; however, if there are some very special requirements for the facility, these must also be carefully considered in the
site selection.

In the case of a laboratory located within another complex or building, the problems of utility services can be very critical. In many
such circumstances, the services available to the laboratory will be
very constrained and the possibilities for augmenting them very limited.

Investigation of the utility services should always be among the
first steps in the determination of any project's feasibility; in the
case of a laboratory, this step is even more critical and should never
be postponed or shortchanged.

F. Security

The owner and the designer need to establish the types of security
which the laboratory facility must have. Most often the security requirements can be met through appropriate planning and design of
the facility's exterior and interior. Nonetheless, in assessing the suitability of a specific location for a laboratory, the owner and the designer should keep in mind whether some feature of the site or location
would be a security problem; for example, if the streets are such that
deliveries to laboratory will have to be made in an isolated, difficult-
to-monitor area, then, if deliveries need to be secure, the facility will
have to include special monitoring devices and/or additional security
staff. The cost implications of such security issues should be considered in the site selection process.

IV. TYPICAL COMPONENTS OF A FACILITY

A. General

Laboratory facility spaces can generally be divided into three classes:
the laboratory spaces and their support spaces, the laboratory staff
spaces and their support spaces, and the overall facility administration
and operation spaces and their support spaces. The following discussion provides a basic listing of these types of spaces and some considerations that should arise in their design and incorporation in a
laboratory facility. It is essential that the specific requirements for
each space to be included in a laboratory facility be established in
the programming phase of the facility's design.

B. Laboratory and Laboratory Support Spaces

Laboratory Spaces

There is no unique, best arrangement for a laboratory. The laboratory's size and layout will depend upon the purposes it is to serve,
the nature of the institution that houses it, and the research style of
the organization. With these facts in mind, one can still describe some
of the more common sizes and types of laboratory layout.
    The choice among laboratory layouts for a specific laboratory facil-
ity has to be made in concert with the owner's understanding of how
the facility is to operate. In some circumstances, large, open laboratory spaces where many lab workers are in a single space will serve
the institution's purposes for the facility; for example, in a university
teaching/research laboratory. In other circumstances, the laboratory
work is to be conducted by smaller groups; in that case, the laboratory
may be laid out as a series of modular laboratories of similar plan and
equal size connected by an interior corridor. This modular layout

allows considerable flexibility in the association of small and large laboratory research groups, allowing both privacy and community. At the other extreme is the small unit laboratory–an individual laboratory that is sized and laid out to accommodate only one or, at most, a few laboratory workers. Figure 1-2 illustrates the basic features and differences among large, open laboratories, modular labs, and single unit labs.

The basic feature that controls their layout is the laboratory casework or benches at which the work of the laboratory is performed. Typically, laboratory benches have a flat work surface, 2-2½ feet deep, set approximately 3 feet off the floor. The bench usually has storage drawers or cabinets below the work surface; fixtures for gas, electric, water, compressed air, and vacuum services; built-in cup or service sinks; and often a raised reagent storage shelf at the back of the work surface. Other storage space is usually provided by wall-hung or suspended cabinets above the work surface. The laboratory benches can be laid out in a laboratory in three, very basic ways: against a wall, as a peninsula projecting from a wall, or as a free-standing island. Invariably, laboratories are composed of some combination of these three basic bench arrangements. In the design phase it is critical that the users check the bench layout to make certain that it provides the necessary work area, storage spaces, sinks, and other services.

Each laboratory worker should be provided with at least 6 feet (preferably more) of clear bench work space; the clear distance between benches where workers are back to back should be a minimum of 5 feet; and a service aisle or corridor in a laboratory should be a minimum of 3 feet 8 inches (44 inches) wide.

Figure 1-3 illustrates these basic dimensional requirements that should be met in the laboratory layout. Figure 1-2 shows the basic bench arrangements–wall, peninsula, and island–are combined to make some typical laboratory layouts.

Each laboratory manager will have his own preference for the type of laboratory benches or casework to be used in the facility. The architect needs to be informed of this preference early in the design so that the facility layout can reflect this decision. Basically, laboratory benches fall into two groups–fixed casework systems and demountable systems. Fixed casework systems are the conventional systems and are composed of a base cabinet supporting a work suface with wall-hujng or ceiling-mounted storage cabinets above. Demountable systems consist of a supporting frame or service core from which the laboratory work surfaces and cabinets are hung.

The fixed casework systems are less flexible, obviously, than demountable systems. In the demountable systems, the cabinets and work surfaces can be rearranged or removed, if necessary, without taking the whole bench apart. It must be remembered that radical rearrangements of the laboratory will, no matter what bench system

LARGE OPEN LABORATORY

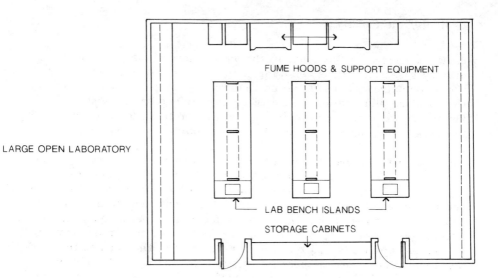

FUME HOODS & SUPPORT EQUIPMENT

LAB BENCH ISLANDS

STORAGE CABINETS

MODULAR LABORATORIES

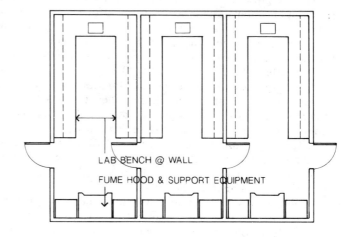

LAB BENCH @ WALL

FUME HOOD & SUPPORT EQUIPMENT

SINGLE UNIT LABORATORY

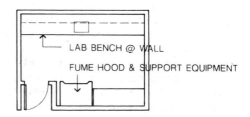

LAB BENCH @ WALL

FUME HOOD & SUPPORT EQUIPMENT

FIGURE 1-2 Laboratory types.

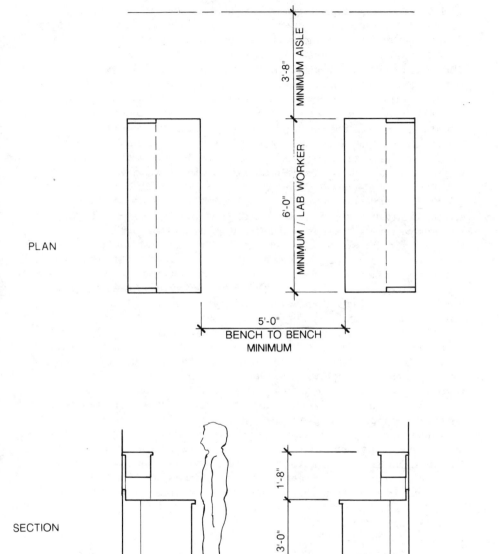

PLAN

SECTION

FIGURE 1-3 Basic laboratory planning dimensions.

is chosen, require rearrangement of the services to the benches. In
many circumstances, the apparent flexibility of a demountable system
is illusory, since rearrangements contemplated would require either
major changes to the bench services or the modifications in layout.
The most flexible feature associated with any facility is the behavior
of the people who use it. Given these facts, it is often best to choose
the conventional bench system and to make the laboratory layout as
convenient as possible for different modes of laboratory work. Demountable systems should be chosen only when the laboratory owner
is certain that the rearrangements they allow will in fact be made in
the facility on a relatively frequent basis. Otherwise the premium
usually paid for the demountable systems will be wasted.

The other choices involved in the selection of laboratory benches
are the materials for the cabinets and work surfaces. The cabinets
are generally available in wood, steel, and plastic laminate finishes.
The choice among these is aesthetic as much as it is practical, since
all types provide serviceable cabinets. The laboratory manager is
likely to have a preference and should be consulted. The architect is
also likely to have a preference, and in this matter it is recommended
that his suggestions be followed, since the overall appearance of the
completed laboratory is his responsibility. The cost difference among
the materials is generally negligible.

The choice of work surface is more a practical matter than an aesthetic one, since the durability of the surfaces in service is affected
directly by the nature of the laboratory reagents it is likely to encounter. Laboratory bench work surfaces are generally available in stone,
resin composites, plastics, and plastic laminates. The laboratory manager and the architect should consult with each other and the laboratory
equipment suppliers before selecting a work surface material. The
suppliers should provide test reports on the performance of their work
surface materials when exposed to the various laboratory chemicals
and to standard wear conditions. The suppliers should also provide
a list of their benchwork installations, which the architect and the
owner can inspect. This list should give the date of installation and
a reference at the facility where they were installed. It is always good
practice to ask the reference for an evaluation of the benchwork's performance and to visit older facilities to assess how the materials and
installations actually stood up.

As part of its basic equipment, almost every laboratory contains
at least one chemical fume hood. Fume hoods provide a controlled exhaust area in which work with volatile materials or aerosols can be
conducted safely. Fume hoods are typically designed to fit within the
usual bench layouts for laboratory spaces. The specific number of
fume hoods required in a lab and their location must be decided by the
owner. The designer should not count the fume hood as part of the
laboratory bench space assigned to a worker but rather as a shared

resource. In many laboratory layouts these hoods are located against a wall and across an aisle from the main bench work areas. This arrangement, which is shown in Figure 1-2, allows easy access to the fume hoods by all laboratory workers and keeps the main bench work areas from becoming congested and from losing valuable work space to the fume hoods.

Laboratory Support Spaces

The nature, number, and size of these support spaces will vary from laboratory to laboratory. The following briefly described the more common ones required in a biomedical (recombinant DNA) research facility.

Reagent and balances area. Space must be provided for the storage of laboratory chemicals. This area, which can be in an alcove off the main lab space or in a separate room, should contain cabinets, shelves, and other required storage equipment or space for the standard reagents in the lab. This area is not intended to serve as bulk storage for all chemical supplies, rather it is to be a working area convenient to the laboratory for the storage and preparations of regularly used lab chemicals. The area should also contain balance tables and balances for the weighing out of required amounts. This area is *not* intended to serve as the storage area for particularly hazardous chemicals—volatiles, explosives, pressurized hazardous gases, etc.-which require special storage precautions.

Glasswashing and autoclaving room. This room provides for the washing and sterilizing of the laboratory glassware and other equipment. It is a wet, sloppy, and sometimes smelly operation. Consequently, it is best arranged as a separate room from all other laboratory spaces and functions. The glasswashing and autoclaving room needs good access from all lab spaces. Typically, dirty glassware will be picked up in carts and wheeled to the glasswashing area. This room needs to have its doors sufficiently wide and easily opened for the movement of the carts. Typical equipment for a glasswashing and autoclaving room would include double-bowled and single-bowled sinks, dishwasher, drying oven, sterilizing oven, autoclaves, and storage and presentation cabinets for the cleaned glassware. The autoclaves, which are pressure vessels for the sterilizing of equipment by superheated steam, require special planning since they need to have a steam source. If the facility is large enough, the autoclaving area can be made into a separate room from the glasswashing room. In designing this space the planner must consider the thermal output of autoclaves in calculating the heat level for this area. This calculation is important to determine the cooling capacity and exhaust-air requirement of the ven-

tilation system. In addition, special exhaust ventilation may be needed
to control flow when autoclave doors are opened, as well as for odor
and humidity control.

The cleaned and sterilized glassware needs to be readily available
for use in the laboratory. This can be accomplished in several different ways. A supply of cleaned glassware can be stored in the laboratory itself. In that case, a schedule of regular delivery of the cleaned
glassware needs to be maintained, and the laboratory has to set aside
storage space for the glassware stock. An alternate and very good
method is to include in the glasswashing room storage and presentation
cabinets for the cleaned glassware. This arrangement allows the laboratory staff to come and get the cleaned glassware on an as-needed
basis. The presentation cabinets can be built into the corridor wall
of the glasswashing room with glass doors on both the corridor and
room sides. With such a set-up, the laboratory staff does not have to
enter the glasswashing room, which is an operational advantage for
the facility.

Media preparation room. This room is used for the preparation
of sterile plates and broth for the laboratory. It needs to have good
access to the glasswashing room and the laboratory. It needs to have
good access to the glasswashing room and the laboratory. It should
be provided with a sink, bench space, and storage for plates, bottles,
tubes, working chemicals, etc. As in the case of the glasswashing
room, this room is also served by carts and needs to accommodate them
in its doorway and space layout.

Heavy equipment room. This room is designed for equipment that
is too large, too noisy, too hot, or otherwise inconvenient for placement in the laboratory. Since this equipment is very important to the
lab work, this room is often made immediately accessible to the laboratory. The room has to be designed so that the noise, heat, vibrations,
drafts, moisture, etc. produced by the equipment are isolated from
the laboratory and other spaces in the facility. The sorts of equipment
that would normally be contained in a heavy equipment room include
ultracentrifuges, large freezers, sonicators, cryostats, ice machines,
and dry-ice cabinets. Typically, the room would also contain some
work space, either a small lab bench or work table.

Warm room. Warm rooms are temperature-controlled spaces which
are usually provided as a package by manufacturers. The package
includes the insulated walls, floor, ceiling and door(s), and the temperature control equipment. These rooms are used for the holding
and cultivation of experimental plates and solutions. This room is a
specialized piece of laboratory equipment and should be directly accessible from the laboratory. It may also have access from a corridor if

it serves more than one laboratory or other support functions in the
facility.

Cold room. The cold room is exactly the same as a warm room,
except it maintains a cold temperature instead of a warm one. Like
warm rooms, they come as a package form a manufacturer. For the
purposes of facility planning, the only major difference between a warm
room and a cold room is the fact that the cold room requires refrigeration equipment. This equipment, which can be noisy and needs ventilation and condensate drainage, has to be carefully planned so that its
operation does not interfere with the laboratory work.

Photography and darkroom. Biomedical researchers use radioactive
chemicals in their work. These low-level radioactive materials are often
traced photographically. The exposure of the photographic negatives
needs a small space separate form the main laboratory where this work
can be done. A small, light-safe alcove off the main laboratory space
usually suffices. The exact layout of this space will depend upon the
specific equipment proposed by the lab operators; however, for the
purposes of illustration, a typical alcove might be approximately 4
feet wide by 8 feet deep, containing a counter for a camera with stand
and lights, a working aisle, and a back counter and storage cabinets.

In order to process the photographic materials, unless they are
all done by instant methods, there will have to be a small darkroom.
This room does not have to be directly connected to the laboratory
since the photographic materials can be brought to it easily. Making
the darkroom separate from the laboratory also offers the advantage
of having it serve broader institutional uses. The size of the darkroom
will depend upon the nature of the equipment it is to house and the
volume of materials it has to process. These decisions need, once again,
to be made during the programming of the facility and to be checked
during its design.

Specialized laboratories. Depending upon the nature of the work
being done in the facility, the laboratory may have associated with it
specialized, smaller laboratories. A typical example would be a tissue
culture laboratory associated with a general recombinant DNA laboratory. Because the work in tissue culture laboratory requires a higher
level of containment than the general laboratory to minimize external
contamination, the tissue culture lab would be provided as a separate
room accessible from the general lab. The nature, number, size, and
equipping of these specialized laboratories is entirely dependent upon
the facility operators and has to be determined in the facility programming.

Distilled water system. The laboratory will require a source of
high-purity water. Currently, reverse osmosis systems are the most

favored methods for providing purified water. Typically, the reverse
osmosis equipment and an associated storage tank and pumps will be
provided by a manufacturer of the equipment. The sizing of the room
to contain the equipment has to be done in close coordination with the
manufacturer's requirements. The water distribution system also has
to be carefully thought through so that all the spaces and benches to
be served are supplied with it. The size of the system is worked out
by the equipment manufacturer and is based on the laboratory's needs.

Vacuum and compressed air systems. Vacuum and compressed
air can be provided either by centralized systems or by bench-mounted
equipment. Due to the noise generated by vacuum and compressed air
equipment, the equipment for centralized systems is frequently isolated
from the rest of the facility. Centralized systems have the advantage
of allowing the provision of vacuum and compressed air throughout
the facility. They have the disadvantages of requiring an isolated
room to house the equipment and of building into the facility distribution systems and equipment that will require additional maintenance.
The provision of vacuum and compressed air by small bench-mounted
equipment eliminates the problems of the centralized systems, but it
does take up work spaceon the lab benches. In many cases the vacuum
drawn from a cup sink attachment may suffice for the laboratory operations. The choice of vacuum and compressed air systems will have to
rest with the laboratory operators, but it should be made in the programming of the facility so that proper allowances can be made in the
design for the system chosen. Whichever systems are used, it is important to make sure that the systems are protected from contamination
during use. Filters or liquid traps on the inlet side and exhaust filters and traps on the outlet side should be considered.

Bulk chemical and supply storage. Storage space must be provided
in the facility for bulk chemicals and other laboratory supplies. The
amount of storage space needed will have to be assessed by the laboratory operators, who should know the volume of materials they will be
using and the frequency of deliveries they expect. Bulk storage
spaces should have easy access to the receiving area or loading dock
of the facility. The movement of the chemicals and supplies to the
laboratory spaces should also be easy.
    There are two types of bulk storage that need to be given special
consideration--volatile solvents and compressed gas cylinders. The
volatile solvents should be stored in a separate, explosion-free (no
source of ignition) room equipped with a special fire-protection/suppres-
sion system. The solvents themselves should be kept in the room in
ventilated cabinets specifically designed for this type of storage. The
compressed gas cylinders should be stored tied to wall-mounted racks
clear of the movement of carts and other equipment. Depending upon

the nature of the gases stored, the cylinders may also have to be in a separate, explosion-free room.

Refuse rooms. Laboratory waste may require separate refuse storage and handling rooms. If low-level radioactive waste is produced by the laboratory, then it must be stored separately from all other materials and held for pick-up and disposal by licensed disposal services. The radioactive waste room will require special ventilation; its requirements must be established in the basic facility program and its layout checked in the design process. These same considerations apply to toxic chemical wastes and waste solvents as well.

Biologically contaminated waste can be disposed of with the regular facility waste if it is autoclaved prior to being placed in the waste bins. The proper handling and disposal of these materials are discussed in other sections of this monography.

Wastewater treatment systems. The wastewater from the laboratory will generally require some treatment prior to its being placed in a community's sewer system. The exact nature of treatment requirements must be established in consultation with local officials. At a minimum, the laboratory facility should allow space for neutralization chambers for the wastewater. These chambers are usually tanks or sumps containing marble chips which neutralize any acids in the wastewater stream in order to protect the community's sewer pipes from corrosion. The chambers need to be accessible for replacement of the neutralizing medium. They do not have special requirements except that they be properly connected to the laboratory sanitary drainage system and the building drain.

If local officials or special laboratory circumstances require additional treatment of the laboratory wastewater, then specialized consultants, including industrial hygienists and sanitary engineers, need to be involved in the programming and design so that the effects on the facility's layout, services and cost can be assessed properly.

Production facilities. If the laboratory is to contain some sort of production facility, such as fermenters, the architect should be informed at the outset of the design process so that the equipment requirements can be included in the design. Since biological production facilities are at present still relatively small in comparison to most chemical production facilities, it is most likely that the inclusion of the production equipment will not radically affect the facility layout. As a rule, the production facilities are best located at the rear of the laboratory facility where they can take advantage of proximity to receiving and shipping areas in the facility.

There is one special issue concerning production facilities in a laboratory that should be resolved at the very outset of the project:

the effect of the inclusion of the production facility on the building
code and zoning code classifications of the laboratory. Most of these
codes make a critical distinction between laboratories and light industries. If the local officials are wont to categorize a biological production facility as a light industry rather than an adjunct to the laboratory, then the entire project might be jeopardized. At the least, the
facility design will have to meet different building code and zoning
code requirements, which can affect the site and building layout, the
materials of construciton, the building system, and the cost of construction. Given these possibilities, the owner and the architect should
bring the matter to the local officials' attention as part of the project's
feasibility study and have the matter resolved one way or the other
as soon as possible.

Animal facilities. Biological laboratories often use animals as part
of their test procedures. If these animals are to be housed in the facility, special preparations need to be made in the facility design to accommodate the animals. The architect and his consultants need to know
the types, sizes, and numbers of the animals to be housed. They also
need to know the type of experimental and naimal facility equipment
that will be needed in this space. A whole host of questions must be
answered in the programming; for example, Is quarantine required?
How will naimal carcasses be disposed of? How much feed storage is
needed? Will small and large animals be housed? What size cages are
needed? Is a cage washer needed? What staff facilities hould be provided for the animal care workers? Will dissection be done in the laboratory spaces, in the animal quarters, or in a separate space? What
will the traffic between the animal quarters and the laboratory spaces
be? How will the building services to the laboratory and the animal
quarters be zoned? Will the animal quarters need emergency power?
    The design of the animal facilities needs to be responsive to the
overall requirements of the laboratory facility design, and it also needs
to conform to the regulations governing the treatment of laboratory
animals. The architect should include in his consultants a specialist
in animal facility design if provisions for the housing of and experimentation on animals are to be made in the laboratory facility.

C. Laboratory Staff Spaces

Laboratory Staff Offices

The senior staff, including the laboratory director, research director,
senior researchers, etc. should be provided with a defined office space.
The size of the office area will vary in accord with the institution's
standards for its staff, but a minimum workable office with a desk,
chair, side chair, files, and book shelves should be at least 80 square
feet.

The supervised laboratory workers will also need some desk space. This space is usually included within the laboratory benches. At a minimum they will need three feet of desk height space (29 inches from the floor). As an option to the desk space in the laboratory, they can be provided with carrel spaces in a room outside the laboratory.

The critical decision for a laboratory layout is the relationship between the senior staff offices and the laboratory itself. Some researchers like to have an office directly associated with the laboratory. Figure 1-4 shows a modular laboratory with a research office attached to it. Such research offices do not have to be the size of a full office, especially if the researchers have other offices somewhere else in the

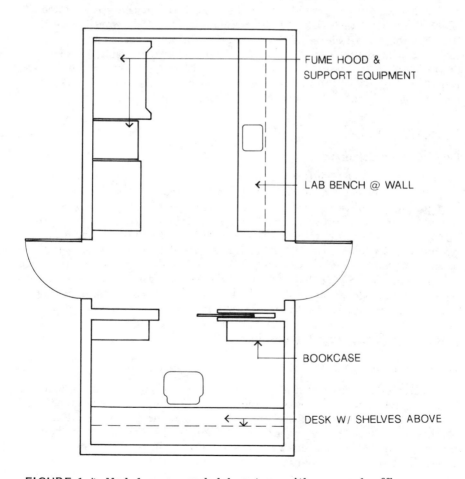

FIGURE 1-4 Modular research laboratory with research office.

facility. On the other hand, some laboratory directors want their offices to be completely separated from the laboratory. As with all the other laboratory planning issues, a decision on the relationship of staff desks and office spaces to the laboratory must be made in the programming of the facility and checked in the design.

Laboratory Staff Support Spaces

The support spaces for laboratory staff, such as conference rooms, lunch rooms, lounges, libraries, locker and shower facilities, etc., are frequently included in the facility plan. The decision of which of these spaces are to be included in the facility is made during programming in accord with the facility's operational needs and the owner's budget. The programmer can establish the sizes of the spaces once their nature and uses are defined by the owner. It is worth remembering that the use of these support functions is not necessarily dedicated to just the laboratory staff. They can serve, equally well, all users of the facility, including administrative staff and visitors. It is important to define who is to have access to the various support spaces in the program, since such decisions will affect the layout of the facility.

D. Facility Spaces

Facility Administration Spaces

In most laboratory facilities there are requirements to house administrative functions not directly related to the laboratory work itself. The nature of these functions will vary from facility to facility. In a laboratory housed within a larger complex, in fact, there may be no need of any additional administrative spaces since they are already provided elsewhere. On the other hand, in a recombinant DNA research laboratory and office facility for a commercial corporation, there may be a need for corporate offices, secretarial areas, mail rooms, copier rooms, reception areas, and visitor's facilities. The exact nature of these sorts of facility requirements hould be established in the facility programming. Their character and design should then be determined in the facility's design.

Facility Support Spaces

The facility as a whole will have support spaces. Such spaces would include toilets, janitor's closets, housekeeping supply storage, refuse rooms, general storage rooms, receiving docks, and mechanical and electrical rooms. These space requirements will be established by the programming and in the design. Obviously, such facility support spaces will serve the laboratory spaces and staff as well.

V. FACILITY LAYOUT

A. Principles of Organization

In every good building design there are two general principles which
are usually operating together to determine the facility's layout. Those
principles are (1) clarity of circulation and (2) maintenance of the
appropriate adjacencies and separations among the facility spaces.
    In its simplest terms, the principle of clarity of circulation means
that whatever has to move into, out of, or through the building should
be able to do it in a simple, unconfusing, and unconfused way without
getting in the way of the movement of others. This principle applies
to people—staff, visitors, vendors, etc.—and to things—deliveries,
supplies, refuse, etc.
    The principle of adjacencies and separations means that uses that
should be near each other are kept that way and those that should
be separated are kept apart.
    In a laboratory facility there is one additional general principle of
organization that comes into play. Because a laboratory facility has
utility service requirements that are significantly greater than those
of normal buildings, the third general principle that applies to laboratories is ganging of the services. This principle means that the water,
wastewater, electricity, gas, and HVAC services should be organized
into groups to service laboratory spaces.

B. Circulation

In general, a laboratory facility needs three distinct entry and exit
points. It needs an entry and exit for people, an entry for supplies
and materials, and an exit for products and refuse. Additional entry
and exit points can be added to the facility if, for some programmatic
reason, there is a desire to separate visitors and staff, or administrative staff and lab staff, or housekeeping supplies and laboratory supplies. Generally, though, such additional entry/exit points are not
necessary to the proper flow of people and materials to and from the
facility.
    Within the facility it is important to have definite entry and exit
points into the laboratory and its support spaces. The laboratory
space should not be used as main circulation corridors to any spaces
other than its own support spaces. For example, one should not have
to enter the laboratory in order for staff or visitors to get to administration areas from the main facility entry.
    Whenever possible the paths of supplies into the facility and refuse
out should be separated from each other and form any main entry/exit
for people. Service entries and exits should be to the side or rear
of the facility, and the supplies and refuse should be able to move
through the facility without having to take circuitous routes.

If a facility has been laid out properly, a diagram showing the
circulation of all major components of the facility's population and all
supplies and refuse can be overlaid on the facility's plans to show that
the circulation patterns are clear, direct, and lacking in conflicts.
In multistory buildings the appropriate separations in circulation may
require the installation of service elevators. Whatever the final configuration of the building or facility, the owner should not accept it
unless its interior circulation and exterior access is very well worked
out.

It should be noted that building codes and fire department regulations require certain features in a building's circulation, for example,
a minimum number of independent paths of egress, a limitation on the
length of a dead-end corridor, or a maximum travel distance to a fire-
safe exitway. All such regulations should be adhered to strictly. Fire
egress, being a provision for an emergency situation, should not be
considered a conflict in circulation.

C. Adjacencies and Separations

The most important separation in a laboratory facility, especially one
that deals with biohazardous materials, is the establishment of a clearly defined laboratory area. The laboratory and its support spaces
should be clearly separated from the rest of the facility. If there are
spaces in the facility that are shared by the laboratory staff and rest
of the facility, for example, toilets, lunchrooms, lounges, or conference rooms, they should be made equally accessible from both parts
of the facility, but they should not be placed within the laboratory
area, nor should the laboratory have to be crossed to get to them from
other areas of the facility.

Adjacencies are less critical. The laboratory and support spaces
should be arranged to meet the laboratory director's view of the best
arrangement to accommodate the flow of work. Laboratory staff support areas should be located so that the staff can get to them readily.
The refuse holding rooms should be adjacent to the refuse pick-up
areas or docks, and the main storage rooms should be near to or easily
accessible from the receiving areas.

D. Services

There are several different ways in which ganged or grouped laboratory services can be distributed in a building. The basic patterns
for such distribution are vertical, horizontal, and interstitial.

In a vertical distribution system, the ganged services rise vertically in service cores and feed laboratory spaces at each level. In
some vertical arrangements, the building is subdivided into modular
components, each served by a separate vertical service core.

In horizontal distribution systems, the ganged services run horizontally at either the perimeter or the center of the building and feed laboratory spaces from the side.

Obviously, a laboratory facility can be designed to be a combination of a vertical and horizontal service distribution system. Ordinarily, the service's cores are designed to meet the clearance requirements for the service pipes, ducts, and conduits and for repair access.

In an interstitial distribution system, the services are provided in spaces above and below the laboratory areas. In fact, an interstitial service arrangement makes a layer cake of laboratory and service levels. In this arrangement the service spaces actually form intermediate floor levels between laboratory floors. The floor to ceiling height in these service spaces is sufficient for workmen to enter the service floor and work on the service pipes, ducts, etc. In a facility where the laboratory arrangements are guaranteed to change frequently, an interstitial service arrangement allows the easiest rearrangement of laboratory spaces. The laboratory floor plan is not tied to a specific set of service entries to the laboratory, and services can be dropped from above or brought up from below in almost any location.

Figure 1-5 illustrates diagrammatically the vertical, horizontal, and interstitial distribution of laboratory services.

E. Change and Expansion

As an additional consideration in the organization of the laboratory facility, the architect needs to be informed of and respond to the owner's expectations about changes in or expansion of the facility.

As part of the programming of the facility, the architect should assess the extent to which the owner expects the facility to change or be expanded during its life. In so far as these eventualities can be determined, they should be accommodated in the initial facility planning and design.

For the purposes of this discussion, "change" will be restricted to mean rearrangements within the confines of the laboratory facility as it already exists in order to distinguish it from "expansion."

A well-designed laboratory facility should be able to accommodate changes in the laboratory operation without radical rearrangements in the facility's layout, since the most flexible aspect of any facility's operations is the behavior of the people using it. There are circumstances, however, where an owner can know that the laboratory's use will change significantly enough to require rearrangement or rebuilding of portions of the facility. Such circumstances include laboratory facilities that are specifically devoted to start-up operations intended to run for a year or so and then to be replaced by another, possibly very different operation. In this sort of circumstance, and in any others where the need for change in the laboratory in the near future

VERTICAL DISTRIBUTON

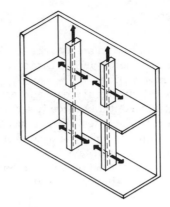

HORIZONTAL DISTRIBUTION

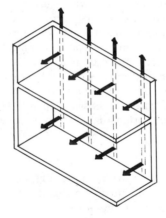

INTERSTITIAL DISTRIBUTION

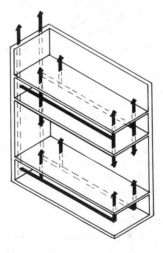

FIGURE 1-5 Service diagrams.

is known before the facility is constructed, it is best to have the facility designed with ganged, easily accessible, and easily modified services. It is also best to have the laboratory spaces clearly separated
from the laboratory staff support spaces and from the general facility
support spaces. This kind of zoning in the design will accommodate
the laboratory changes in a way that minimizes both the costs of the
changes and the disruptions they impose on the rest of the facility.

An expansion to the laboratory imposes different kinds of con-
strains on a design than an internal change. The initial facility design
should, if at all possible, not look like and operate as if it were incomplete without the expansion. There may be considerable time between
completion of the initial laboratory facility and its expansion, and during that time the laboratory should be a complete facility in its own
right. The owner needs to determine whether it is worthwhile including in the initial construction the services needed to accommodate the
expansion. These accommodations can be quite expensive; for example,
the provision of a larger electrical service can involve substantial sums
of money. The owner also needs to be prepared to spend money on
construction features in the initial facility which will be essential to
the expansion; for example, the building's foundation and structure
should be designed for the expected expansion, if it is known before-
hand, since later changes to such basic construction elements can be
prohibitive.

If an expansion of the facility is known about in the initial facility
planning, the owner's best course of action is to have the expanded
facility designed, at least through the schematic phase, as part of the
architect's and his consultants' design services. In this way both
the initial and the expanded facility can be laid out properly, and the
basic decisions affecting the design of the one can inform the design
of the other. The architect and his consultants will be able to design
the laboratory construction on a pahsed basis, which will ensure, within the limits of their design abilities and the owner's knowledge of the
facility's future, that the laboratory will look and operate properly
during all phases of its life. This principle applies equally in cases
of a single expansion or in cases of a series of contemplated expansions.

VI. SAFETY CONSIDERATIONS AND OTHER PRECAUTIONS

A. Types of Hazards and Risks

In a biological laboratory facility there are a fairly wide range of hazards and risks to be dealt with. There is the normal building hazard
of fire, which is made more critical by the storage of volatile chemicals
and compressed gases. There are the hazards of laboratory accidents.
There are the hazards of low-level radioactive materials. There are
the hazards of biologically contaminated refuse. There are the hazards

associated with biological materials. Finally, there are the risks of
damage to the experimental materials and risks of interruptions to fermentation processes.

Some of the hazards are not that difficult to deal with; good housekeeping and laboratory procedures and the proper use of the dishwash-
ing, autoclaving, and refuse disposal equipment will provide safety
in the handling and disposal of the low-level radioactive wastes and
the biologically contaminated materials. The other hazards and risks,
however, require more careful consideration and planning in the facility's design.

B. Fire Safety

Wherever possible, it is good practice to protect any building against
fire by the provision of a complete fire-detection and -suppression
system. Fire detection means smoke and/or heat detectors located
throughout the facility. Fire suppression means automatic sprinklers
or other fire suppression means throughout the facility. In general,
a sprinkler system is the cheapest, easiest to install and maintain,
and the most reliable fire-suppression system available. There are
many varieties of sprinkler systems—dry pipe, wet pipe, preaction—and
the choice among them should be made on the advice of the designer
and his fire-protection engineering consultant.

In some areas of the facility a sprinkler will not be appropriate,
for example in electrical rooms, computer rooms, and chemical storage
rooms. Electrical rooms should be equipped with fire detectors, but
the provision of a suppression system is usually not worthwhile since
a fire in the electrical equipment will not be easily suppressed without
cutting the electrical supply.

In comptuer equipment rooms, if there is one in the facility, a
halogenated fire-suppression system can be provided. These systems
use a gas to interrupt the combustion process. They work in enclsoed
areas; however, they are expensive to install and to recharge. The
halon gas is also itself a hazard if its concentration is too great, consequently the halon fire-suppression system has to be designed to
provide the necessary concentration to suppress the fire without being
a hazard to humans.

In a volatile chemicals storage room, a halon system is the best
fire suppression system available. These rooms are usually small and
unoccupied so that the problems of the system's cost and its potential
hazards are limited. It should be noted that a fire-suppression system
will not prevent an explosion in the volatile storage room. Explosion
protection is provided by storing the chemicals in specially designed
ventilating cabinets which prevent the build-up of explosive vapors.
To prevent damage to the building and injury to humans, pressure
relief panels or vents should be provided as part of the construction

of the volatile chemicals storage room. If an explosion occurs these
panels or vents blow out in a safe direciton, relieving the explosive
pressure and directing them away from protected areas.

The alarms on the fire detection and suppression systems should
sound both locally in the facility, outside it, and to the fire department.

The requirements for the fire-deteciton and -suppression systems
should be thoroughly checked with and reviewed by local fire department and code envorcement officials. Their requirements for the systems should be adhered to strictly and their suggestions listed to seri-
ously. The design of the systems should be done in accord with all
governing codes and standards by engineering specialists in fire protection.

In those cases where an automatic sprinkler system cannot be provided in the facility, for example, where there is an inadequate water
supply or a very constrained budget, fire extinguishers msut be provided. As a general rule the extinguishers should be located wherever
the hazard of fire is significant—in the laboratory, in storage rooms,
in refuse rooms—and in sufficient numbers to provide ready accessibil-
ity throughout the facility. The extinguishers should be visible in
common areas such as corridors. It is good practice to make the extinguishers small enough to be handled easily, not more than 20 pounds
in capacity, and to choose extinguishers that are rated for all classes
of fire, type ABC extinguishers.

Even in facilities that have automatic sprinkler systems, it is a
good safety precaution to provide small extinguishers in the laboratory,
no more than 5 pounds in capacity. These extinguishers will allow
lab workers to suppress accidental fires without having the detection
system and sprinkler system activated. This limits unnecessary alarms
to the fire department and unnecessary wetting of laboratory equipment.

C. Laboratory Accidents

In laboratory accidents there are three types of hazards—those to the
building and its physical contents, those to humans inside and outside
the facility, and those to the materials used in the laboratory. The
hazards to the building and its physical contents stem principally from
the threats of fire and explosion that might result from a laboratory
accident; the means to deal with these threats have already been discussed.

The threats to humans inside the laboratory facility from laboratory accidents are dealt with in three ways—the maintenance of good
laboratory operational practices, the proper design of containment
systems, and the proper selection and placement of safety equipment.
The basic laboratory safety equipment items are deluge showers, eyewash units, fire blankets, and first aid equipment. As part of the
laboratory design careful consideration must be given to the proper
location of all of this safety equipment.

Deluge showers provide a deluge of water to neutralize chemical spills or splashes on lab workers. The showers should be located within the laboratory spaces in conveniently accessible places; an excellent, typical location is at the end of a laboratory bench in an aisle or corridor. Deluge showers are for use in circumstances of panic, therefore a simple pull chain with a large ring is the best device for their activation. There is some disagreement whether deluge showers should be located over a floor drain. Some plumbing inspectors insist on having a drain, in which case it must be provided. Unfortunately, drains may be more problematic than useful. The deluge shower produces so much water that a floor drain directly below it will not prevent water from spreading over the floor. In addition, operation of the shower should be very rare in a well-managed lab, hence the drain trap may dry out unless it is recharged periodically by pouring water down the drain. For these reasons, it is often more reasonable to omit a drain at the deluge shower and to be prepared to mop up the water when, and if, the shower is used. The deluge shower should always be supplied with an alarm which is activated when the shower is used. The alarm should be heard inside and outside the laboratory space to alert any occupants of the building that someone is in trouble in the laboratory.

Eyewash units are small deluges designed to be directed into the eyes to flush out chemicals. There are two basic types of plumbed eyewashes—fixed faucet arrangements and a hand-held wash unit. The fixed faucet types require the lab worker to hit a panic button or lever and place his face between the water streams. The hand-held units are a hose with a nozzle and a hand-activated lever. The hand-held units have the advantage of allowing direction of the water spray to locations other than the face and eyes. On the other hand, the fixed types are very effective at deluging the eyes and face without much coordinated movement by the affected lab worker. The choice of type of plumbed eyewash should be made by the laboratory manager based upon his own experience with laboratory accidents and his own preference. As in the case of deluge showers, the eyewash units should be located within the laboratory in very convenient and accessible places.

The eyewash and deluge shower can be combined into a single safety cabinet unit. Such units are available from laboratory bench and equipment manufacturers. The advantage of buying a combined safety cabinet unit is that it can be coordinated with the other laboratory benches in the lab layout. The disadvantage of such a unit is that it takes up lab bench or equipment space that separated deluge showers and eyewashes do not.

Eyewashes can also be provided in the laboratory without the need for plumbing them. In that case, they are bought as a piece of laboratory safety equipment, like a first-aid kit. These sorts of eyewashes are usually some sort of bottle or canister. Obviously, both the plumbed and bottle/canister eyewashes can be provided in a laboratory for additional worker safety.

A fire blanket is a fire-resistant blanket used to smother a fire on a worker's clothing. The blankets come in wall-mounted cases, either folded or rolled. The rolled type of blanket is designed so that a person can grab the blanket and wrap themselves in it in a simple motion. The folded type requires a more coordinated action to cover the person. The advantage of the folded type is that it can be carried to a fire more easily. This advantage is slight, and the rolled type is generally preferable in the laboratory.

The laboratory should also include first-aid equipment. Such equipment can range from simple first-aid kits to very well-equipped crash carts. The choice of such equipment should be made by the laboratory manager based upon his assessment of the type of accidents to be dealt with and the level of training of the laboratory workers in first aid. The owner's insurance carrier should also be consulted before elaborate first-aid equipment is included in the laboratory since untrained or improper use of such equipment can lead to the owner's exposure to unwanted liability.

Laboratory accidents involving the spread of biologically hazardous materials are discussed in the next section.

D. Containment Considerations

The containment of biologically hazardous materials concerns the prevention of spread of the materials both within the laboratory and from the laboratory to the outside environs. The principal design features of the laboratory facility that affect containment are the plumbing and HVAC systems. The proper design of these systems is dealt with in other sections of this chapter. It is the architect's responsibility to make certain that the facility's design and construction is done in a manner that is entirely consistent with the operational requirements of the plumbing and HVAC containment measures. The nature and extent of the containment measures should be established by a specialist in biohazard containment; this specialist may be either a consultant to the architect or an independent consultant to the owner. The communication between the architect and safety consultant should be very direct and full. The architect needs to be informed about the types of separation that must be maintained between spaces so that the doors, partitions, ceilings, floors and penetrations in them are properly designed, specified, and constructed to meet the consultant's containment requirements. Similarly, the architect's plumbing and HVAC consultants need to be directed by the safety consultant as to the specific features that must be incorporated in their designs to meet the containment requirements.

The containment requirements will direct the architect on matters such as the inclusion of operable sash in windows in laboratories (if there are any windows), the provision of air-lock vestibules as room entries or as separations between rooms, the specification of hardware on doors such as automatic closers and air-seals, the choice of finishes for floors, walls, and ceilings, and the nature and extent of the caulking systems used in the facility's interior.

These architectural design and detail issues deal generally with two aspects of biohazard containment—maintaining the proper air-movement patterns in the facility and facilitating the proper housekeeping and maintenance of the laboratory. The maintenance of the proper air movements affects the architectural work in that the design must provide the necessary separations of space and avoid the introduction of short circuiting air-movement pathways. These sorts of provisions are not difficult to achieve if the architect is properly advised of the facility's requirements. The appropriate vestibules, air-locks, door hardware, and caulking will be included in the design and construction contract documents and the architect will be on the lookout for their inclusion in the contractor's work.

The choice of finish materials is also not a difficult matter if the architect is properly informed. In the lower levels of containment, the floor and wall finishes need only to be nonabsorptive and easily cleaned so that in the event of a laboratory spill or accident the biologically hazardous materials can be removed with simple clean-up procedures. Walls can be made of gypsum board so long as they are caulked at the bottom, have a vinyl or rubber tile base, and are painted properly; they can also be tiled or receive special plastic seamless or sheet finishes if the laboratory budget allows it and the owner so prefers. If the walls are made of concrete block or other absorptive or creviced materials, they should either be finished with an impervious paint system or covered with another finish material, such as tile or painted gypsum board. The laboratory floors can be painted concrete, vinyl, or rubber tiles, PVC or other seamless sheet goods, ceramic or quarry tiles, terrazzo, or other seamless toppings. As with the walls, the essential requirements are that the floor finish not be absorptive and that it be easily cleaned. If a tile system is used rather than a seamless one, then the tile joints need to be sealed; liquid acrylic floor finishes are usually very effective with almost all these materials.

The exposed ceilings in the laboratory space can be lay-in acoustic tiles in the lowest level of containment so long as the underside of the floor or roof above is sealed and penetrations caulked. The lay-in tiles are an inexpensive way to provide acoustic treatment in the laboratory and allow for an inexpensive installation of the lighting and other laboratory systems. In laboratory spaces that require a higher level of containment, the ceiling exposed in the laboratory space itself can be painted gypsum board or some other seamless, nonabsorptive material.

In the highest levels of containment, the biohazard containment consultant should direct the architect on the selection of finishes so that the finished spaces meet all the regulatory requirements for the laboratory's licensing and operation. Since the containment measures at these highest levels often require that the laboratory work be conducted within specialized containment equipment, the architect and his consultants must also be fully informed on the space needs and service requirements of this equipment.

Since the containment measures will likely be very dependent upon the proper operation of the facility's HVAC system, the facility should be equipped with sensors and alarms to warn when the system is not functioning properly. It should also be designed so that it provides the proper containment when the HVAC system is malfunctioning or not operating; for example, in the event of a power failure, automatic dampers should shut to prevent the improper venting of laboratory air. The owner should also make certain that the contract for construction of the facility includes professional balancing and testing of the facility's HVAC system prior to his acceptance of the completed facility.

E. Risks to Biological Materials and Products

Since most of the biological processes, materials, and products are temperature-sensitive, they can be damaged or lost if the facility's HVAC or electrical systems fail. The principal threat to the facility that can lead to such losses is a power failure. The extent to which a power failure will compromise the facility's operation must be decided by the owner. In some circumstances the threat is not great; for example, in a small research laboratory without production capacity, refrigeration equipment or cryostats may be able to maintain critical materials for long enough to be reasonably assured that the power failure can be remedied. On the other hand, the threat may be critical; for example, in a production facility the processes may have to be maintained so rigidly that even short power interruptions can compromise the operation.

The answer to such threats to a facility is the provision of some means of emergency power. The electrical consultant for the facility's design should assess the alternate means that might be employed to provide such emergency power after the owner has decided what services and equipment should be maintained in the event of a power failure. These decisions should be weighed thoroughly since emergency power systems can be very expensive and their inclusion in the facility will affect the design.

VII. AMENITY AND IMAGE

A. The Architect's Role

The architect for a laboratory facility has a basic responsibility to ensure that the facility meets the owner's needs for the operation of the
facility, that is, that it accommodates his operations and staff in a ma-
terially and structurally sound building which provides the necessary
mechanical and electrical services and which is safe and conforms to
all the regulations and codes which apply to it. Beyond this basic
design responsibility, the architect also has another set of responsibilities which deal more with the pleasures and art of architecture.
The laboratory facility should, if the architect is capable, be a pleasure
to its owner and its users and be an object of some interest and en-
lightenment to the public at large. The ways in which these other
aims can be met in the design is the province of the architect's art.
The following secitons deal briefly with a few of the domains within
this province.

B. Light and Views

As a general rule, people prefer to live and work in spaces that are
well lit by daylight and offer interesting or diverting exterior views.
In a laboratory facility, the provision of daylight can be prohibited
in some spaces by the detrimental effects of sunlight on sensitive biological materials. In these circumstances, the architect should consider
the provision of daylight in all those areas where it can be tolerated,
either with windows, clerestories, or skylights. In those cases where
exterior views are limited by the facility's site or its location within
another complex, the architect should consider the use of courtyards
or atria to provide visual relief.

C. Exposition

It is possible in the facility's design to have the building expose its
use and operation to the public at large. The simplest, most·direct
method to accomplish this is to build visitor's amenities into the facility.
Where it is appropriate to the laboratory's functions, the laboratory
work can be displayed to visitors. This can be done with glass walls,
a tour path, or even a separate visitor's pavilion in larger laboratory
complexes. The inclusion of such features in a laboratory facility
needs to be carefully assessed by the owner, especially during the
facility programming and schematic design. In some cases, of course,
exposition of this sort will be directly contrary to the owner's intentions for the facility. In that circumstance, for example, in the case
of proprietary product development facilities, the architect may choose
to express the building's functions and uses without exposing them.

D. Expression

Expression is the most abstract architectural method. It encompasses any and all of the ways in which the architect uses the building's shape, materials, spaces, decorations, or other visible features to explain the building itself; or to reveal one or more of the aspects of its operation, use, or construction; or to explore some independent or larger theme.

A few rudimentary examples of architectural expression will help to explain the idea. In Gothic cathedrals the vaults are often constructed with ribs visible from the sanctuary below; these ribs are a visible expression of the ways in which the vaults were constructed and of the directions that the compressive forces in the vaults follow to the vault supports. In many skyscrapers of the early twenthieth century the window and exterior wall-surface patterns are arranged to enhance the building's vertical appearance. In some airport terminals, the building's roof is made to appear as a wing or airfoil section.

The direction that the architectural expression takes is dependent upon the architect's imagination and interest. Some architects even eschew expression. Wherever possible, though, an attempt at expression should be made, since it is the means through which construction becomes art.

There are some obvious opportunities for expression in the design of biological laboratories. Since the basic additional principle of organization for laboratories is the ganging of services, the expression of the relationship of the services to the whole building can be exploited as a theme in the laboratory design. This was done by Louis Kahn, a later master of modern architecture, in his designs for the Richards Medical Laboratory in Philadelphia and for the Salk Institute Laboratories in La Jolla. Kahn's design device was to make a visible organized difference between the spaces that provided services and those spaces (the laboratories) that were being served.

The double helix of the DNA molecule also offers itself as an expressive motif in laboratory design. Viewed two-dimensionally, the DNA molecule flattens into a diaper pattern, which can be used as part of a decorative brickwork scheme for a laboratory building. This, in fact, was recently done by the contemporary architect Robert Venturi in his design for the Lewis Thomas Laboratory at Princeton.

The owner should understand that architectural expression does not require the expenditure of extra money to be included in the facility's design. It is often part and parcel of the facility's organization and fabric and not a superfluous addition which can be stripped away. One of the reasons the owner is hiring an architect for the facility's design is to derive the benefits of his artistic abilities as well as his planning and technical knowledge.

VIII. CONCLUSION

Based on all of the previous information, the following points should be obvious.

There are a few simple principles that should be mastered and adhered to in the design of a laboratory facility; these are:

1. The facility should be programmed very carefully.
2. The facility's proposed site should be investigated extremely thoroughly for its suitability to the project.
3. A competent architect should be engaged to lead the project design team.
4. There should be close and complete communication between the architect and the owner during all phases of the facility's design and construction.
5. Where special expertise is required, such as in cost estimating, biohazard safety assessment, or animal facility design, the architect should recommend and the owner should insist on the engagement of experienced, specialized consultants.

2

Ventilation for Biomedical Research, Biotechnology, and Diagnostic Facilities

MELVIN W. FIRST   *Harvard School of Public Health, Boston, Massachusetts*

I. INTRODUCTION

Until recently, biological controls have been the principal method of
hazard management for microbiological agents. Although biological
controls alone have been reasonably successful for laboratory and small-
batch industrial operations, many failures have been documented. By
1976, the recorded number of laboratory-associated infections had
reached a total of 3921 (1), with fewer than 20% associated with a known
accident (2). With the increasing number and type of microbiological
agents in use and the steady increase in batch size, prudence dictates
that all reasonable safety measures be employed. This means that engineering controls should *not* be considered a substitute for biological
controls, or vice versa. In fact, they should be used to supplement
each other, thereby providing the extra factor of safety that occupational safety and health practitioners strive for when they are uncer-
tain about the safety of a process or a product. At the very least, it
will always be necessary to prepare agents whose effects are equivocal,
solely because of limited information about their toxic and infective
nature.

II. VENTILATION AIR

A. Ventilation as an Engineering Control Method

Ventilation is a well-established and widely used engineering method
of hazard control in the chemical, metallurgical, and mineral industries.

It is recommended for engineering control of biohazards also, because
of its (1) easy adaptability to most situations involving exposure to
biological agents, (2) resistance to deterioration and sudden failure,
and (3) avoidance of interference with normal operations. Ventilation
for microbiological control can be divided into two distinct categories:
(1) *general ventilation* to satisfy legally mandated air exchange rates,
to provide temperature and humidity control, and, from a safety standpoint, to maintain appropriate pressure differentials in rooms where
hazardous materials are being used as a means of preventing airborne
contamination from escaping, and (2) *local exhaust ventilation* to provide controlled air motion to trap evolving aerosols and draw them away
from the breathing zone of workers. Local exhaust ventilation facilities
include biological safety cabinets in which to conduct operations that
require work, operator, and environmental protection. Other frequently used local exhaust ventilation facilities include chemical fume hoods,
flexible exhaust hoses (sometimes called elephant trunks), and canopy
hoods. Because energy conservation is an important requirement in
addition to effective hazard control, the ventilation engineer must cope
successfully with both requirements simultaneously.

B. Supply Air for General Room Ventilation

Building codes and public health regulations mandate minimum air exchange rates for workplaces. They are designed to assure an adequate
supply of healthful breathing air for the occupants. The air exchange
rate requirements are variable, but usually modest for workplaces that
do not contain processes that generate airborne contaminants. Air exchange rates are customarily expressed as a certain number of complete
air changes per hour, a number of cubic feet per minute per square
foot of floor area, or a number of cubic feet of fresh air per minute
per occupant. A typical requirement for an office is 5 cubic feet per
minute (cfm) of outside air for each person who normally would occupy
the space with a total of 15 cfm circulated per occupant to all portions
of the building during the time it is occupied (3). In crowded areas
where smoking is permitted, the minimum required outside air exchange
rates are unlikely to please non-smokers and outside air rates of 30
cfm per active smoker will be needed to maintain a satisfactory environment (4).

C. Supply Air Requirements for Heating,
Cooling, and Humidifying

When building heating and cooling is accomplished with ventilation air
systems, much larger amounts of circulating air than are provided to
meet the legal minimum outside air exchange rates are needed to satisfy
the heating and cooling loads imposed by external weather conditions

plus internal machinery, lighting, and process vessel heat emissions.
This is due to the narrow human comfort limits on the maximum (for
heating) and minimum (for cooling) air temperatures that can be introduced into occupied spaces. Therefore, it is necessary to circulate
an adequate air volume to cope with the heating and cooling loads imposed on the building without exceeding human comfort limitations.
This means that larger volumes of circulating air are needed to heat a
building located in a cold climate than an identical building sited in a
warmer climate, whereas the reverse is true for hot weather cooling.
Building heating and cooling loads are calculated on the basis of heat
losses (cold weather) or gains (hot weather) by solar radiation and
heat transmission through the building structure, plus the tempering
(heating or cooling) of outside ventilation air and makeup air for all
exhaust systems. In offices and similar workplaces, heat transmission
through the building structure by conduction and radiation represents
the major source of heat gains or losses, but for laboratories as well
as biotechnology pilot plant and manufacturing facilities, tempering,
cleaning, and otherwise conditioning outside air to replace that lost
through exhaust systems, e.g., hoods, will represent, by far, the
major requirement. For example, a standard 4-foot laboratory hood
discharges approximately 1000 cfm of conditioned air to the out-of-doors
when operated at an average face velocity of 100 fpm. When operated
continuously, it costs approximately $3600 per year for air heating
and cooling at the 1981 cost of electricity in the Boston area (5). Although it is possible to increase the amount of tempered outside air
brought into a building to satisfy all heating and cooling requirement,
it is energy conservative in situations where no airborne contaminants
are released to recirculate a major fraction of the building air through
an air conditioning unit and return it to the occupied spaces. Thus,
the volume of building air being discharged on each cycle is equal to
the legal minimum outside air quantity drawn in as a replacement. Although restricting the inflow of outside air to the minimum legally permitted is energy conservative, it often leads to complaints by occupants
who sense a less than invigorating environment.

D. Air Conditioning

Treatment of outside and recycled supply air usually includes some or
all of these steps: heating, cooling, humidifying, dehumidifying, and
air purification by filtration (to remove aerosol particles) and activated
charcoal adsorption (to remove malodorous or otherwise objectionable
gases and vapors). In old buildings, heating by radiators located in
each space may be the only air-treatment step in addition to introducing
the minimum amount of outside air required by local building codes.
In modern, tightly sealed buildings, all the air-treatment steps are
likely to be utilized at appropriate seasons, and all the operations may

be conducted in a central system with the cleaned, conditioned air
being piped to each location requiring ventilation. When temperature
regulation of individual workrooms is desirable (as it often is), it is
usually accomplished by simultaneously conducting a cold air and a
warm air supply through ducts to a mixing box located at the room
supply air inlet and there proportioning the two airstreams in response
to a thermostat located in the workroom. A second way of accomplish-
ing individual workroom temperature regulation is to pipe all of the
cleaned, conditioned air needed for ventilation purposes from a central
system and to provide individual room temperature modulation by means
of total recirculation units provided with heating and cooling coils that
are located within the space itself. The cooling coils of room total recirculation units often perform an important dehumidification function
during hot, humid weather and must be provided with suitable drains.
Close control of relative humidity on a year-round basis is the most
technically demanding and energy-consuming air-treatment step as it
often requires air reheating after first cooling to a very low temperature to condense excess moisture.

E. General Room Ventilation for Contaminant Control

When airborne contaminants are generated from multiple sources that
are more or less uniformly distributed throughout the workplace, there
will be a need to introduce quantities of outside air that are much
greater than the minimum prescribed by law to replace contaminated
air that must be evacuated to maintain a safe indoor environment. A
high number of air changes per hour (20-40) can be effective for reducing airborne contamination by rapid dilution and evacuation of hazardous gases and aerosols. Dilution ventilation is especially effective
for reducing airborne contamination levels when all the air is introduced
at low velocity over a large area of perforated ceiling and removed at
floor level, producing a unidirectional flow of uncontaminated air that
passes over the workers' breathing zone before encountering the sources
of contamination. Ideally, airborne contaminants exit through floor
vents without exposing workers. This airflow pattern reflects the
cleanroom concept of controlling contamination generated inside the
workplace. It is effective but requires very high air rates that are
costly unless most of the air can be purified and recycled. This principle is employed for the protection of transplant and immunosuppressed
patients because it makes it possible to completely envelope them in a
steady flow of air containing less than an infective dose. The same
air is usually recirculated through high efficiency particulate air
(HEPA) filters capable of removing all viable particles. For less demanding applications, multiple ceiling diffuses are used to introduce
supply air into a workplace in a draft-free manner. Comfort ventilation air (heating, cooling, etc) may be supplied by one system and

makeup air for health and safety exhaust systems by another, or the two supply systems may be combined.

III. EXHAUST AIR SYSTEMS

A. General

Supply air is exhausted from the workplace through three pathways: (1) To the atmosphere with the aid of a mechanical exhaust blower. The minimum volume of air exhausted by this means is intended to satisfy regulatory requirements for fresh air circulation. It also includes all air from exhaust systems that may not be recirculated for health and safety reasons. (2) To the intake of a supply air heating, ventilating, and air conditioning (HVAC) system for recycling to the workplace to conserve energy. This fraction of recirculated air is referred to as return air and must be purified before reuse, especially when there is a possibility that hazardous airborne contaminants may be generated in the workplace. (3) To the atmosphere and to adjacent interior spaces through open doors, windows, and gaps in the building structure, including interior partitions. The latter losses are caused by air pressure differentials. Infiltration is also possible but exfiltration is the more usual situation in a controlled-air building. The air supplied must equal the sum of exhaust air, return air and exfiltration. If there is a deficiency of supply air, uncontrolled infiltration will make up the difference.

B. Local Exhaust Ventilation

Substantial air economy can be obtained relative to the quantities required to control airborne contaminants with increased room ventilation rates by the judicious use of process enclosure and local exhaust ventilation. When a process or device can be mechanized and automated, it can be totally enclosed and provided with only enough exhaust airflow (by the use of mechanical exhaust blowers) to ensure that all air leakage through the enclosure will be inward, thereby avoiding worker exposure. Even when a worker-machine interface must be maintained, large benefits in worker safety and economy of operations can be obtained by the use of ventilated partial enclosures relative to the alternative of open bench work and reliance on enhanced general workroom ventilation. Ventilated partial enclosures are referred to as local exhaust hoods. Even when it is not possible to place a contaminant-producing operation within a partial enclosure, safety and economy can be obtained by locating it in front of an exhaust air opening that draws in a sufficient volume to product an inward air flow adequate to capture contaminants at the point they become airborne and convey them without loss into the opening of the exhaust hood; referred to as

an exterior capture hood. Open-end flexible exhaust hoses work on
this principle. Local hoods that do not enclose must exhaust larger
air volumes than those that do, and they are more susceptible to disruption by cross drafts. Guides to the design and operation of local
exhaust systems are contained in *Industrial Ventilation—A Manual of
Recommended Practice* (6). The guiding principles of local exhaust
ventilation are (1) the more complete the enclosure, the greater the
economy of operation, and (2) there exists a minimum inward hood
face velocity (control velocity) that prevents the escape of contaminants
while providing an adequate margin of safety. Control velocities range
from 100-2000 feet per minute through the open face, depending on the
amount of air turbulence created by the operations conducted inside
and outside the local exhaust hood. The lower value is appropriate for
a laboratory fume hood, whereas the upper value is appropriate for
creating a control velocity for a high-speed centrifugal machine. The
face velocity cited for fume hoods is based on average conditions, but
disturbing air currents can originate from high-velocity ceiling air
supply diffusers, rapid foot traffic past the hood face, rapid opening
and closing of doors, proximity of open windows, and intense heat
sources inside or outside the hood enclosure. Whereas mild disturbances of this nature can be overcome by increasing the face velocity of
laboratory fume hoods to 125 feet per minute, caution is needed because
face velocities higher than 125 feet per minute create their own turbulence effects around the edges of the open face and defeat the purposes
of the hood. As a rule, the correct remedy is to eliminate disturbances
to smooth inflow patterns by locating laboratory fume hoods away from
access doors, open windows, traffic lanes, and air supply grilles and
diffusers. Although comfort air exhaust systems for an entire building,
or major parts of very large structures, are usually combined into
central systems, the exhaust systems for hoods and all other health
and safety equipment should be vented through individual, dedicated
exhaust ducts and fans. The principal reason for the separation is
safety, to avoid backflows of contaminated air into unprotected areas
that may be caused by system malfunctions. A second reason is related
to engineering considerations—the air volumes and pressure requirements of comfort and safety systems are often incompatible.

C. Sterile Work Environments

A conventional laboratory fume hood is adequate to protect the operator
working with hazardous biological materials, but many biological operations must be conducted in a controlled environment to avoid contamination by airborne organisms. When the biological materials being processed are innocuous, contamination can be avoided by conducting the
operations in a clean bench, also called a laminar flow clean work station. It is a partially enclosed work area continuously enveloped in a

smooth, unidirectional flow of filtered air from above or from the rear of the work surface as shown in Figure 2-1. When correctly designed, installed, and operated, clean benches protect the work from microbiological contamination in a convenient and economical manner.

When, however, the work involves microorganisms or their metabolic products that may be hazardous to man, or when the microbiological agents are treated with toxic chemicals, worker protection as well as work protection is required. A number of attempts have been made to provide the dual requirements economically by a laboratory hood known as a balanced laminar airflow cabinet, shown in Figure 2-2. The feature that distinguishes it from a horizontal laminar flow clean work station is a slotted air intake grille located on the work surface at the front of the work opening and extending the entire width of the opening. In theory, all the filtered air directed toward the work opening is captured by the front intake grille and recirculated, thereby maintaining the open front as a neutral airflow zone and avoiding discharge of hood air toward the operator or taking unfiltered air into the hood from the laboratory. In practice, 100% of the air discharged into the interior of the cabinet through the filters is not re-entrained into the front grille, and the remainder is discharged into the operator's breathing zone through the front access opening of the

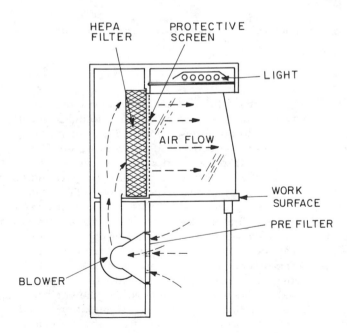

FIGURE 2-1 Component diagram of a laminar airflow clean workbench.

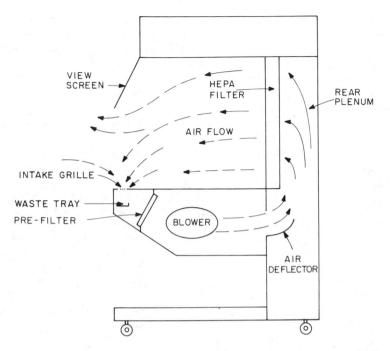

FIGURE 2-2 Component diagram of a balanced laminar airflow work-bench (7).

cabinet. In fact, the design concept for the operation of a balanced laminar air flow hood is unsound in theory and has been found severely wanting in practice with respect to operatory safety (7). Escape of materials from the cabinet can only be prevented by drawing air from the workroom, *past the operator*, into the hood or bench. The vector of the air at the open face must be inward and, preferably, perpendicular to the face to provide satisfactory operator protection. This means that a bench that promises to provide operator and work protection simultaneously must have two distinct airflow patterns: inflow through the open face to protect the operator and a downflow of clean air inside the enclosure to protect the work area from outside contamination. To ensure operator and work safety, as well as environmental safety when working with hazardous biological agents or products, biological safety cabinets must be used. The revised standard for the certification of biological safety cabinets, National Sanitation Foundation Standard No. 49 (8), contains the following definitions of the several classes of this type of equipment that are commercially available:

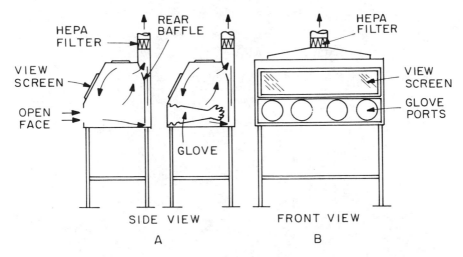

FIGURE 2-3 Component diagram of Class I biological safety cabinets (2): (A) partial containment open-faced model; (B) total containment gloved model.

Class I. A ventilated cabinet for personnel and environmental protection with an unrecirculated inward airflow away from the operator. The cabinet exhaust air is treated to protect the environment before it is discharged to the outside atmosphere. (See Figure 2-3)

The air passing over the work surface is untreated room air and does not provide a sterile environment for the work:

Class II. A ventilated cabinet for personnel, product, and environmental protection having an open front with inward airflow for personnel protection, HEPA filtered laminar airflow for product protection, and HEPA filtered exhaust air for environmental protection. (See Figure 2-4).

Because Class II cabinets have an open front, they provide only partial containment.

Class III. A totally enclosed, ventilated cabinet of gas-tight construction. Operations in the cabinet are conducted through attached rubber gloves. The cabinet is maintained under negative air pressure of at least 0.5 inches water gauge (in. wg.). Supply air is drawn into the cabinet through HEPA filters. The exhaust air is treated by double HEPA filtration, or by HEPA filtration and incineration. (see Figure 2-5)

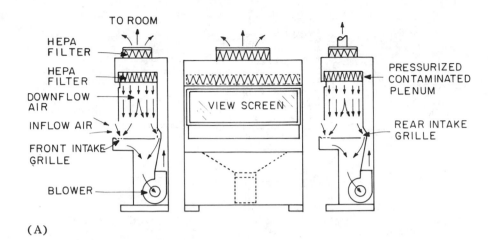

(A)

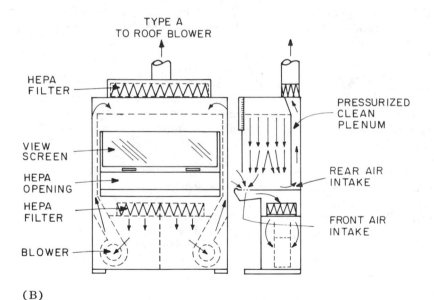

(B)

FIGURE 2–4 Component diagrams of four Class II biological Safety cabinets: (A) Type A (2); (B) Type B1 (2); (C) Type B2 (courtesy Germfree, Miami, FL); (D) Type B3 (courtesy Labconco Corp., Kansas City, MO).

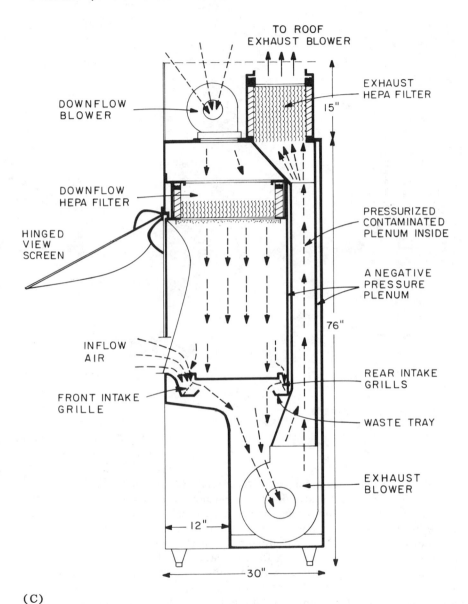

(C)

FIGURE 2-4 (continued)

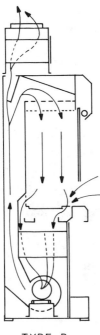

TYPE B$_3$

(D)

FIGURE 2-4 (continued)

It will be clear form the definitions that Class I biological safety
cabinets are essentially standard laboratory fume hoods with exhaust
air purification to avoid atmospheric contamination.   Class III cabinets
are total encapsulation devices that are appropriate for the most dangerous microorganisms.   Class II biological safety cabinets are the work-
horse biological safety devices for laboratory and small pilot-scale operations, they provide product, worker, and environmental protection
at thousands of locations in the United States.   There is virtually no
market for Class II biological safety cabinets that do not qualify for
certification by the National Sanitation Foundation[*] and all cabinets
are supposed to be field recertified upon installation and at intervals
not less than annually, thereafter.

---

[*]Ann Arbor, MI 48106

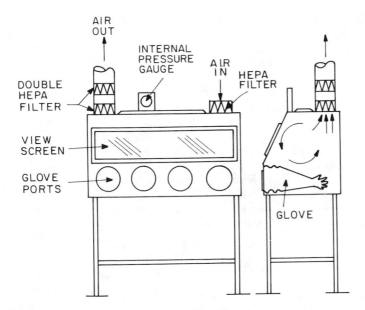

FIGURE 2-5 Component diagram of a Class III biological safety cabinet (2).

Class II biological safety cabinets are available in several different types to satisfy a wide variety of applications. They are defined in NSF Standard No. 49 as follows (8):

Type A cabinets

(1) maintain a minimum calculated average inflow velocity of 75 feet per minute through the work area access opening (2) have HEPA filtered downflow air from a common plenum (i.e., plenum from which a portion of the air is exhausted from the cabinet and the remainder supplied to the work area (3) may exhaust HEPA filtered air back into laboratory and (4) may have positive pressure contaminated ducts and plenums. (see Figure 2-4a)

Type B1 cabinets

(1) maintain a minimum (calculated or measured) average inflow velocity of 100 feet per minute through the work area access opening (2) have HEPA filtered downflow air composed largely of uncontaminated recirculated inflow air (3) exhaust most of the contaminated downflow air through a dedicated duct exhausted to the atmosphere after passing through a HEPA filter and (4) have all biologically contaminated ducts and plenums under negative pres-

sure, or surrounded by negative pressure ducts and plenums.
(see Figure 2-4b)

Type B2 cabinets (sometimes referred to as "total exhaust cabinets")

(1) maintain a minimum (calculated or measured) average inflow
velocity of 100 feet per minute through the work area access opening (2) have HEPA filtered downflow air drawn from the laboratory
or the outside air (i.e., downflow air is not recirculated from the
cabinet exhaust air) (3) exhaust all inflow and downflow air to
the atmosphere after filtration through a HEPA filter without recirculation in the cabinet or return to the laboratory room air and
(4) have all contaminated ducts and plenums under negative pressure, or surrounded by directly exhausted (nonrecirculated
through the work area) negative pressure ducts and plenums.
(see Figure 2-4c)

Type B3 cabinets (sometimes referred to as "convertible cabinets")

(1) maintain a minimum (calculated or measured) average inflow
velocity of 100 feet per minute through the work access opening
(2) have HEPA filtered downflow air that is a portion of the mixed
downflow and inflow air from a common exhaust plenum (3) discharge all exhaust air to the outdoor atmosphere after HEPA filtration and (4) have all biologically contaminated ducts and plenums
under negative pressure, or surrounded by negative pressure
ducts and plenums. (see Figure 2-4d)

All of the air discharged from each of the Type B cabinets goes
to the outdoors rather than back into the workroom, as may be the
case with Type A cabinets. Therefore, Type B cabinets are safe for
biological use with small amounts of toxic volatile products, whereas
Type A cabinets are safe only for biological agents and nonvolatile
particulate products of biological or chemical origin. Although Type
A cabinets, that discharge inflow air directly back into the workroom
after thorough filtration, are maximally conservative of energy, Type
B cabinets are favored for safety reasons, and Type B2 total exhaust
cabinets are favored for maximal protection of the work as well as the
worker.

IV. VENTILATION SYSTEM CONTROLS

A. Fixed Flow vs. Modulated Flow Systems

When buildings have been made as airtight as possible for energy conservation purposes, unbalances between the volumes of mechanically

supplied and exhausted air cannot be corrected by air infiltration
through window casings, building joints, and single door entries.
When the unbalanced condition results in a marked deficiency of supply
air, the exhaust air volume may be reduced to such a degree that a
serious reduction in worker safety results. Therefore, it is essential
that industrial ventilation systems be designed, installed, and operated
so as to maintain at all times a predetermined balance between supply
and exhaust air volumes. This may be accomplished in either of two
ways: one is to use a fixed volume system in which supply and exhaust
airflow volumes are balanced and maintained in this invariant condition
regardless of work load or outside temperature variation. Continuous
operation of all exhaust air systems is an absolute requirement for this
ventilation mode. Comfort conditions are maintained by modulating
temperature and humidity of the supply air. This type of air system
is simple to design, requires a minimum of sensors and controllers, and
is likely to require a minimum of maintenance. It is, however, very
wasteful of air (and the heating, cooling, and humidity control energy
associated with the wasted air) whenever operations are less than highly routinized and continuous. The second way to maintain a balanced
air system is to modulate the supply volume to exactly match the exhaust air volume requirements throughout the daily and weekly work
cycle. This type of system requires a fairly complex control system to
satisfy heating, ventilating, and air conditioning requirements simultaneously with exhaust air requirements, but it eliminates energy losses
associated with exhausting excess supply air that has been conditioned
for internal use. Modulating systems become even more complex when
it is necessary to maintain a constant differential pressure between
adjacent rooms for safety reasons. In spite of their greater complexity
and initial cost, modulating systems are usually preferred because
energy savings over the life of the system are substantial. An exception is when the severity of the hazard is sufficiently great that safety
considerations dominate and more than offset the attraction of energy
savings. Under the imperative of assuring safety, simpler systems
are usually preferred because they operate with greater reliability.

B. Pressure Regulation

Biological safety requires a careful balance between supply and exhaust air volumes. Requirements will vary depending on the hazard
rating of the biological agents or products being used, the quantity
being handled, and the nature of the operations. Frequently, odor
control from such facilities as animal rooms will dictate the use of pressure-controlled zones even when the hazard potential is nonexistant.
Containment of airborne contamination is accomplished by establishing
pressure boundaries that prevent airflow out of potentially contaminated
areas. This means that highly contaminated areas must be maintained

at a lower pressure than less contaminated areas and that public or uncontrolled areas must be maintained at the highest pressure. The pressure differentials between the three zones—maximally contaminated, less contaminated, and public access areas—must not be large for practical reasons. For example, a modest pressure differential of 0.5 in. wg. against a door measuring 7 × 3.5 ft equals a total pressure of 65 pounds that must be overcome to close it against inrushing air, or to open it in the opposite direction. A pressure differential of 0.5 in. wg. on an 8 × 12 ft partition is equivalent to a distributed load of 255 pounds, and when a door in such a partition is wrenched open, the sudden pressure pulse puts a severe momentary dynamic load on the wall that it may not have been built to withstand. Pressure differentials of this magnitude can, and do, move interior room partitions.

For reasons of structural integrity as well as to reduce infiltration through cracks in doors, windows, partition walls, and around utility penetrations, differentials across pressure boundaries seldom exceed 0.05 in. wg., reducing the pressure loading on the door cited earlier from 65 to 6.5 pounds, a much more manageable value. Although values as low as 0.02 in. wg. have been employed, they are not recommended because small pressure differentials make it difficult to maintain the integrity of the pressure boundary during periods when people and equipment are entering or leaving. Therefore, it is necessary to introduce at each portal a pressure-controlled anteroom, a double door entry or lock, that is at a pressure 0.05 in. wg. above the pressure inside the potentially contaminated area. When the interconnecting door is opened with this arrangement, air will flow from the higher pressure anteroom into the lower pressure contaminated zone. Because there is still a small chance the anteroom can become partially contaminated when the interconnecting door is open to the contaminated area, it is necessary to maintain the anteroom at a lesser pressure than all uncontrolled areas adjacent to it, so that when the door between them is opened, air will flow into the anteroom rather than the reverse. It is sometimes necessary, but less usual, to have more than the three controlled pressure zones that have been described.

Continuous maintenance of the defined pressure zones with a high degree of constancy and reliability is essential for the safety of the workers directly involved and for others in the vicinity. Extremely sensitive and accurate pressure sensors and controllers are required to regulate the small pressure differentials that are called for in typical ventilation plans. As an aid in continuously maintaining the design pressure differentials, it is strongly recommended that two highly visible liquid-filled draft gauges be installed in each anteroom: one to monitor the pressure differential between the anteroom air-lock and the access corridor and the other to monitor the pressure differential between anteroom air-lock and the workrooms containing potentially hazardous microbiological agents. Optimum safety is achieved when the workers who are directly involved in the potentially hazardous operations share responsibility with the maintenance staff for monitoring that design pressure conditions are being met.

The desired pressure differentials are established and maintained by regulation of supply and exhaust airflow volumes in the controlled pressure workrooms and in all contiguous spaces. When a contaminated workroom is to be kept under negative pressure relative to all connecting interior spaces, the exhaust air volume from that workroom must be adjusted so as to be greater than the supply volume by 200-400 cfm depending on the size of the workroom. A small entry and egress lock might have an excess exhaust volume of only 50-75 cfm to maintain the lock under a significant but lesser negative pressure than the potentially contaminated workroom. The lock itself can be maintained at a more negative pressure than connecting corridors and other free access spaces by similar airflow relationships or by slightly pressurizing all public access spaces relative to the out-of-doors and the access locks through the introduction of an excess of supply over exhaust air for all such areas. Either method will work, but as a practical matter it is preferable to maintain the contaminated space at minus 0.05 in. wg. than to keep the contaminated workroom at a negative pressure of 0.15 in. wg. relative to atmospheric pressure, and the other spaces in an appropriate pressure cascade.

Sometimes it is necessary to prevent contamination from entering a workroom rather than the reverse. To create this situation, the supply air volume would have to be 200-400 cfm greater than the exhaust volume, but otherwise the same principles apply.

It is a relatively simple matter to establish the airflow patterns that will produce the desired pressure relationship between workrooms when using a fixed volume ventilation systems as it is only necessary to turn on all the exhaust systems in the pressure controlled space, adjust the supply air volume until the desired pressures are read on a gage, and then fix the settings of supply and exhaust systems so they remain constant thereafter. Nevertheless, even fixed volume supply and exhaust ventilation systems, when chosen for safety reasons, can, at times, involve considerable complexity. Consider a biotechnology facility devoted to the production of large quantities of a dangerous organism such as an AIDS virus. The fermentation room must be maintained at a constant temperature of 68 ± 1°F and at a relative humidity between 40 and 60%. For safety reasons, the fermentation room must have an air exchange rate equivalent to 30 air changes per hour and be maintained as the most negative space in a 4-zone pressure-controlled cascade that includes robing and disrobing rooms, preparation rooms, and interlocks. Disrobing rooms are of particular concern as they have a potential for dispersing organisms from contaminated clothing. They should be maintained under negative pressure relative to surroundings and all exhaust air should be HEPA-filtered before release from the disrobing room. As a special precaution, all air exhausted from the facility must pass through two stages of HEPA filters before release (9). In the biotechnology industry, fermenters are growing in size as demand for products increases and operations become established on a firm basis. A vertical, 20-foot-high fermenter for a 10,000 gallon batch requires a room 35-40 feet high, and this places special demands on the design and operation of HVAC as well as health and safety exhaust air systems.

As already noted, a constant flow ventilation system is safe and simple, but may be an excessive energy consumer. Modulating systems are preferred for economy of operation, but such systems can become very complex in theory and difficult to operate in practice. Consider a workroom containing three adjustable-sash fume hoods that must be maintained under a negative pressure of 0.05 in. wg. at all times for contamination control reasons. To save energy and maintain a constant face velocity into each hood regardless of sash position (a highly desirable hood feature), the exhaust volume from each hood must be modulated so that the exhaust flow rate will be exactly proportional to sash height. At one extreme, when all three hoods are wide open simultaneously, they will be exhausting the maximum volume of room air, and supply air must be equal to this maximum exhaust volume minus 300 cfm to maintain the required negative room pressure. At the other extreme, all three hoods may have their sash in the down position, but the exhaust volume must still be 300 cfm greater than the supply to maintain the predetermined constant pressure differential with contiguous spaces. Most of the time, however, the system will be exhausting a variable volume of air through the three hoods that is between the two extremes, and the supply system instrumentation must be able to sense the position of each of the hood sashes, add the air requirements for each, subtract 300 cfm from the total, and provide that precise volume of air to the workroom while, at the same time, making certain that the temperature and humidity in the room remain constant and the minimum volume of outside air exchange required by the local building codes is adhered to. Instrumentation is now available to accomplish all these tasks, but there is not yet a large body of experience in their application to hazard control.

V. AIR-CLEANING TECHNOLOGY FOR BIOHAZARD CONTROL

A. High-Efficiency Air-Cleaning Elements

When working with hazardous microbiological agents and when toxic chemicals are used in conjunction with this activity, air cleaning is required before ventilation air can be recirculated back to the work space or before it can be discharged to the environment. When only

hazardous microbiological agents are employed (without the use of
hazardous volatile chemicals), absolute air filtration is an adequate
air-cleaning step. High-efficiency particulate air (HEPA) filters are
used universally for this purpose. HEPA filters for nuclear and military applications are characterized by a minimum efficiency of 99.97%
for a test aerosol containing 0.3 μm particles (10). This size was selected during World War II as the least filterable particle, meaning that
efficiency increases for smaller as well as larger particles. The technical explanation for this is related to the different separation mechanisms responsible for the filtration of small and large particles. For
particles smaller than 0.3 μm, diffusional migration of particles to filter fibers is the principal separation force, and it increases in effectiveness as particles become smaller. For particles larger than 0.3
μm, inertial forces are the principal separating mechanisms. Inertial
force causes particles to deviated from the airflow path and impact on
filter fibers. The larger the particle, the greater the inertial force
exerted, and the more easily it is filtered. Particles in the 0.1-0.3
μm size range possess little intertia and are not very diffusive, hence
they represent a least filterable size (11).

HEPA filter media are made on a Fourdrinier paper-making machine
from graded glass fibers ranging in diameter from less than 0.25 μm
to 3-4 μm. The smallest fibers provide for the high-efficiency removal
of submicrometer particles, and the larger fibers provide strength.
The thin glass fiber web is extensively pleated to provide a larger
surface area in a small volume and then sealed inside a rigid wooden
or metal box-like frame for strength and to provide surfaces for mount-
ing gaskets (see Figure 2-6). Most manufacturers of HEPA filters
are able to provide products that substantially exceed 99.99% efficiency
for 0.3-μm particles. When tested with aerosols containing bacteria
and viruses, retention efficiencies of 10^5 and greater have been recorded (12). This means that the decontamination capability of HEPA
filters is at least as great as the microbiological retention capability
of Class I and Class II biological safety cabinets. In fact, Class II,
Type A biological safety cabinets customarily recirculate purified cabinet air back to the workroom atmosphere. This is considered acceptable for all but the most hazardous microorganisms, provided the recycled air does not contain toxic volatile chemical substances.

Volatile toxic chemicals penetrate HEPA filters. When present,
other air purification means are required. The universal collector of
volatile organic chemical vapors is gas adsorption activated charcoal.
Adsorption is a mass-transfer process wherein certain components of
a fluid are deposited on the surfaces of a solid by surface forces. Ad-
sorbents are characterized by their extremely porous structures, which
provide internal surface areas many times larger than the external
surface. Vapors diffuse into the pores and are bound to the internal
surface in various ways. Before the adsorbent becomes incapable of

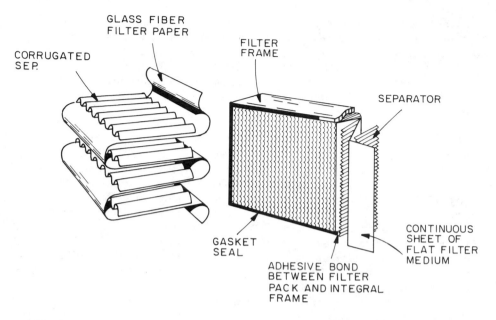

GLASS FIBER
FILTER PAPER

FILTER
FRAME

CORRUGATED
SEP.

SEPARATOR

CONTINUOUS
SHEET OF
FLAT FILTER
MEDIUM

GASKET
SEAL

ADHESIVE BOND
BETWEEN FILTER
PACK AND INTEGRAL
FRAME

FIGURE 2-6 Component diagram of a HEPA filter (courtesy Flanders Filters, Inc., Washington, NC).

removing the dispersed material at the required rate, it is either re-placed or regenerated for further use. Activated charcoal is the most widely used adsorbent for gas treatment. It is produced by first mak-ing a charcoal from coconut and other nut shells. Next, the charcoal is activated by controlled heating in a steam atmosphere to drive off organic matter and to generate large internal surfaces on which adsorp-tion can take place. The surface areas of activated charcoals range from 1100 to 1600 m^2/g. Although activated charcoal can simultaneous-ly adsorb several substances with widely differing retentivities, sub-stances of greater retentivity can displace (desorb) those of lesser retentivity when the charcoal nears saturation. The formation of an adsorbed surface layer of gas molecules by adsorption can be likened to the condensation of a vapor to form a liquid film. In both processes, the heat emitted is a specific value for each substance and of the same order of magnitude. Because physical adsorption is related to liquifi-cation, it usually occurs only at temperatures and pressures close to those for condensation. Similarly, the physically adsorbed layer (ad-sorbate) can be removed (desorbed) by applying a specific amount of heat. Adsorption, like condensation, takes place more readily at lower temperatures and higher pressures (13).

An all-purpose air-purifying system for air recirculation or for air exhausted to the environment contains several elements in series: (a) a particulate prefilter to remove large particles so that (b) the small-particle-removal HEPA filter will not be prematurely overloaded and (c) a 2-4-inch-deep gas adsorption activated charcoal bed to remove toxic volatile chemicals that penetrate the particulate filters.

B. In-Place Testing of Installed High-Efficiency Air-Cleaning Elements

Both types of air purifiers need to be tested after installation to provide assurance that they have no leaks and are capable of performing up to the factory efficiency ratings of the individual air-purification elements. This is an essential step whenever hazardous materials are being processed, whether all exhaust air is being discharged or partially recirculated. It is futile to expect these two ultra-high-efficiency air-purifying devices to perform up to their rated efficiency unless the installations are checked routinely for leaks.

C. Bag-in, Bag-out Air-Cleaning Element Change-out Systems

When dangerous microbiological agents, volatile chemicals, and radioactive substances are used, changing of HEPA filters and gas adsorption beds is considered to be a hazardous operation. To reduce the exposure of personnel assigned to this task, specially constructed caissons are available that make it possible to remove contaminated cartridges and replace them with new ones in a totally closed system that effectively separates the operator from the contamination by an impenetrable transparent plastic barrier. This device, widely used in the nuclear industry, is known as a bag-in, bag-out system (see Figure 2-7). Stand-alone caissons for HEPA filter and activated charcoal air-cleaning elements that contain the bag-in, bag-out feature are available commercially in any desired capacity. They are easily incorporated into unit and central exhaust air systems whenever either or both air-purification elements present handling dangers because of the materials they collect from the air passing through. Hazardous operations requiring use of a cabinet containing a bag-in, bag-out system for air-cleaning element change-out would not be considered safe for conduct in a Class II, Type A biological safety cabinet. At least one manufacturer of NSF-certified Class II, Type B biological safety cabinets offers models containing this feature already built into the cabinet structure.

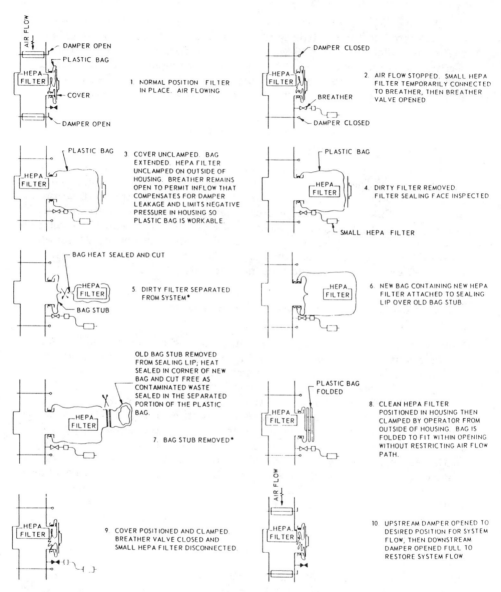

FIGURE 2-7 Procedure for bag-in, bag-out filter change (10).

VI. VENTILATION SYSTEM DESIGN AND CONSTRUCTION

A. Balanced Systems

In spite of the fact that HVAC facilities represent a sizable percentage
of the total building construction costs for a biotechnology laboratory
or production facility (as high as 20-25%), the systems are frequently
poorly designed and, to compensate for this lack, installed with large
numbers of "trimming dampers" placed in all duct runs for the purpose
of correcting design errors. Instead of designing a balanced system
(without trimming dampers) and making certain it is constructed and
installed in strict accordance with design specifications, it has become
a common practice to engage ventilation system "balancing specialists"
to place the systems into compliance with design airflow values after
the rest of the bulding is completed. The difficulties associated with
this substitute for engaging in a sound engineering design effort are:
(1) It may not be physically possible to balance the system adequately
in its "as-built" condition, and expensive remedial work will be required. (2) After balancing large combined systems with numerous
trimming dampers, any alteration in the damper positions will throw
the system permanently out of balance until a balancing specialist is
called in to redo the work. (3) When ventilation systems are to be
balanced after construction has been completed, the duct systems and
other vital parts will already be concealed above false ceilings, in utility chases, and behind cabinets, hoods, and similar large pieces of
permanently installed equipment and furniture. There is no cheap
or easy way to make a poorly designed and carelessly installed HVAC
system perform satisfactorily after the building has been accepted and
occupied.

B. Duct Construction

The design and construction requirements of health and safety exhaust
ventilation systems are very different from those employed for comfort
air conditioning. Guidelines are contained in *Guidelines for Laboratory
Design—Health and Safety Considerations* (14), and details of systems
in *Industrial Ventilation—A Manual of Recommended Practice* (6). Among
important differences are the construction methods and materials used
for health and safety ducts and plenums. Leak tightness is an essential requirement. This calls for soldered or welded pipe seams plus
flanged and gasketed connections between parts instead of conventional
lock seams and telescoping pipe joints. Many systems will carry saturated atmospheres containing corrosive chemicals, and the ducts and
plenums must be constructed of stainless steel or plastic piping and
all joints must be sealed to avoid dripping from condensation. Similar
corrosion-avoidance measures are required for exhaust blowers and
all other interior parts of such systems and, most especially, all sensors and controllers.

C. Exhaust and Supply Air Points

Another important difference between comfort ventilation and exhaust
hood systems is that all health and safety systems should be designed
to place their exhaust blowers on the roof of the building to maintain
all parts of the system under negative pressure so that in the event
of leakage there will be a flow of air into the system rather than a leak
of contaminated air somewhere inside the building.

An additional reason for placing the discharge point of all exhaust
systems on the building roof is to avoid bringing potentially contaminated air into the building fresh air intakes. This calls for careful planning and a detailed examination of likely contaminant sources before
outside air intakes are selected. A minimum distance of 30 feet from
air discharge openings and stack outlets is recommended to reduce fume
reentry problems, but it is good practice to design for the maximum
feasible separation. Outside air intakes located at ground level are subject to contamination from automobile and truck fumes, whereas air intakes at roof level are subject to contamination from laboratory exhaust
stacks or high stacks serving off-site facilities in the vicinity. When
buildings contain more than 10 stores, it is advisable to locate the air
intakes at the midpoint of the building. Difficult sites may require
wind tunnel tests to investigate the fume reentry problem under simulated conditions. All roof-mounted exhaust installations should have
the stack on the positive side of the fan, and it should extend at least
10 feet above the roof parapet and all other prominent roof structures
in order to discharge the exhaust fumes above the layer of air that
clings to the roof surface and prevents contaminants from displacing
upward. This arrangement will help to avoid reentry through nearby
air intake points. To further assist the exhaust air to escape the roof
boundary layer, the exhaust velocity should be at least 2500 fpm and
there should be no weather-cap or other obstruction to prevent the exhaust discharge from rising straight upward (14). An ideal roof exhaust air installation is shown in Figure 2-8. It incorporates a weatherproof motor and direct-connected fan to eliminate often troublesome belt
drives and weatherproof drive housings. People worry needlessly about
precipitation entering the stack. When the fan is running, precipitation
cannot enter against the upward flow of air, and when the fan is off,
the weep-hole at the low point of the fan scroll casing permits precipitation to drain, harmlessly, to the roof.

VII. SAFETY CRITERIA FOR WORKING WITH
HAZARDOUS BIOLOGICAL MATERIALS

A. The Place of Ventilation in the Regulation
of Containment

The discussion that follows will be confined to a summary of ventilation
requirements for operations at each biosafety level, although it must

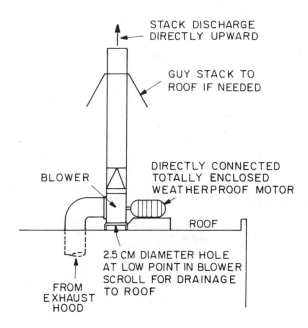

STACK DISCHARGE DIRECTLY UPWARD

GUY STACK TO ROOF IF NEEDED

BLOWER

DIRECTLY CONNECTED TOTALLY ENCLOSED WEATHERPROOF MOTOR

ROOF

2.5 CM DIAMETER HOLE AT LOW POINT IN BLOWER SCROLL FOR DRAINAGE TO ROOF

FROM EXHAUST HOOD

FIGURE 2-8 Exhaust stack and weather proof blower (15).

be kept in mind that ventilation facilities represent only a part of the total management program needed to handle hazardous biological agents in a safe manner (16-18). The biological levels referred to are discussed in detail in Chapter X. For biosafety levels 1 and 2, the only reference to ventilation facilities is a statement that "If the laboratory has windows that open, they are fitted with fly screens" (18).

Facilities for biosafety level 3 work must include the following ventilation arrangements: (1) Access through two sets of doors, the space between functioning as an airlock and provided with the special ventilation arrangements that have been described earlier. (2) All windows are to be closed and sealed, necessitating a mechanical supply and exhaust air system that maintains prescribed air-pressure relationships with connecting access airlocks, adjacent workrooms, and public access areas. (3) No air exhausted from a biosafety level 3 facility may be recirculated, although HEPA-filtered air discharge from a Class II cerfified biological safety cabinet may be recirculated within the same workroom or laboratory.

Ventilation facilities required for biosafety level 4 include: (1) individual supply and exhaust air ventilation systems with interlocked controls and malfunction alarms to maintain pressure differentials that assure inward flow from areas outside the facility toward areas of highest risk, (2) decontamination of all exhaust air by HEPA filters and discharge to the environment well away from other buildings and build-ing air intakes, (3) exhaust air filter housings provided with in-place filter-testing facilities and an in situ decontamination capability, (4) HEPA-filtered supply air, (5) double HEPA filtration for all air ex-hausted from total containment biological safety cabinets, and (6) a redundant, double HEPA filtration system for all air exhausted from an entire decontamination suite maintained for personnel who must work in a one-piece positive pressure suit ventilated by a life support system.

Biosafety level 3 ventilation requirements are likely to be of most interest and utility as they apply, as well, for preparing large-scale cultures of biosafety level 2 organisms. Mechanical ventilation sys-tems suitable for biosafety level 3 and 4 facilities are complex and must be designed, installed, and vetted by knowledgeable and experienced professionals.

B. Regulatory Concerns

In a recent study of the international biotechnology industry by the Congressional Office of Technology Assessment (19), the sections con-cerned with worker safety stated that the U.S. Occupational Safety and Health Act imposes general duties on employers to maintain safe workplaces and to eliminate or control hazardous substances. Of all countries surveyed, U.S. governmental controls directed specifically toward biotechnology were considered to be the least restrictive. The report does, however, contain the important caveat that the long-term effects on workers' health from exposure to novel organisms and prod-ucts is currently unknown. A recognition that these concerns are truly of an international nature has motivated the Organization for Economic Cooperation and Development (OECD) to develop a set of principles to regulate biotechnology. It proposed "to classify organisms according to their pathogenicity so that countries do not regulate a particular organism differently" (20).

Worker safety is important in biotechnology laboratories and produc-tion facilities because many of the products (including microorganisms) can be potent allergens, even when they are denatured and no longer dangerous from the infection standpoint. Therefore, concern does not cease after the products have been processed. The dangers may, in fact, be enhanced when the products are converted from a liquid slurry into a finely divided dry powder. In the realm of real hazards, engi-neering controls, including numerous methods of applying ventilation and air cleaning technology, are certain to play a prominent role.

REFERENCES

1. Pike, R. M. (1976). Laboratory-associated infections: Summary and analysis of 3,921 cases. *Health Lab. Sci.* 13: 105-114.

2. CDC. (1984). *Biosafety in Microbiological and Biomedical Laboratories*. Centers for Disease Control and National Institutes of Health, U.S. Government Printing Office, Washington, DC.

3. International Conference of Building Code Officials. (1982). *Uniform Building Code*. ICBCO, Whittier, CA.

4. Yaglow, C. P. (1955). Ventilation requirements for cigarette smoke. *ASHRE Jour. Sect., Heating, Piping, and Air Conditioning* (Jan. 1955):1-4.

5. DiBerardinis, L., First, M. W., Price, J. M., and Martin, K. 1983). Storage cabinet for volatile toxic chemicals. *Am. Ind. Hyg. Assoc. J. 44*: 583-588.

6. ACGIH. (1986). *Industrial Ventilation—A Manual of Recommended Practice*, 19th ed. American Conference of Governmental Industrial Hygienists, Cincinnati, OH.

7. Ivany, R. E., First, M. W., and DiBerardinis, L. (1986). Performance testing of a laminar airflow workbench designed for animal studies. Presented at The 29th Biological Safety Conference, American Biological Safety Assoc., Lexington, KY, October 5-9, 1986.

8. NSF. (1983). *Standard No. 49, Class II (Laminar Flow) Biohazard Cabinetry, Revised*. The National Sanitation Foundation, Ann Arbor, MI.

9. Baum, J. (1987). Personal Communication.

10. Burchsted, C. E., Fuller, A. B., and Kahn, J. E. (1976). *Nuclear Air Cleaning Handbook*, 2nd ed. Oak Ridge National Laboratory, available from National Technical Information Service (NTIS), Springfield, VA.

11. First, M. W. (1968). Filters, prefilters, high capacity filters, and high efficiency filters; review and projection. In *Proceedings Tenth AEC Air Cleaning Conference*, CONF-680821, M. W. First and J. M. Morgan, Jr. (Eds.). NTIS, Springfield, VA, p. 65.

12. Harstad, J. B., Decker, H. M., Buchanan, L. M., and Filler, M. E. (1967). In *Proceedings Ninth AEC Air Cleaning Conference*, Vol. 2, AEC-660904, J. M. Morgan, Jr. and M. W. First (Eds.). NTIS, Springfield, VA.

13. Jorgensen, R. (1983). *Fan Engineering*, 8th ed. Buffalo Forge Co., Buffalo, NY.

14. DiBerardinis, L., Baum, J., First, M. W., Gatwood, G., Groden, E., and Seth, A. (1987). *Guidelines for Laboratory Design—Health and Safety*. John Wiley & Sons, New York.

15. First, M. W. (1977). Control of systems, processes, and operations. In *Air Pollution*, A. Stern (Ed.). Academic Press, New York.

16. CDC, Office of Biosafety. (1974). *Classification of Etiologic Agents on the Basis of Hazard*, 4th ed. U.S. Department of Health, Education, and Welfare, Centers for Disease Control, Office of Biosafety, Atlanta, GA.

17. NIH. (1976). Guidelines for Research Involving Recombinant
     DNA Molecules. *Federal Register 41*: 27902-27943.
18. NIH. (1984). Guidelines for Research Involving Recombinant
     DNA Molecules. *Federal Register 41*: 46266-46291.
19. OTA. (1984). *Commercial Biotechnology—An International Assess-
     ment.* Congressional Office of Technology Assessment, Washington,
     DC.
20. Sun, M. (1985). Administration drafts biotech plan for OECD.
     Science 230: 925.

3

Certification of Biosafety Cabinets

DAVID LUPO   *B & V Testing, Inc., Waltham, Massachusetts*

I. INTRODUCTION

The biological safety cabinet has long been recognized and accepted
as the principle device used in the laboratory to provide physical control and containment of aerosols. The biological safety cabinet, when
properly used and maintained, offers personnel protection, environmental protection, and product protection.

II. STANDARDS

The initial specifications prepared by the National Institutes of Health
(NIH) and the National Cancer Institute (NCI) were intended to assist
their staff in procurement of acceptable cabinets for use at NIH. It
was evident that a system that would encourage innovation and allow
all units to be tested and validated in an independent fashion was needed. In response, NIH and NCI asked the National Sanitation Foundation (NSF) to prepare a standard for Class II cabinets to provide facilities and personnel to test units to the standard and to publish lists
of cabinets that met the standard.
    In response to this request, NSF formed a committee composed of
personnel from NIH, NCI, the Centers for Disease Control (CDC),
and manufacturers and potential users of cabinets from research laboratories. After long deliberation, NSF Standard #49 resulted. This
guideline has been generally accepted as "state of the art" for manufacturing, performance, testing, and periodic evaluation and certifica-

tion of biological safety cabinets. Standard #49 has superseded previous standards. A number of cabinets have been listed by NSF as meeting this standard, and only those that pass a number of performance tests are permitted to display the NSF seal. Although NSF Standard #49 is the guideline for Class II biological safety cabinets, it does reference the three types (Class I, II, and III) of cabinets presently available or in use in microbiological laboratories.

III. DESCRIPTION OF CABINETS

Open-fronted Class I and Class II biological safety cabinets are partial containment cabinets, which offer significant levels of containment to laboratory personnel and to the environment when used in conjunction with good microbiological technique. The Class III cabinet, a closed gas-tight biological safety cabinet, provides the highest attainable level of proteciton. NSF Standard #49 describes these cabinet types as follows:

A. Types of Cabinets

Class I

A ventilated cabinet for personnel and environmental protection, with an unrecirculated inward airflow away from the operator.

   Note: The cabinet exhaust air is treated to protect the environment before it is discharged to the outside atmosphere. This cabinet is suitable for work with low- and moderate-risk biological agents, where no product protection is required.

Class II

A ventilated cabinet for personnel, product, and environmental protection having an open front with inward airflow for personnel protection, HEPA filtered laminar airflow for product protection, and HEPA filtered exhaust air for environmental proteciton.

   Note: When toxic chemicals or radionuclides are used as adjuncts to biological studies or pharmaecutical work, Class II cabinets designed and constructed for this purpose should be used.

   Type A (formerly designated Type 1). Cabinets that (1) maintain a minimum calculated average inflow velocity of 75 fpm through the work area access opening; (2) have HEPA filtered downflow air from a common plenum (i.e., plenum from which a portion of the air is exhausted from the cabinet and the remainder supplied to the work area); (3) may exhaust HEPA filtered air back into the laboratory; and (4) may have positive pressure contaminated ducts and plenums. Type A cabinets are suitable for work with low- to moderate-risk biological agents in the absence of volatile toxic chemicals and volatile radionuclides.

Type B1 (formerly designated Type 2). Cabinets that (1) maintain a minimum (calculated or measured) averaged inflow velocity of 100 fpm through the work area access opening; (2) have HEPA filtered downflow air composed largely of uncontaminated recirculated inflow air; (3) exhaust most of the contaminated downflow air through a dedicated duct exhausted to the atmosphere after passing through a HEPA filter; and (4) have all biologically contaminated ducts and plenums under negative pressure or surrounded by negative pressure ducts and plenums. Type B1 cabinets are suitable for work with low- to moderate-risk biological agents. They may also be used with biological agents treated with minute quantities of toxic chemicals and trace amounts of radionuclides required as an adjunct to microbiological studies if work is done in the direct exhausted portion of the cabinet or if the chemicals or radionuclides will not interfere with the work when recirculated in the downflow air.

Type B2 (sometimes referred to as "total exhaust"). Cabinets that (1) maintain a minimum (calculated or measured) average inflow velocity of 100 fpm through the work area access opening; (2) have HEPA filtered downflow air drawn from the laboratory or the outside air (i.e., downflow air is not recirculated from the cabinet exhaust air); (3) exhaust all inflow and downflow air to the atmosphere after filtration through a HEPA filter without recirculation in the cabinet or return to the laboratory room air; and (4) have all contaminated ducts and plenums under negative pressure, or surrounded by directly exhausted (nonrecirculated through the work area) negative pressure ducts and plenums. Type B2 cabinets are suitable for work with low-to moderate-risk biological agents. They may also be used with biological agents treated with toxic chemicals and radionuclides required as an adjunct to microbiological studies.

Type B3 (sometimes referred to as "convertible cabinets"). Cabinets that: (1) maintain a minimum (calculated or measured) average inflow velocity of 100 fpm through the work access opening; (2) have HEPA filtered downflow air that is a portion of the mixed downflow and inflow air from a common exhaust plenum; (3) discharge all exhaust air to the outdoor atmosphere after HEPA filtration; and (4) have all biologically contaminated ducts and plenums under negative pressure, or surrounded by negative pressure ducts and plenums. Type B3 cabinets are suitable for work with low- to moderate-risk biological agents treated with minute quantities of toxic chemicals and trace quantities of radionuclides that will not interfere with the work if recirculated in the downflow air.

Other types.   Other cabinets may be considered Class II if they
meet these requirements for performance, durability, reliability, safety,
operational integrity, and cleanability.

Class III

A totally enclosed, ventilated cabinet of gas-tight construction. Operations in the cabinet are conducted through attached rubber gloves.
The cabinet is maintained under negative air pressure of at least 0.5
inches water gauge (in. wg.). Supply air is drawn into the cabinet
through HEPA filters. The exhaust air is treated by double HEPA
filtration, or by HEPA filtration and incineration (see Laboratory Safety
Monograph, January 1979, NIH Supplement to Guidelines).

IV. EQUIPMENT EVALUATION

A. Manufacturer's Responsibility

The above-mentioned Class II biological safety cabinets are to be tested
in accordance with NSF Standard #49 at the manufacturer's facility.
All equipment is shipped to the end user with a performance evaluation indicating that after assembly, the equipment meets manufacturer's
specification and the appropriate related standard NSF Standard #49.

B. User Responsibility

Upon arrival the user should inspect each piece of equipment for physical damage which may have occurred during transportation (broken
glass, dents, scratches, etc.). Before the biological safety cabinet
is uncrated the laboratory must be inspected to determine its final
location. The positioning of the cabinet in the laboratory is critical
and may effect the operation of the cabinet. The location of the cabinet
should be in an area where personnel travel is at a minimum. It should
not be placed in an area where the front opening of the cabinet may be
affected by doors opening or closing (access doors, incubator doors,
refrigerator doors, etc.) or where laboratory supply air registers diffuse air toward the front opening of the cabinet. The cabinet should
not be placed in an area where air turbulence created by air conditioners or fans may have an effect on the front opening of the biological
safety cabinet.

   Consideraiton should be given to the laboratory electrical system
to determine if it meets manufacturer's specifications for voltage and
amperage. Depending on the requirements of the facility and the end
user, it may be desirable for the electrical circuit to be part of the
facility's emergency power system to prevent failure in the event of
a power outage.

If it is determined that the cabinet will be exhausted out of the building, all duct work should be in place and the system checked to determine if the system meets manufacturer's requirements for exhaust.

The biological safety cabinet may be equipped with accessories such as petcocks for gas, air vacuum, etc. These plumbing requirements should be met and in place.

After all of these requirements have been completed, the equipment can be positioned in the lab and appropriate systems connected.

V. CERTIFICATION

A. In-Place Testing

Once the biological safety cabinet is installed on site, it is to be certified in accordance to meet the appropriate standard (NSF Standard #49) as well as the manufacturer's specifications. As of this writing, companies that test and certify biological safety cabinets are not required to meet any particular criteria with regards to competency or professionalism. Several groups are investigating the possibility of licensing or accrediting the certifiers of biological safety cabinets. Until such a program is implemented, the selection of the certification company is left to the facility or individual investigator. Some of the qualifications that must be considered carefully are as follows:

1. The company that is selected should not be affiliated with the manufacturer of the equipment, its representative, or subsidiary. It should not be affiliated with the manufacturer of components used in fabrication or assembly (HEPA filters) or their representatives or subsidiaries.
2. Although formal training for this particular type of work is limited, several avenues are available to certifiers of biological safety cabinets. These include training at a manufacturer's facility, courses available at various educational or governmental facilities, educational seminars, and experience with the related product.
3. Due to the portability of the testing equipment involved, all of the testing equipment should be calibrated by qualified personnel at or before manufacturer's recommended intervals with documentation available to appropriate agencies.

VI. RECOMMENDED FIELD TESTS

A Class II biohazard cabinet, properly used, is acknowledged as an effective primary barrier complementing good laboratory practices and procedures. It is essential that each cabinet be tested and performance evaluated on-site, assuring that all physical containment criteria are met at the time of installation, prior to use, and periodically thereafter. The ongoing recertification should be performed at least annually, and when HEPA filters are changed, maintenance repairs are required, or a cabinet is relocated. More frequent recertification may be considered for particularly hazardous or critical applications, or workloads. The following physical tests should be performed on site to qualify a Class II cabinet for certification:

Primary Tests
1. Velocity Profile Test
2. Work Access Opening Airflow (Face Velocity)
3. HEPA Filter Leak Test
4. Cabinet Integrity Test
5. Airflow Smoke Patterns

Secondary Tests
1. Vibration Test
2. Electrical Leakage and Ground Circuit Resistance
3. Noise Level Test
4. Lighting Intensity Test

VII. FIELD TEST PROCEDURES

A. HEPA Filter Leak Test

1. The DOP concentration upstream of the HEPA filter may be determined by using a repeatable reference source, such as an upstream DOP concentration or calibrated internal reference system.
2. If the cabinet is ducted so the exhaust filter cannot be scanned, drill a 3/8-in. (9.5-mm) hole in the duct (if necessary) and insert a sampling line into the hole. DOP penetration shall not exceed 0.005%. After testing, seal the hole with an expandable plug.

B. Electrical Leakage and Ground Circuit Resistance Test

1. This test is performed to determine the electrical leakage and ground circuit resistance to the cabinet gorund connection and dedtermine if a potential shock hazard exists. The cabinet shall be set to operate at the manufacturer's recommended airflow velocities.
2. Apparatus: electrical safety tester with 1 input impedence.

Decontamination should be performed on equipment prior to the replacement or repair of any component exposed to contamination, or before cabinet is to be moved.

4

Integrated Pest Management Program for Research Facilities

GARY D. ALPERT   *Harvard University, Cambridge, Massachusetts*

I. INTRODUCTION

When a fly approaches, instead of reaching for a fly swatter, our society reaches for an aerosol can. We have become conditioned to use chemical warfare against pests and expect to receive immediate results. Today, the use of chemicals to control urban pests has increased to the point where human contact with pesticides in an urban setting may exceed exposure to agricultural pesticides (1). Insecticides are now applied in private homes, public buildings, trains, planes, buses, food storage facilities, movie houses, grocery stores, universities, restaurants, apartments, offices, nursing homes, hospitals, mortuaries, and research facilities, to name just a few.

Public concern has increased over the use of pesticides and their potential risks in contributing to cancer, birth defects, and environmental contamination. Some urban insecticides (Chlordane, Lindane, Heptachlor) in common use five years ago are now banned or restricted because of their toxic effects. At the same time, the public is becoming less tolerant of insects and rodents in food, fiber, and homes. The problem is how to selectively eradicate pests that have invaded sensitive urban areas without creating a hazard to humans and the environment.

There is increasing evidence of the limitations of a strictly chemical approach to controlling urban pests. Insecticides commonly used to control pests in research institutions were originally developed for agricultural use, and thus are designed for short-term control requir-

ing frequent reapplication. Insect repellency, where sublethal concentrations of chemicals are detected and avoided, often inadvertently
serves to disperse pests into nearby uninfested areas (2,3). Insecticide resistance, where pests genetically pass on detoxification enzymes
to their numerous offspring, results in gradually deteriorating control.
As a result, many organophosphate and carbamate insecticides are
no longer effective in controlling urban pests (4). When pest populations rebound, they may swing to even higher levels than existed
prior to treatment.

In a research setting, special precautions must be taken when
using pesticides. Chemical drift onto research equipment and experiments in progress must be prevented. When compressed air sprayers
are used to apply pesticides, drift is unavoidable. Solvents and insecticide carriers as well as the active ingredients will contaminate contact surfaces. Contamination also occurs when chemicals are applied
onto radiators or into airflow systems.

At the same time, insect and rodent pests in research laboratories
cannot be tolerated. Breeding mice will shred paper to make a nest,
including paper containing original results. Cockroaches and mice
can contaminate surfaces and research materials with their urine and
fecal pellets. Cockroaches often breed in laboratory drawers near
sinks, especially when these drawers contain pipettes and infrequently
washed glassware. Insects are excellent mechanical vectors (they
are covered with numerous tiny hairs and they have pads on their
feet) and may contaminate tissue culture, rearing media, gels, and
other substances. Insects may fall into research containers, fly into
petri dishes to lay eggs, or crawl across sensitive surfaces. Time
lost in repeating experiments due to contamination or questionable
results can be minimized with an effective pest control program.

II. INTEGRATED PEST MANAGEMENT

A pest control program designed for research areas should use a "least-
chemical means" approach. When warranted, less toxic pesticides should
be used in a manner that minimizes exposure to humans and the environment. The use of least-chemical means in conjunction with other nonchemical measures is known as Integrated Pest Management (IPM).
IPM is a cooperative effort that can best achieve the goal of eliminating
pests in a research institution. This chapter should provide sufficient
guidance so that building personnel can proceed with confidence in
using IPM.

The principles of IPM require monitoring the pest population, using
available nonchemical control measures, evaluating results, using least-
chemical means if necessary, reevaluating results, and maintaining
a monitoring program. Each research facility should have an IPM program designed to meet its specific needs. For example, research with

human pathogens requires zero pest levels, tissue culture and media prep rooms require a totally nonchemical approach, and recombinant DNA research must meet NIH pest control guidelines (5).

Recent advances in nonchemical pest control methods can be applied to the research setting (6). Solid melt adhesive traps are a long-term survey and control tool (7). Barriers can be applied to surfaces to prevent insects from climbing. Caulking compounds can eliminate pest breeding areas. Least-chemical means such as pheromones, attractants, repellants, dessicants, baits, dusts, and hormones show promise. Many plant products have properties of insect repellency or mortality. An extract from orange peel, alpha-limonene, is effective in controlling adult fleas.

Other products have been proven ineffective in spite of advertising claims. Ultrasonic devices do not control insects, and their effect on rodents is limited. The effectiveness of an "electro-gun" or nematodes to control subterranean termites has been unreliable in the field. Electromagnetic devices are unproven. Recently advertised electric grid cockroach traps costing close to $300 catch no more insects than a 50 cent sticky trap. The reader can refer to Table 1 for a list of effective IPM strategies for common pests of research laboratories.

The goal of an IPM program is safer, more effective pest control. An IPM approach uses long-range planning to identify and control the source of a pest infestation. Harvard University and Massachusetts Institute of Technology have been conducting IPM programs for their research laboratories successfully for many years. Conventional carbamate and organophosphate insecticides have not been allowed in these facilities. Instead, chronic cockroach and ant infestations have been controlled using IPM strategies (8).

III. GOVERNMENT REGULATIONS

A. Federal

Pesticides include agents developed for control of insects, fungi, weeds, and rodents. The Environmental Protection Agency (EPA) regulates pesticides to protect the public health and the environment in accordance with the Federal Insecticide, Fungicide and Rodenticide Act (FIFRA). FIFRA was first established as public law in 1947 and has been revised and expanded in 1972, 1975, and 1978 (9-11). All pesticides must have an EPA-approved label, indicating product contents and directions for safe, effective use. A copy of the label should be on file for every pesticide used in a facility.

The National Institutes of Health (NIH) developed regulations governing recombinant DNA research in 1976. "An effective insect and rodent control program shall be maintained" as stated in the Federal Register (1986). A general description of a pest control program de-

TABLE 1 Common Pests of Research Laboratories and Methods of Control

Pest	IPM methods of control
Cockroaches	
German	Baits, boric acid, sticky traps, insect growth regulators, caulking, repellents, housekeeping
American	Baits, exclusion (check plumbing and pipes from below), sticky traps, boric acid
Oriental	Sticky traps, exclusion (check drains, plumbing pipes from below), boric acid
Brown-banded	Sticky traps, caulking, boric acid, insect growth regulators, housekeeping, baits
Ants	
pavement	Sticky traps, baits, caulking, housekeeping, insect growth regulators
pharoah	Sticky traps, baits, insect growth regulators
carpenter	Sticky traps, boric acid into nest, baits
thief	Baits, sticky traps, insect growth regulators
Silverfish	Exclusion (check radiator pipes), sticky traps, boric acid(?)
Flies	
drain	Clean drains, sticky traps, cover drains
house	Housekeeping, fly lights, sticky traps, fly swatter, screens
Fleas	Vacuum, glueboards, insect growth regulators boric acid
Mites	Vacuum, glueboards, boric acid, sticky traps, repellents
Spiders	Sticky traps, housekeeping
Mice	
house/lab	Snap traps, automatic live traps, glueboards, exclusion, housekeeping
field	Exclusion, snap traps, glueboards, automatic live traps
Rats	
Norway	Exclusion, housekeeping, snap traps, glueboards
roof	Housekeeping, snap traps, glueboard, exclusion

signed for research laboratories that meets NIH guidelines is presented
in this chapter.

The Occupational Safety and Health Act (OSHA) sets standards
for protective clothing, respirators, eye protection, and safe exposure
levels. A series of pamphlets is available from the Government Printing Office on the safe handling of pesticides (12).

B. State

Each state further regulates the use of pesticides through its own laws.
These laws supersede federal regulations and are generally more restrictive. Frequently it is the Department of Agriculture that regulates
the labeling, distribution, sale, storage, application, and disposal
of pesticides. State laws are often revised and their list of registered
chemicals changes frequently.

Some states have adopted right-to-know laws. In this case, a
material safety data sheet is required for all pesticides in use. These
data sheets must be on file for distribution upon request. Important
data on combustion and safety protection are included.

C. Local

Some cities and towns have ordinances governing recombinant DNA
research. Wording of one such city ordinance is as follows: "The
premises on which all P2 and P3 recombinant DNA experiments are
carried out must be effectively free of rodent and insect infestation,
in accordance with all laws governing such matters." Cities that have
tried to supersede state laws governing the use of pesticides are sometimes required to support their position in court.

IV. DESIGN OF FACILITIES

A. Building Design

There are important pest control issues that should be considered during the building design stage. By being aware of factors that increase
the attractiveness of a building to pests, architects and builders can
avoid creating conditions that will have to be modified at a later date.
An example is the excessive presence of ledges, overhangs, and niches
that provide an ideal roosting area for pigeons and other birds. The
excrement from birds can be a health hazard due to the many pathogenic organisms present (13). When fecal material decomposes, it can
become airborne and can be drawn into the building through air intake
vents and window-unit air conditioners. Acidic bird droppings can
deface the building and immediate surroundings, and cleanup can be
difficult and hazardous. There are several different types of barriers

that can be installed to prevent birds from roosting. Strips of pronged metal needles projecting outward can be attached permanently to most ledges. Sticky repellents applied to ledges may last for a year before being covered with feathers and debris. The installation of fine screening which cannot be seen from below is another alternative. Advice from a bird control expert is helpful.

The use of window-unit air conditioners can create unusual pest problems. Birds and their numerous parasites frequently nest under these units. When the birds die or leave, mites migrate in enormous numbers through openings in search of a host. The bites from these parasites cause intense itching and discomfort. The mites are unable to survive and reproduce without a bird host, but experience has shown that they can contaminate research projects.

The shape of a building is often a factor in trapping insects. An example is the presence of cluster flies (*Pollenia rudis*), which develop in earthworms during the summer and in the fall enter through cracks in tall buildings and aggregate in large numbers in attic spaces. On warmer days they become active and drop below where they become trapped on the inside of windows.

B. Inspection of Exterior

An inspection should be made of the building exterior at least annually. A checklist similar to the one below can be used.

1. *Rodent entry.* All doors, screens, and windows should be tight fitting. Gaps greater than 1/4 inch allow mice to enter.
2. *Window-unit air conditioners.* Units should be tightly caulked when installed to prevent insect and mite entry.
3. *Lighting.* Sodium vapor lights of low wattage can be used near entrances to reduce insect attraction. Mercury vapor lights of higher wattage can be placed further from the facility to draw insects away. Electric fly grids can be a very effective control measure when placed near animal labs or food preparation areas. Seek professional assistance in selecting and placing units for maximum effectiveness. Indoor units located low on the wall capture more flies.
4. *Air intake vents.* Make sure all air filters are in place, otherwise insects such as flies and wasps can be drawn into ventilation systems and deposited into laboratories.
5. *Waste disposal.* Outside garbage cans and other waste receptacles should be rat-proof. Rats are able to enter most dumpsters and can push past the piston on most compactor units. However, single unit compactor-dumpsters that are watertight and rat-proof are now available. Garbage bags containing food, needles, and other materials are frequently broken by the piston and accumulate on

the ground outside of the dumpster. Waste from kitchens and
animal laboratories can support large numbers of rats. Make sure
that potential rat harborages such as pallets and construction
materials are removed. An effective system is to locate waste disposal units away from the building or in a rodent-proof structure.
6. *Landscaping.* Shrubbery should not be too thick near a building.
   A buffer zone of grass or gravel will discourage most pests. Prune
   away any tree branches that touch the building.

C. Inspection of Interior

A checklist similar to the one below should be developed for annual
interior inspections.

1. *Water leaks.* The detection and repair of water leaks is critical.
   Many insect pests feed on fungi associated with wet materials.
2. *Plumbing.* Restroom fixtures should be caulked where the plumbing enters the wall. Floor sinks in janitor closets should also be
   sealed tightly.
3. *Waste disposal.* All garbage cans should have a disposable liner.
   Periodic examination and cleaning of the can under the lining is
   usually necessary. Trash should be emptied at the end of the
   work day so that there is less food for insects and rodents over-
   night and on weekends.
4. *Drains.* Clogged, defective, or unused drains can contribute
   significantly to cockroach and other insect infestations. Unused
   drains that are not capped before new equipment is installed become
   infested.
5. *Locker rooms.* Cockroaches may first become established in locker
   rooms and later spread throughout a facility. Lockers resting on
   raised hollow supports are frequently infested. Make sure all
   plumbing in the area is tightly sealed.
6. *Sinks.* Sinks with built-in lower cabinets create ideal conditions
   for pests. Free-standing sinks are best in animal rooms and other
   areas that are frequently washed.
7. *Research laboratories.* Many cockroach infestations occur near
   laboratory sinks. The upper drawers in the cabinet next to the
   sink as well as the central void space behind the sink and between
   the bank of opposing drawers is often infested. One should pull
   out an entire bank of drawers to inspect. When pipette drawers
   are lined with paper, cockroaches are even more numerous.
   Splash boards behind sinks should be of sufficient quality that
   they do not become delaminated over time, creating a cockroach
   harborage site.

D. Construction

During construction or a major renovation, pest control materials can
be placed between the walls and into other voids. Boric acid powder
(99%) is recommended because of its long-term control properties (14,
15). Boric acid stays effective for years over a range of temperature
and humidity conditions. Data from California indicates that those
buildings pretreated with boric acid powder experienced significantly
fewer pest problems over time. Wall voids containing boric acid
powder serve as traps and barriers to the dispersal of insects within
a structure (16).

Frequently, rodents, especially mice, enter a building during the
construction stage. Workers leave food debris on site and mice can
enter in large numbers before the building is occupied. The best
time to control mice is near the end of the construction process. All
mice should be trapped and removed. Poison baiting is not recommended since mice will die in the building and create odor and fly
problems. An inspection should be made to ensure that all door
thresholds are mouse-proof.

V. PEST CONTROL SERVICE PERSONNEL

A. In-House Program

Many institutions look into the possibility of assigning someone on
their own staff responsibility for pest control. This may occur because
of dissatisfaction with pest control vendors, security reasons, sensitivity
of research, or because of cost savings. Advantages of using in-
house staff include detailed familiarity with the facility and a quick
response to complaints.

There are several disadvantages to using in-house staff. Pesticides and regulations constantly change, requiring diligent efforts
to stay abreast of new developments. Special knowledge and experience in solving difficult pest problems such as termite and bird control
are usually found in larger pest control companies. Vacations, sick
leave, and employee turnover must be considered. Pest control activities may be set aside if the staff is too busy with other duties. Professional pest control companies receive substantial discounts on
chemicals and supplies from distributors.

Unless the institution is very large, it is usually better to contract
for outside pest control service. However, there should be at least
one staff person from each facility that is assigned to work with the
pest control contractor. This person's duties should include sending
out an annual survey of the building to measure the degree and extent
of any pest infestations, coordination of contractor service schedules,
preparation of areas for treatment, and ensuring that work orders
for building repairs are completed.

B. Pest Control Contractors

A competitive bid package outlining IPM specifications should be used to select a qualified pest control contractor. Qualifications should include insurance coverage, state license, references, and years in business.

Specifications should be drawn up in advance and discussed prior to contract negotiations. Refer to the section in this chapter on design of an IPM program for general specifications in research laboratories. The number of hours of actual service should be specified.

C. Pest Control Consultants

Consultants can assist in writing the contract specifications, evaluate and select the pest control vendor, identify pests, and solve difficult pest problems. Consultants are also useful in training in-house staff and even pest control technicians. Since each facility has its unique needs, a consultant can make sure the program is implemented effectively. Do-it-yourself pest control without guidance is seldom successful.

VI. DESIGN OF AN IPM PROGRAM

A. Inspection

Pest control technicians should have an inspection form for recommendations at each facility. When conditions are discovered that promote the presence of pests, management should be notified. Loading dock doors that do not close tightly and therefore allow mice to enter is one example. Recommendations should be made regarding housekeeping and building policies that affect pest control measures. The storage and eating of food in research laboratories should be prohibited.

B. Survey Traps

Sticky traps are an ideal means for monitoring a large research building for pests. Potential pest sites such as laboratory sinks, janitor closets, floor kitchens, and storage rooms should be surveyed using these traps. It is a good idea to date the traps and replace them when they are old or damaged. Once each month is an adequate frequency for checking traps. During this check, staff should be questioned regarding any pest activity seen. A record should be kept of the survey results. To survey for rodents, snap traps, glueboards, and automatic live traps are effective tools. Although survey traps are not normally intended as a control measure, they often eliminate small and incipient infestations. The major value of traps is early detection and location of pests along with an evaluation of control efforts.

C. Control Measures

First consideration should be given to nonchemical control measures.
Can caulking and sealing cracks in a slab foundation prevent pest
entry? Can the rodents be sealed out by installing door sweeps? Can
electric light traps intercept incoming flies? Will repairing the leak
or sealing off the drain eliminate drain flies?

When chemical control measures are necessary, prior approval of
the principal investigator and safety department should be required.
Building personnel should always be notified in advance of treatment
dates and asked to vacate the immediate area during application of
volatile chemicals. (Common pests found in research institutions and
effective nonchemical and least-chemical methods of control are listed
in Table 1.)

D. Documentation

A logbook should be kept in a central location. This log should
contain material safety data sheets for all approved pesticides,
sample EPA pesticide labels, contract specifications, and a record
of in-house or vendor service activity. A sign-in and -out sheet
can be helpful. Numbers of pests captured in monitor traps and
pests reported by building personnel should be recorded in the
logbook.

Analysis of control results can be obtained from the data compiled
in this logbook. The decrease in number of reports or number of
pests over time can be quantified. The central logbook improves
communication between the service technician and research staff
regarding pest sightings. The technician should record for every
pesticide application the following: date, time, pesticide name,
amount, concentration, target pest, exact location, and method of
application.

VII. DISCUSSION

In an effective IPM program, building personnel should be responsible
for reporting pests in a timely and precise manner. Waiting until
the pests are numerous and then reporting that they are everywhere
is not a good policy. Try to be specific regarding location of pest
(e.g., sink or desk or location on the floor). This is the single most
helpful piece of information you can provide the vendor.

A commitment to good housekeeping and building repairs by the
institution will produce the best pest control results. An annual inspection of the building and a buildingwide clean up in preparation
for the inspection would motivate many laboratories to improve housekeeping in their area.

The future advancement of IPM in research institutions is encouraging. There is recent evidence using hot or cold temperatures effectively controls urban pests. Keeping some rooms below normal temperature will prevent pests from breeding.

Most current research in the pest control field is compatible with a reduction in conventional pesticide use. There is an ever-increasing number and variety of sticky traps available. Improvements in automatic live traps for rodents and new attractant baits for cockroaches have been made. The time has come for research institutions to take advantage of the many advancements they have made possible and apply them in their own laboratories.

REFERENCES

1. U.S. Environmental Protection Agency. (1979). National Household Pesticide Usage Study. 1976-1977, Report No. EPA/540/9-80-002, 126p.
2. Ebeling, W., Reierson, D. A., and Wagner, R. E. (1967). Influence of repellency on the efficacy of Blatticides. II. Laboratory experiments with German cockroaches. *J. Econ. Entomol.* 60: 1375-1390.
3. Ebeling, W., and Reierson, D. A. (1969). The cockroach learns to avoid insecticides. *California Agriculture 23*: 12-15.
4. Nelson, J. O., and Wood, F. E. (1982). Multiple and cross-resistance in a field-collected strain of the German cockroach (Orthoptera:Blattellidae). *J. Econ. Entomol.* 75: 1052-1054.
5. Federal Register 51(88), Part III, Department of Health and Human Services (N.I.H.). Guidelines for Research Involving Recombinant DNA Molecules; Notice. May 7, 1986.
6. Bennett, G. W., and Owens, J. M. (1986). *Advances in Urban Pest Management.* Van Nostrand Reinhold Company, New York.
7. Moore, W. S., and Granovsky, T. (1983). Laboratory comparisons of sticky traps to detect and control five species of cockroaches (Orthoptera:Blattidae and Blatellidae). *J. Econ. Entomol.* 76: 845-849.
8. Alpert, G., Liberman, D., and R. Fink. (1987). Control of German cockroaches in bio-medical research facilities using a systems approach (in preparation).
9. Public Law 80-104. The Federal Insecticide, Fungicide and Rodenticide Act. June 25, 1947.
10. Public Law 92-516. The Federal Environmental Pesticide Control Act of 1972. October 21, 1972.
11. Public Law 95-396. The Federal Pesticide Act of 1978. September 30, 1978.

12. Occupational Safety and Health Administration. Permissible exposure limits for toxic chemicals. Safety and Health Standards 29 CFR 1910.1047.

13. Weber, W. J. (1979). *Health Hazards from Pigeons, Starlings, and Sparrows.* Thompson Publications, California.

14. Ebeling, W., and Wagner, R. E. (1964). Built-in pest control for wall and cabinet voids in houses and other buildings under construction. *California Agriculture 8:* 8-12.

15. Ebeling, W., Wagner, R., and Reierson, D. A. (1969). Insect-proofing during building construction. *California Agriculture* (May): 4-7.

16. Slater, A. J., McIntosh, L., Coleman, R. B., and Hurlbert, M. J. (1979). German cockroach management in student housing. *J. Environmental Health 42:* 21-24.

5

Hazardous Chemical Management

MARTIN A. LEVITT* *Boston College, Chestnut Hill, Massachusetts*

I. INTRODUCTION

This chapter cannot properly succeed in its designated task unless a clear definition of the term "hazardous" can be established. All chemicals have the potential to be defined as hazardous if they are used inappropriately. Water, one of the compounds necessary to sustain human life, becomes a hazardous material when it comes into contact with an alkali metal such as potassium or sodium, causing it to burst into flames.

In general, all chemicals should be treated with a very healthy measure of respect. Because of the tremendous number of chemicals available in today's marketplace, sufficient information cannot possibly be gathered on all of the short- and long-term effects of exposure to these chemicals. Another factor worth considering is that chemicals thought to be safe today may, indeed, be declared extremely hazardous when enough long-term evidence is collected to implicate them as a causal factor in disease. Asbestos became an extremely popular material because of its fireproof properties, and it was used as a major component in hundreds of different types of building materials. Asbestos has since been found to be responsible for asbestosis and mesothelioma, and is heavily implicated as a major contributing factor to lung cancer. The Environmental Protection Agency has recently lowered what is considered to be a safe limit for asbestos and has concluded that no exposure to this material can be thought to be a safe exposure.

The EPA, OSHA, the American Conference of Governmental Industrial Hygienists, NIOSH, U.S. Dept. of Transportation, State Environ-

---

Current affiliation: Director of Industrial Hygiene and Training, NAACO, North Andover, Massachusetts

mental Quality Divisions, NFPA, associations of physicians, and the Cancer Society all contribute information to help delineate which chemicals are hazardous, carcinogenic, extremely flammable, etc. Some of these agencies set current limits as to how much exposure to any one of these chemicals may be considered safe, but a careful reading of a number of sources for information on a particular chemical will reveal that these expert agencies cannot agree on what constitutes a safe exposure level.

For our purposes, a hazardous chemical may be considered to be a chemical which is highly flammable, toxic, corrosive, carcinogenic, explosive, reactive, or is a solvent.

The proper management of hazardous chemicals must include all of the following considerations:

1. identification-hazards associated with chemical
2. purchasing
3. storage—centralized vs. decentralized
4. protection measures—facilities and equipment
5. emergency procedures—protect personnel, facility, environment

II. IDENTIFICATION—OR WHAT IS A HAZARDOUS CHEMICAL?

A. Definition

By definition, a hazardous chemical is one that present a danger to the user. Most chemicals in use today may be considered hazardous depending on their amount, concentration, and application.

The hazards associated with hazardous chemicals can be delegated to two different classifications, which determine how they should be handled from receipt to disposal: (1) physical hazards, and (2) physiological hazards.

Physical hazards include fire, explosion, and reactivity based on the chemical and on the particular type of containment equipment necessary to maintain the chemical in an inert state.

Physiological hazards include both acute and chronic toxic effects resulting from overexposure to a particular hazardous chemical. Acute effects include burns, allergic responses, and damage to mucous membranes, lungs, and the nervous system. Chronic effects may include cumulative damage to organs over a long period of time and may be carcinogenic.

B. Information

Appropriate management of chemical hazards must first consider the particular chemical to be used: its particular properties, particular storage criteria, protective equipment, etc.

One source for this type of data is the Material Safety Data Sheets (MSDS) supplied by manufacturers on request to comply with the OSHA Hazard Communication Requirements and state right-to-know laws. Although this information may not be complete, it at least offers a starting point and most often includes the type of criteria needed to make appropriate decisions. Information required to be included is:

1. Identification information—name of manufacturer, emergency phone numbers, chemical name, and formula
2. Hazardous ingredients
3. Physical characteristics
4. Fire and explosion hazard data
5. Health hazards
6. Reactivity data
7. Spill information
8. Protection equipment necessary
9. Handling and storage procedures
10. Special precautions

MSDSs must be provided by a chemical manufacturer after a chemical is purchased, but this information should be available before purchase so that appropriate handling measures may be considered.

III. MANAGEMENT OF HAZARDOUS CHEMICALS

A. Substitution

The most obvious and safest way to manage hazardous chemicals is not to use them. Although on the surface this may seem to be both a facetious and simplistic statement, it has a particular relevance for today's modern chemical laboratory.

Careful consideration of the experiment or procedure to be undertaken may indicate certain choices to be made in the selection of chemicals, especially solvents, which, if safety is in mind, could result in a less hazardous product being used. This option should result in a twofold benefit—a safer experiment and a less hazardous, and often less costly, waste to be disposed of when the process is complete.

B. Reduction

Purchasing Strategies

Obviously, if not using any hazardous chemicals is a desirable but unreasonable policy, then "less is better." *Less is Better* is the title of a 16-page pamphlet published by the American Chemical Society

(1) which challenges our supermarket mentality in buying larger-size containers because they are more economical. Indeed, in a very careful analysis, it is proven that buying more than is needed for the procedure contemplated is a very costly decision for a number of reasons, two of the most important being safety and the cost of disposal.

The scenario is one we are all familiar with. A researcher needs a particular amount of a chemical, looks it up in a catalog, and finds that if he buys the larger size container, his cost per unit volume is significantly reduced (Table 1). The larger amount is purchased because of its economy, a small portion is used, and the remainder sits on a shelf in the lab. Time goes by, and if the material is not forgotten, or if the label has not fallen off, the material is considered too old to use because the chances of contamination could jeopardize valuable research results and the material is set aside to be disposed of at considerable cost.

Purchasing the minimum quantity of material necessary to do a procedure has considerable added benefits to the overall safety of the whole operation. If the chemical is used up, there is no remainder left

TABLE 1 Purchasing-Disposal Economics (in 1984 dollars)

	Size of package	
	500 ml	2500 ml
Purchase price	$15.50	$52.00
Unit purchase price	3.1¢/ml	2.1¢/ml
When 1000 ml are used, the cost will be:		
Unit purchase cost	3.1¢/ml	5.2¢/ml
Package disposal cost	0	$22.67
Total cost		
(Purchase and disposal)	$31.00 (purchase only)	$74.67
Unit total cost	3.1¢/ml	7.5¢/ml
When 1677 ml are used, the cost will be:		
Unit purchase cost	3.7¢/ml	3.1¢/ml
Package disposal cost	$6.54	$22.67
Total cost		
(Purchase and disposal)	$68.54	$74.67
Unit total cost	4.1¢/ml	4.5¢/ml

for disposal, the individual lab does not become a ministockroom, and the associated safety liability and disposal costs are reduced. Recognizing this situation, at least one chemical supplier (Pfaltz and Bauer, Inc., Waterbury, CT) is offering a service where they will ship the exact quantity of chemical needed for the research. Other companies may soon follow suit and offer repackaging service. It may be to the institution's advantage to find out if this service will be made available by its largest supplier.

Purchasing Procedures

Once the decision has been made to purchase a hazardous chemical, the next question to ponder should not be how much of a hazardous chemical to buy, but how little? Strict controls must be put in place by both the persons responsible for the purchasing function and those responsible for the general safety of the laboratory and facility.

A request form should be used by the would-be purchaser of a chemical that provides at least the following amount of information to the person responsible for the purchase:

1. Amount of chemical used in past—month and days
2. Projected amount to be used in future
3. Estimated shelf life
4. Adequacy of storage space for this chemical for time necessary to completely use up containers
5. Appropriate equipment available for transport, storage, dispensing
6. Appropriate personnel
7. Cost in bulk
8. Cost in smallest usable quantities
9. Estimated amount to be disposed of—if bought in bulk or small quantities

This usage information should be checked against past purchases for the following reasons:

1. To see if the information is accurate
2. To see if others use the same chemical
3. To determine the actual quantity to be purchased

Strict control over chemical purchases does not have to result in long delays. Blanket purchase orders can be used to allocate a specific dollar amount to a "low bid" vendor with the arrangement that a chemical ordered will be shipped immediately. This type of arrangement reduces the amount of inventory in storage as chemicals are not present until they are needed, frees valuable space, and ensures that chemicals are fresh and have not been sitting for a long time on a shelf in an institution deteriorating.

These procedures give greater control to those exercising the purchasing function rather than to those controlling the laboratories and
could result in the purchase of smaller quantities, leading to loss of
discounts from both purchases and increased freight charges. Careful
consideration and good control could result in a system where chemicals
could be ordered on a weekly or monthly basis predicated on a particular rate of use. Monitoring this rate by checking with the stockroom
or the person responsible for hazardous waste reception could provide
feedback to indicate if the system was working as designed.

Proper planning is a vital part of the safety program of any facility
that buys and uses chemicals, and sufficient emphasis at this stage
will produce positive benefits as the chemicals come into use.

IV. STORAGE

One of the most crucial problems in dealing with hazardous chemicals
is what to do with them until they are actually used.

A. Centralized vs. Decentralized Storage

There are two options to consider when it comes to storage for hazardous chemicals—centralized and decentralized storage.

The nature of the chemicals and their associated hazards will usually dictate a preference for a centralized location. Often, however,
the actual physical configuration of the facility may preclude a large
central storage area that will meet appropriate regulatory criteria.
Assuming that necessary regulatory guidelines could be met in either
situation, consideration should be given as to the most desirable
approach.

A chemical storage area in each laboratory certainly offers conven-
ience for the researcher but would preclude bulk purchases and pose
additional hazards in the laboratory. A centralized storage area, on
the other hand, would be less convenient but would offer the added
option of bulk purchases and avoid the necessity of each lab having
its own inventory of seldom used chemicals. Although different grants
used in the purchase of chemicals complicate matters, a controlled purchasing plan along with an inventory control record for each chemical
could result in a very efficient way to save both valuable shelf space
and hard-to-come-by research grant money, as chemicals could be
purchased in larger quantities if they were to be used by more than
one researcher.

B. Design of a Storage Facility

The expense of creating an appropriate storage area will usually require that most, if not all, of the hazardous chemicals within the build-

ing of institution be placed there. The safety of this controlled area should be at least as great as would be found if the chemicals were present in a lab area with the greater association of fire, explosion, and spills because persons are actively working with the chemicals.

Significant attention to the design of an adequate storage space should begin with an analysis of the types of hazards to be protected against. Many chemicals have several different hazards associated with them, and close attention to one aspect should not obscure others.

In chemical storage, one of the most serious consequences of an accident is fire with its associated release of toxic gases and vapor, and associated physical destruction of the facility, economic loss, and potential for loss of life. Consequently, protection from fire must be a prime consideration in the design of a hazardous chemical storage area.

Fortunately, storage for flammable materials has been extensively considered by the NFPA in its Flammable and Combustible Liquids Code-NFPA 30 (2,3).

NFPA 30 defines flammable liquids as those having flash points below 100°F and combustible liquids as those having flash points above 100°F. The flash point is defined as the minimum temperature at which a liquid gives off sufficient vapor to form an ignitable mixture with air near the surface of the liquid. This concept is used because in a flammable liquid fire it is the vapor rather than the liquid that actually burns.

In order to contain these chemicals in a storage area adequately protected from fire, the NFPA calls for 2-hr fire resistant construction, adequate ventilation, leak-proof joints between walls and floors, special curbs or drains if containers larger than five gallons are to be stored, and sets maximum limits on the amount of flammable and combustible material to be stored based on size of the room, adequacy of fire resistant construction, and the availability of an automatic fire extinguishing system. Most recommended is a separate building for flammable storage complete with explosion-proof blowout panels on roof or walls. The code also recognizes a "cut-off" storage room, which is a room containing at least one exterior wall to allow access for firefighters. Flammable liquids with flash points below 100°F, also referred to as class I, are specifically excluded from basement storage. Also allowed according to the code are inside storage rooms, but their size and storage capacity is expressly limited.

A less expensive alternative to a separate building or costly interior renovations in already crowded buildings may be the purchase of hazardous chemical storage containers, which combine the advantages of a separate building with low cost and portability. These containers as manufactured by Safety Storage Inc. and others offer different sizes and options depending on the chemicals to be stored. They may be equipped with blow-out panels, explosion-proof lighting, chemical fire extinguishing systems, ventilation, heat, alarms indicating leaking containers, and other options to expressly design the container to match the chemicals to be stored.

In a laboratory setting, flammable chemicals should be stored inside a flammable storage cabinet, which is essentially a metal or wooden cabinet assembled in such a way as to contain vapors. The design of the cabinet itself is discussed in NFPA 30, and the amount of flammable materials to be stored in a laboratory both inside and outside of flammable storage cabinets is prescribed by NFPA 45 for laboratories using chemicals.

C. Storage Strategies

As the storage of chemicals is considered within a flammable storage area, emphasis should then be placed on particular classes and combustibilities. If the flammable storage area has been designed with a water sprinkler system, then chemicals that are incompatible with water should be excluded. A CO_2 or halon fire extinguishing system will probably be a more appropriate choice because of its increased ability to put out flammable liquid fires.

Storage of chemicals should depend on the characteristics of the individual chemicals involved. It is most appropriate to store chemicals that react similarly together. Flammables should be stored with flammables and combustibles, corrosives should be stored together, acids should not be stored with bases, and the list could go on. One of the problems encountered will be the fact that every further division and classification will require a separate storage area. The J. T. Baker Chemical Co. recommends 16 different classifications, the U.S. Coast Guard classifies chemicals into 43 division, and the resources of an institution could be completely consumed by trying to provide an isolated area for each possible different type of chemical. The question now involves whether separation of chemicals is sufficient as opposed to isolation.

Practically, common sense and adherence to a reasonable scheme will allow an institution to provide for a safe storage area, albeit a compact one.

Leslie Bretherick, in "Incompatible Chemicals in the Storeroom: Identification and Segregation" (4), provides a reasonable solution to the storage problem that will be outlined here (see Figure 5-1).

Suggested specifically are eight categories of storage:

1. Flammables that are compatible with water
2. Flammables that are incompatible with water
3. Nonflammables that are compatible with water
4. Nonflammables that are incompatible with water
5. Materials that become unstable above room temperature—keep away from heat
6. Materials that are unstable at room temperature requiring refrigeration in an explosion-proof refrigerator
7. Pyrophoric materials, which will burst into flame when their container is broken

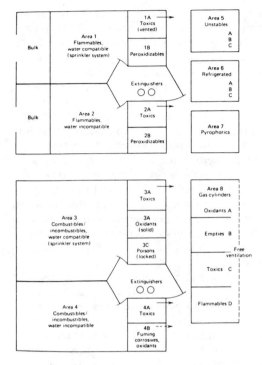

FIGURE 5-1 Outline of schematic arrangements for segregated storage of a wide range of laboratory chemicals (Note: This is not intended as a working ground plan.)

8. Compressed gases in cylinders—isolation from sparks or heat and oxidants separated from flammables

Additional consideration must be paid to toxics, peroxides, oxidants, poisons, fuming corrosives, etc. to see where they fit into the basic first four categories. Chemicals that belong in the same storage area but that are incompatible with each other, such as acids and bases, should be separated as much as possible. This may be accomplished with compartmentalized or modular storage cabinets or by placing them on remote shelves in the same storage area.

The storage area should be well lit, may contain flammable storage cabinets for further segregation, should have aisles at least 3 feet wide and contain no dead-ends. It should also have an adequate amount of corrosion-resistant shelving with raised edges, braced for extra strength, attached securely to the walls or floor. The shelves should not be overcrowded and should not exceed the design weight limits. Every item placed on a shelf should be clearly labeled with the chemical name, precautions to be taken in handling and the date received. If a chemical has an expiration date this should be highlighted. Most chemical companies voluntarily do an adequate job of labeling their chemicals but care must be taken so that a neophyte user is protected from unknown hazards. Chemicals should be received by someone in the organization who has a clear understanding of the basic classifications, sufficient knowledge, and resources to look up chemicals to classify them according to institutional guidelines and label them appropriately for the user. Depending on the size of the chemical container, labeling may consist of DOT symbols (NFPA 704 Diamond System) classifying flammability, reactivity and health hazards from 0 as least hazardous to 4 as most hazardous, to simple pictograms identifying protective equipment to be used. Care should be taken so that chemicals are rotated using the oldest stock first. Inventory records should be kept, indicating who purchased the chemical, date purchased, amount used, and amount to be disposed of. This can be done adequately by using a file card system or more efficiently with any basic data base management system and a personal computer.

Additional areas will be necessary for the storage of hazardous waste and for the dispensing of flammable liquids. Hazardous waste should be segregated from new chemicals to prevent corrosion as well as confusion. It should also be segregated as much as possible using the previously mentioned classification scheme as well as arranging the chemicals according to DOT categories for final disposal. All hazardous waste should be very clearly labeled with a tag of sufficient size indicating the name of the chemical, name of researcher, lab room number, date collection was started, and date container was filled. Careful attention should be paid to overfilling of containers. While it is not economically sound to place into lab packs minimally filled containers, containers that are filled to the very top have a nasty habit of spilling over. Flammable solvents expand during warm weather and may cause the container to leak or rupture. A lab pack is a way of shipping hazardous waste to another facility or to a landfill and consists of a 30 or 55 gallon drum containing vermiculite and 10-15 gallons of compatible chemicals according to Department of Transportation guidelines.

Dispensing areas should be separated from general storage areas because of the increased risk of fire, explosion, and exposure to toxic vapors. Volatile, flammable solvents in large containers can be dispensed through a number of methods: pouring, gravity flow, pressure from inert gas, or pumping. Pumping using a hand pump is the safest method available but is often overlooked by persons in a hurry.

Flammable liquids being transferred from large metal containers to smaller ones should at least be connected to each other by a bonding wire which will prevent a difference in electrical potential from creating a spark and igniting the vapors. It is also good practice to have the larger metal container connected to an earth ground so that there will be no electrical difference from the container to the ground. Ideally all metal containers will be safety cans having flame arrester vents, and class 1A flammable liquids may not be stored in containers larger than 1 gallon outside of the storage area except in a 2-gallon safety can (NFPA 30). Transferring of chemicals using any of the previous methods exposes chemicals, which will be inhaled if proper respirator protection is not used.

V. PROTECTION OF PERSONNEL

In all matters relating to hazardous chemicals, the best solution to limit exposures is an engineering solution. Harmful vapors can be ducted away using several engineering techniques such as transferring in a hood, using a slotted hood, placing a portable duct at the point of transfer, etc. Controlling the exposure to the user is always the best way to solve the problem. Relying on personal protective devices to protect the user from harm presupposes cooperation from the user which cannot always be guaranteed. The only antidote to potential self-destructive behavior placing the person at risk is constant, repeated training and supervision. Personnel involved in working with hazardous chemicals must be sufficiently trained so that they know the risks involved and the appropriate methods to avoid them. It is the responsibility of all persons involved in the program, from the president of the institution to the safety official, the supervisor, and the student, to assure that all proper precautions are taken to avoid exposure to hazardous and toxic materials.

Protection from exposure and avoidance of fire, spills, and accidents requires the installation in the storage area in the laboratory of certain equipment and devices to protect the facility and the user.

Fire protection equipment ranging from water sprinklers, automatic fire extinguishing systems using CO_2, foam, halon, and dry chemicals, to simple hand-operated fire extinguishers should be present in these areas according to the type of chemical used or stored. Obviously water sprinklers are not indicated in an area of flammable, water-incompatible chemicals.

Hand-held fire extinguishers are capable of reacting to three different types of fires, and it is important that those involved understand the difference in fires and utilize the right type of extinguisher. Class A fires are of ordinary combustibles such as wood and paper. Class B refers to flammable liquids such as gasoline and other solvents. Class C fires are electrical fires. A pressurized water fire extinguisher is rated only for A fires. It will be ineffective against a Class B fire and add to the problem in a Class C fire because water is a conductor of electricity. Certain dry chemical and halon fire extinguishers, while effective against Class B and C fires, will be rendered useless when applied to a Class A fire. All fire extinguishers are clearly marked as to their effectiveness for types of fires, and the operating instructions should be read and understood long before the opportunity to use the extinguisher presents itself.

The presence of drench showers, eyewashes, fire blankets, protective clothing including coveralls, aprons, gloves, respirators, goggles and safety glasses, as well as the availability of self-contained breathing apparatus and access to a first aid kit may all be considered to be vital components that should be present based upon the particular nature of the chemicals to be used. Careful attention to the MSDS should give guidance as to what types of precautions are necessary to be taken before one works with any potentially hazardous chemical.

Unfortunately and invariably, accidents will happen. Of primary importance is to protect the user of the chemicals. When a fire starts, if it cannot be put out with a fire extinguisher quickly, it is best to evacuate quickly, closing gas valves and doors and shutting off electricity if possible and ringing the general alarm alterting others in the building. If clothes are burning, instruct the person not to run to the fire blanket but to practice the NFPA refrain that they should already be familiar with—STOP, DROP, AND ROLL—until the flames are out. Once a person is on the floor it is alright to try to cover them with a fire blanket, but putting it on them while they are standing will cause the blanket to act as a chimney and duct smoke and flames to the face.

A person overcome by smoke or fumes should be immediately transported to fresh air, administered oxygen, cardiopulmonary resuscitation applied if necessary, and emergency personnel notified.

In any type of emergency laboratory personnel should try to remain calm and do whatever is necessary to preserve life.

When chemicals are splashed on the skin or clothing or in the eyes, immediate action is necessary—immediately douse with water for at least 15 minutes using the drench shower, remove clothing if necessary (the potential for serious burns taking precedence over modesty), and, if in the eyes, hold lids open and wash for at least 15 minutes and have someone summon emergency help.

When chemicals are ingested, have the person drink large quantities of water while on the way to medical assistance. Try to clearly identify what chemical has been ingested to medical authorities.

Inherent in this basic information is the fact that an emergency plan should already have been prepared taking into account most probable emergencies. All personnel in the area should have some basic understanding of first aid and CPR and should be familiar enough with the emergency plan so that they will take appropriate action as necessary and not panic.

VI. PROTECTION OF THE ENVIRONMENT

Protection of the environment, while not as immediately important as the protection of the individual, should also be considered a vital part of any emergency plan. Environmental incidents, if not handled quickly and properly, have the potential to be injurious to those involved in the incident and to others along the chain in the ecological cycle.

The spill control plan should take into account the following criteria before it can be considered adequate: (1) accurate information about the exact nature of the material spilled, (2) quick availability of information about the chemical spilled—manufacturer's MSDS, NIOSH pocket guide to chemical hazards, Emergency Response Guidebook published by the Department of Transportation Materials Transportation Bureau as well as others providing handling and hazards data for a large selection of chemicals—and (3) adequate and appropriate equipment available.

The nature of the chemical spill obviously dictates the nature of the material to be used in its cleanup, but previous considerations must be given so those materials will be on hand and ready when an accidental spill occurs. The most important matter to be considered is naturally the protection of the person responding to the spill. The worst case should always be considered, and personal protective equipment including but not limited to self-contained breathing apparatus, full coveralls, face shields, goggles, cartridge respirators, and different types of gloves should be available in close proximity to the storage area and the laboratory.

A commercial spill cart may be used, but it is more appropriate to design your own based on the types of incidents that may occur at your facility.

Equipment used in containing and cleaning up the spill can be anything from paper towels to booms and dikes and other specially prepared materials created expressly for dealing with spills.

At the very minimum, materials should be available to absorb a spill from the largest single container in the area. Generally, material should be available for dealing with solvents, acids, corrosives, and mercury.

A. Organic Solvent Spills

Organic and flammable liquids present additional problems when spilled. As the size of the spill increases, so does the surface area and the amount of vapors available to be inhaled or to be involved in a fire or explosion. Step 1 should include turning off all sources of ignition (flames or sparks). Step 2 should include putting on the appropriate respiratory protection, because the person cleaning up the spill is the most likely to be injured. Several commercial products are available made of expanded silicate, cellulose, polypropylene, polyethylene, and other materials and are sold as pillows, dikes, pads, rugs, and granules. These materials absorb better than clay or vermiculite and can usually be incinerated in the diposal process instead of landfilled. Their lighter weight and different forms make them quite suitable to contain and clean up most spills.

After the respirator is in place and nonconductive shoe coverings are put on, the spill should be surrounded or diked. Once the spill has stopped spreading, additional material should be added to absorb all of the liquid from the outside in. Special care should be taken so that the respondent does not step into the spill. When all of the liquid is absorbed, the material should be picked up using nonsparking (plastic) tools, placed into a plastic bag, tied and put into a 55-gallon drum for disposal, keeping in mind that the disposal material is still quite flammable.

B. Acid and Caustic Spills

Procedures for responding to spills of acids or caustic solutions are generally the same but differ in the type of protective equipment necessary and in the application of the absorbent on reactive material. Acids may cause serious burns to the face, eyes, and hands, and it is extremely important to wear protective gloves and eye goggles before spill cleanup is begun. Concentrated acid spills will also require the use of an approved respirator.

Commercial products that change color when neutralized are available to neutralize the acid or base so that it can be disposed of safely. It is essential to know whether an acid or base has been spilled so that the right material may be used in the cleanup. Once this has been determined, apply the chemical (usually available in a spray bottle) to the perimeter of the spill and then inward. Expect this mixture to foam and produce a lot of heat. When foaming has stopped, continue to add the chemical until the indicated color change has occurred. Once the material has cooled, absorb it with appropriate materials and place into a plastic bag for later disposal.

C. Mercury Spills

Mercury gives off toxic vapors and, because it is present in many types of scientific apparatus and instruments, especially thermometers, spills should never be ignored. The threshold limit value (TLV) for organic mercury compounds is 0.001 parts per million (ppm), indicating that it takes an extremely small amount to be toxic.

When mercury is spilled it may be cleaned up in a number of ways depending on the amount spilled. Careful attention should be paid to the production of vapors. Droplets may be picked up by suction using an aspirator bulb or a specially designed mercury vacuum cleaner which vacuums up the spill and then uses centrifugal force to separate the mercury from the other materials also picked up. Several commercial preparations are available that use powdered sulfur to precipitate mercuric sulfide. Sponges are available to absorb the mercury and activated carbon granules are used to keep mercury vapors under control.

In the spill area shoe covers should be used and care should be taken to not step in the spill area. After the spill is cleaned up, mercury levels should be checked with appropriate instruments or specially impregnated mercury monitor paper, which changes color in the presence of mercury.

REFERENCES

1. The American Chemical Society. (1985). *Less Is Better.* Dept. of Government Relations and Science Policy, Washington, DC.
2. The National Fire Protection Association. (1985). NFPA 30 Flammable & combustible liquids code. Quincy, MA.
3. The National Fire Protection Association. (1985). NFPA 45 Laboratories using chemicals. Quincy, MA.
4. Pipitone, D. A. (Ed.). (1984). *Safe Storage of Laboratory Chemicals.* John Wiley & Sons, New York.

6

Hazard Control in the
Animal Research Facility

NEIL S. LIPMAN *Massachusetts Institute of Technology,
Cambridge, Massachusetts*

CHRISTIAN E. NEWCOMER *Tufts University, School of Veterinary
Medicine, Boston, Massachusetts*

I. INTRODUCTION

The sources of hazard for research personnel and animal care staff in
the animal research facility are diverse and numerous. Activities associated with the care and use of laboratory animals lend themselves to
physical, chemical and/or biological injury. Many of the hazards present in the animal research facility are the same as in any general laboratory, however, certain procedures unique to the maintenance of laboratory animals create additional risks for animal care and laboratory personnel. Animals may be a source of injury. They may harbor infectious agents that can cause disease in man, can inflict physical harm
through bites and scratches, or can precipitate cutaneous and inhalant
allergic reactions through their dander and urine. Animals may pose
additional risks if they have been inoculated experimentally with infectious agents or administered toxic or radioactive substances. It is important to recognize the existence of these potential hazards in order
to establish safety programs that limit their effects. Safety programs
should be designed to be commensurate with the level of hazard. Insufficient precautions can result in serious consequences, but unneces-
sarily restrictive programs can be costly and time-consuming, inviting
noncompliance by participating personnel.
    A number of regulations and guidelines have been established that
directly or indirectly deal with the control of hazards when utilizing

Supported in part by grant RR01046 from the Division of Research
Resources, National Institutes of Health.

animal subjects or maintaining animal facilities. These include the Animal Welfare Act (1), the Good Laboratory Practices Act (2), the Occupational Safety and Health Act (3), and the *Guide for the Care and Use of Laboratory Animals* (4). These documents are concerned with a variety of areas ranging from architectural and engineering specifications to adequate veterinary care, quarantine requirements, and occupational safety for personnel. Although many of these guidelines are concerned with the safety of the laboratory animal subject, safety of animals and research personnel are interdependent, thus safety procedures beneficial for the animal most often benefit personnel. A number of other federal regulations may guide specific types of experimental animal utilization according to the nature of the compound or test material under study. In addition to protection of the environment and personnel from direct exposure to experimental agents, the disposition of animal carcasses containing these materials requires special considerations. The purpose of this chapter is to define and address the potential hazards found in the animal research facility and provide recommendations for their prevention or control.

II.   DEFINING THE HAZARDS

A.   Animal-Related Hazards

Trauma

Animals may be a source of physical and biological injury. Animals can inflict physical harm to their handlers through bites and scratches. Animal bites can cause severe mechanical damage to tissues. In addition to physical trauma, the microbiological flora of the oral cavity and the nailbed may be a source of wound contamination, which can lead to a localized or systemic infection, e.g., rat bite fever. Prevention of animal-induced injury is based on the knowledge and practice of proper animal restraint by personnel. Familiarity with the behavioral characteristics of the species or of individual animals may be helpful and is conducive to humane care and handling of laboratory animals. Most laboratory animals are extremely docile, however, the possibility of unpredictable events should not be overlooked when handling animals. This is particularly applicable to animals that are frightened or in pain. Cats and rabbits are two common laboratory animals likely to scratch their handlers as a defense mechanism. Proper handling and restraint of these species will prevent the infliction of trauma. There are a number of restraining devices available for many of the commonly used laboratory species. Chemical restraint, i.e., utilization of a sedative, tranquilizer or immobilizing agent, may also to aid in the restraint of an animal.

Zoonotic Diseases

Animals may harbor infectious microorganisms that are zoonotic. Zoonoses are infections or infestations shared by man and lower vertebrate animals. Zoonoses may be caused by viruses, bacteria, fungi, chlamydia, ricketssia, or parasites (see Tables 1, 2, and 3). These agents may cause clinical disease in both their animal and human hosts or may be subclinical or inapparent in one and cause clinical disease in the other. One of the most problematic situations occurs when the animal host is an asymptomatic carrier of a zoonotic agent. Animals have been implicated as a major cause of laboratory-acquired infections. Studies indicate that 17-40% of occupationally acquired infections in the laboratory can be attributed to animals (5,6).

The Process of Infection. An understanding of the natural history of a disease in animals and man is required in order to formulate a disease control program. Transfer of an agent from animal to man or vice versa requires escape of the agent from the reservoir, a means of transmission from the reservoir to a new host, and the ability of the agent to gain entry into the new host, animal or man. Prevention of disease spread requires breaking one of these links. Agents will differ in their pathogenesis, virulence, infectivity, and stability, all of which determine how readily the agent will spread within animal populations and between animal and man (7).

The virulence of an agent is an important characteristic to assess in risk or hazard evaluation. Microbial agents that are capable of causing disease by exposure to a limited number of organisms are of greater concern than those requiring large numbers of the organism. Agents causing high mortality or permanent disability are of obvious concern. The availability of therapeutics should also be considered when assessing the risks involved in containing an agent. Risks are reduced when vaccines or other immunoprophylactic agents are available that prevent infection.

The environmental stability of an agent is another important factor of consideration in the risk assessment of agents present in the animal research facility. Agents capable of surviving in a wide range of temperatures and humidity, or resistant to disinfectants, sunlight, and other adverse environmental conditions pose a greater threat than those that are susceptible. Control of such agents is much more difficult due to their wider dispersion and persistance in the environment.

Mechanisms of Escape. Agents can escape from their reservoir naturally via excretion or secretion from body orifices and/or release from external lesions, mechanically by investigators removing contaminated bodily fluids or tissues, during surgery or necropsy, or by ectoparasites that contaminate the animals' environment. Transmission

TABLE 1 Zoonotic Viral, Bacterial, Rickettsial, Chlamydial, Fungal, and Protozoal Agents in the Laboratory

Disease/Etiology	Natural host(s)	Comments
I. Viral		
Poxviruses		
Monkey Pox	Primarily Old World nonhuman primates (NHP)	Produces systemic disease; important to differentiate from smallpox.
Benign epidermal Monkey Pox/Tanapox Virus(?)	Old World nonhuman primates	Cutaneous lesions which regress in 2-3 weeks in humans.
Yaba Virus	Old World nonhuman primates	Has been demonstrated only by experimental human infection to be zoonotic.
Vaccinia Virus	Numerous mammalian hosts	Generally important to consider only when animals are being infected experimentally.
Contagious Ecthyma (ORF)/Parapoxvirus	Sheep and goats	Most common zoonotic pox virus in laboratory environment. Inoculation of skin abrasions or bites produces large painful nodules which regress in 1-2 months.
Hemorrhagic Fever Viruses		
Yellow Fever	Nonhuman primates	Mosquito vector required; animal quarantine measures are essential to control.
Korean Hemorrhagic Fever/Hantaan Virus	Rat and several species of wild rodents	Systemic disease has been recorded in several laboratory animal handlers. Laboratory rats have subclinical disease.
Lassa Fever/Lassa Virus	Multimammate rat	Representative of several other Arenaviruses which occur in the laboratory under unusual circumstances.

Disease	Source	Comments
Marburg Virus Disease	African green monkeys, other NHP	Focal epidemics have occurred with lab exposure.

Miscellaneous Viruses

Disease	Source	Comments
Rat Bite Fever / *Streptobacillus moniliformis*, *Spirillum minus*	Rats, mice	Appropriate handling is essential to prevention.
Cat Scratch Fever	Cats	Appropriate training in handling cats is essential for protection.
Rabies	Dogs, cats, NHP, and other random source animals	Vaccinations are available and recommended for personnel at high risk.
Lymphocytic Choriomeningitis Virus	Mice and hamsters	Can be introduced into mouse colonies by tumor/cell line passage. Humans develop flulike illness with a potential for central nervous system involvement.
Measles (Rubeola)	Nonhuman primates	Humans are the ultimate reservoir. Transmission from NHP to man is conceivable in the laboratory animal facility.
Hepatitis A Virus	Many NHP, especially, chimpanzees	Immunoprophylaxis recommended in high risk personnel in laboratory animal facility.
Herpesvirus simiae (Herpesvirus B)	Macaques spp. of NHP	Endemic in most colonies of macaques. Recent fatalities due to this agent have been reported in laboratory animal personnel.

II. Rickettsial/Chlamydial

Disease	Source	Comments
Chlamydiosis (Psittacosis)/*Chlamydia psittaci*	Birds, mammals, and Xenopus frogs	Avian strains are usually implicated in human cases of infection, however, mammalian strains are known to be transmissible to humans.

TABLE 1 (continued)

Disease/Etiology	Natural host(s)	Comments
Q Fever/ *Coxiella burnetti*	Sheep and other domestic ungulates, wild animals	Recent outbreaks in numerous laboratory animal facilities linked to the experimental use of pregnant sheep.
Rocky Mountain Spotted Fever/ *Rickettsia rickettsia*	Dogs, cats, wild rodents	Control of the ectoparasite vector is key to preventing infection in the laboratory animal facility.
Rickettsial Pox/ *Rickettsia akari*	Mice	Elimination of feral mice from animal facility provides control.
III. Bacterial		
Brucellosis/ *Brucella canis*	Dogs	Produces bacteremia and systemic disease in humans. Serological detection useful in health profiling dogs.
Leptospirosis/ *Leptospira spp.*	Large domestics, dogs, rats, mice, and wild caught mammals	Monitoring for infection in animals other than the large domestics is impractical.
Plague/ *Yersinia pestis*	Wild rodents and rabbits, cats, and dogs	Control of feral rodents and ectoparasite. Control on random source animals is preventive.
Tuberculosis/ *Mycobacterium tuberculosis*	NHP	Routine surveillance of nonhuman primates and personnel are necessary to prevent this serious zoonotic disease.
Campylobacteriosis/ *Campylobacter jejuni*	Dogs, cats, nonhuman primates	Human diarrheal disease has been associated with infected laboratory animals.

Disease/Agent	Host	Comments
Salmonellosis/ *Salmonella spp.*	Mammals, birds, reptiles	The most universal zoonotic disease; should be included in the health profiling of all species.
Shigellosis/ *Shigella spp.*	Nonhuman primates	Produces acute diarrheal disease in humans; included in health screen of NHP.
Yersiniosis/ *Yersinia spp.*	Many mammals	Most human cases presumed to be due to food contaminated by animals.
Dermatophilosis/ *Dermatophilus congolensis*	Mammals and reptiles	Cutaneous lesions produced in man; rare in laboratory animal facility.
IV. Fungal		
Ringworm	Cats, dogs, laboratory rodents, other mammals	Examination of newly arrived dogs and cats particularly important to control in laboratory animal facility.
V. Protozoal		
Toxoplasmosis/ *Toxoplasma gondii*	Cat	Control through rigorous sanitation and identification of infected cats; pregnant women should be reassigned to different duty.
Amebiasis/ *Entamoeba histolytica*	Nonhuman primate	Detection and treatment of NHP in quarantine and personnel practices key to control.
Cryptosporidiosis	Mammals, birds, reptiles	Zoonotic transmission in the laboratory animal facility has not been reported; agent produces chronic diarrhea.
Giardiasis/ *Giardia spp.*	Dogs, cats, nonhuman primates, and other laboratory animal species	Detection and treatment of infected animal during quarantine is recommended.
Pneumosystis carinii	Numerous mammals	Animals are not known to be infectious to humans.

TABLE 2 Zoonotic Helminth Parasites in the Laboratory Environment

Disease	Etiology	Natural host(s)	Aberrant hosts	Comments
Cestodiasis	*Hymenolepis nana*	Rats, mice, hamsters, nonhuman primates	Humans	Intermediate host is not essential to the life cycle of this cestode. Direct infection and internal autoinfection can occur also. Heavy infections result in abdominal distress, enteritis, anal pruritis, anorexia, and headache
Strongyloidiasis	*Strongyloides stercoralis, S. fulleborni*	Nonhuman primates, dogs, cats, humans, Old World nonhuman primates	Humans	Oral and transcutaneous infections can occur in animals and humans. Heavy infections can produce dermatitis, verminous pneumonitis, enteritis. Internal autoinfection can occur
Ternidens infection	*Ternidens deminutus*	Old World primates	Humans	Rare and asymptomatic
Ancylostomaisis	*Ancylostoma duodenale*	Humans	Nonhuman primates, pigs	Oral and transcutaneous routes of infection occur. Heavy infections produce transient respiratory signs during larval migration followed by anemia due to gastrointestinal blood loss
	Necator americanus	Humans	Nonhuman primates	
Trichostrongylosis	*Trichostrongylus colubriformis. T. axei*	Ruminants, pigs, dogs, rabbits, Old World nonhuman primates	Humans	Heavy infections produce diarrhea
Oesophagostomum	*Oesophagostomum* spp.	Old World primates	Humans	Heavy infections result in anemia. Encapsulated parasitic granulomas are usually an inocuous sequela to infection

Ascariasis	*Ascaris lumbricoides*	Old World primates	Humans	Infection occurs by ingestion of embryonated eggs only. Embryonation, requiring 2 or more weeks, ordinarily would not occur in laboratory. Heavy infections can produce severe respiratory and gastrointestinal tract disease
Enterobiasis	*Enterobius vermicularis*	Humans	Old World primates	Oral and inhalational infection can occur. Disease in humans characterized by perianal pruritis, irritability and disturbed sleep
Trichuriasis	*Trichuris trichiura*	Humans	Old World primates	Three-week embryonation makes laboratory infection highly unlikely. Heavy infection in humans results in intermittent abdominal pain, bloody stools, diarrhea, and occasionally rectal prolapse
Larval migrans (viscera)	*Toxacara canis* *Toxacara cati* *Toxacara leonina*	Dogs and other canids Cats and other felids Dogs, cats, wild canids, and felids	Humans Humans Humans	Chronic eosinophilic granulomatous lesions distributed throughout various organs. Should not be encountered in laboratory
Larval migrans (cutaneous)	*Ancylostoma caninum* *Ancylostoma braziliense* *Ancylostoma duodenale* *Uncinaria stenocephala* *N. americanus*	Dogs Dogs, cats Dogs, cats Dogs, cats Dogs, cats	Humans Humans Humans Humans Humans	Transcutaneous infection causes a parasitic dermatitis called "creeping eruption"

Source: Reprinted from Fox, J. G., Newcomer, C. E., and Rozmiarek, H. (1984). Selected zoonoses and other health hazards, in Fox, J. G., Cohen, B. J., and Loew, F. M. (eds.), *Laboratory Animal Medicine*. Academic Press, Oralndo, pp. 611-648.

TABLE 3 Zoonotic Ectoparasites in the Laboratory Environment[a]

Species	Disease in humans	Laboratory host	Agent
Mites			
Obligate skin mites			
Sarcoptes scabiei subspecies	Scabies	Mammals	
Notoedres cati	Mange	Cats, dogs, rabbits	
Nest inhabiting parasites			
Ornithonyssus bacoti	Dermatitis, murine typhus	Rodents and other vertebrates, including birds	WEE[b], SLE[c] virus *Rickettsia mooseri*
Ornithonyssus bursa	Dermatitis	Birds	WEE, EEE[d] SLE viruses
Ornithonyssus sylviarum	Dermatitis, encephalitis	Birds	
Dermanyssus gallinae	Dermatitis, encephalitis	Birds	
Allodermanyssus sanguineus	Dermatitis, rickettsialpox	Rodents, particularly *Mus musculus*	*Rickettsia akari*
Ophionyssus natricis	Dermatitis	Reptiles	
Haemogamasus pontiger	Dermatitis	Rodents, insectivores, straw bedding	
Haemolaelaps casalis	Dermatitis	Birds, mammals, straw, hay	
Eulaelaps stabularis	Dermatitis, tularemia	Small mammals, straw bedding	*Francisella tularensis*
Glycyphagus cadaverum	Dermatitis, psittacosis	Birds	*Chlamydia psittaci*
Acaropsis docta	Dermatitis, psittacosis	Birds	*Chlamydia psittaci*
Trixacarus caviae	Dermatitis	Guinea pigs	
Facultative mites			
Cheyletiella spp.	Dermatitis	Cats, dogs, rabbits, bedding	
Dermatophagoides scheremtewskyi	Dermatitis, urinary infections and pulmonary acariasis	Feathers, animal feed, bird nests	
Eutrombicula spp.	Human pest (chiggers) local pruritis	Chickens, occasional mammals obtained from natural habitat	

Organism	Condition/Disease	Host	Agent/Notes
Laelaps echidnirus			Potential Argentinian hemorrhagic fever
Ixodids (ticks)			
Rhipicephalus sanguineus	Irritation, RMSF[e], tularemia, other diseases	Dogs	*Rickettsia rickettsia, F. tularensis*
Dermacentor variabilis	Irritation, RMSF[e], tularemia, tick paralysis, other diseases	Wild rodents, cottontail rabbits, dogs from endemic areas	See above
Dermacentor andersoni	Irritation, Colorado tick fever, Q fever, RMSF[e] other diseases	Small mammals, uncommon on dogs	See above Ungrouped rhabdoviruses
Dermacentor occidentalis	Irritation, Colorado tick fever, RMSF[e] tularemia	Small mammals, uncommon on dogs	See above
Amblyomma americanum	Irritation, RMSF[e] tularemia	Wild rodents, dogs	
Ixodes scapularis	Irritation, possible tularemia	Dogs, wild rodents	
Ixodes spp.			
Ornithodoros spp.	Irritation, relapsing fever	Captive reptiles, wild animals, pigs	*Borelia recurrentis*
Argas persicus	Irritation, seldom bites humans, but can transmit anthrax, Q fever	Domestic fowl	
Fleas			
Ctenocephalides felis	Dermatitis, vector of *Hymenolepis diminuta, Dipylidium caninum*	Dogs, cats	
Ctenocephalides canis (cat and dog fleas)			
Xenopsylla cheopsis	Dermatitis, plague vector, *H. nana, H. diminuta*	Mouse, rat, wild rodents	
Nasopsyllus fasciatus	Dermatitis, plague vector, *H. nana, H. diminuta* murine typhus	Mouse, rat, wild rodents	
Leptopsylla segnis	*H. diminuta, H. nana,* murine typhus vector	Rat	Harbors salmonella
Echidnophaga gallinacea (stick tight flea)	Potential plague vector	Poultry	

TABLE 3 (continued)

Species	Disease in humans	Laboratory host	Agent
Pulex irritans	Irritation	Domestic animals (esp. pigs) and humans	

[a]Found in laboratory animals which cause allergic dermatitis or from which zoonotic agents have been recovered in nature.
[b]WEE = western equine encephalitis.
[c]SLE = St. Louis encephalitis.
[d]EEE = eastern equine encephalitis.
[e]RNSF = Rocky Mountain spotted fever.

Source: Reprinted from Fox, J. G., Newcomer, C. E., and Rozmiarek, H. (1984). Selected zoonoses and other health hazards, in Fox, J. G., Cohen, B. J., and Loew, F. M. (eds.), *Laboratory Animal Medicine.* Academic Press, Oralndo, pp. 611-648.

of the agent from the reservoir to the new host can occur either directly or indirectly. Direct transmission occurs through close association between the reservoir and host. Indirect transmission can occur if the agent is able to survive outside the host, allowing for transmission by either an animate or inanimate object, e.g., vermin and contaminated equipment. Indirect transmission can be controlled by eliminating the vector. Direct transmission is more difficult to control, requiring animal isolation and restricting personnel contact. Isolation techniques instituted may range from space dilution to the use of specialized laminar flow caging equipment and facilities.

Mechanisms of Transmission. Transmission is often accomplished through the formation of aerosols. Aerosols are small particles of liquid or solid material which, due to their size, have prolonged airborne suspension and travel a great distance to expose the new host. The smaller the particle, the greater its ability to settle deep within the respiratory tract avoiding natural defense mechanisms. Particles 10 µm or less in diameter are capable of forming stable suspensions in air and can be disseminated throughout the facility and contaminate personnel and equipment (8). Particles 5 µm or less are capable of bypassing host defenses and settling deep within pulmonary alveolae (9). Aerosols can be created by any forceful activity. Routine husbandry procedures such as the changing of bedding or pressure cleaning animal pens, experimental manipulations such as tracheal instillation of a liquid, or animal activities like coughing or sneezing are likely to create aerosols. Aerosols may contain infectious microorganism(s) or biohazardous materials that were utilized experimentally. Agents infectious in low doses and spread by aerosols should be considered most dangerous (10).

Mechanisms of Exposure. Finally, the agent must gain entrance to its new host. The most common routes of exposure include ingestion, inhalation, direct contact with the skin or mucous membranes, and parenteral inoculation. The susceptibility of the host depends on many interactive factors including physiological and nutritional status, mucosal defense mechanisms, immunological preparedness, and intercurrent disease. The immune system is particularly important to the effective elimination of the infectious agent, and employee vaccination is recommended when available for agents of known hazard to personnel.

Zoonotic Disease Acquisition. Zoonotic agents are usually acquired by the animal in its natural environment. Therefore, animals from random sources, e.g., dogs purchased from pounds, have a higher incidence of zoonotic disease than do purpose-bred animals, which are bred specifically for research utilization. Experimental manipulation can precipitate organism shedding by carrier animals. Because they are often unsuspected and remain undetected until they result in a laboratory-acquired infection, natural infections often pose a greater risk than experimentally induced infections. The vast majority of rodents are purpose-bred. These animals have been surgically re-derived, a process in which fetuses are removed aseptically from the uterus. These animals are free of all potential pathogens except those transmitted vertically, i.e., agents transmitted from mother to off-spring *in utero*, effectively eliminating most of the potential zoonoses.

Disease Classification and Handling. The center for Disease Control (CDC) and the National Institutes of Health (NIH) have provided in their manual *Biosafety in Microbiological and Biomedical Laboratories* (11) guidelines for handling animals known to or suspected of harboring an infectious agent. Zoonotic agents have been placed in one of four biosafety levels dependent upon the potential hazard of the agent and of the associated laboratory function or activity.

Animal diseases listed in Biosafety Level 1 (BL 1) are considered species-specific and as such do not fit the definition of a zoonoses. These agents are not associated with disease in healthy adult humans. Mouse hepatitis and canine distemper viruses are examples of agents in this group. Most infectious zoonotic agents are classified as Bio-safety Level 2 (BL 2). This group includes agents such as *Salmonella spp.* and *Shigella spp.* The primary hazards of these diseases are associated with parenteral inoculation or mucous membrane exposure. Aerosols are not a common means of exposure to agents in this class. Animal diseases in Biosafety Level 3 (BL 3) are in the high risk category. Humans are highly susceptible to these infections, and disease can result form transmission of a low number of organisms. Aerosol transmission of these agents is common. Infections with the diseases in this level may be assymptomatic in the animal host. Common labora-tory-acquired agents in this group include *Mycobacterium tuberculosis, Coxiella burnetti, Herpesvirus simiae*, and Lymphocytic chorio-meningitis virus (LCM). Extremely high risk agents are classified in Biosafety Level 4 (BL 4). Agents in this category such as *Marburg* and *Lassa* viruses are highly pathogenic, and their investigation requires substantial containment facilities and procedures.

Allergic Sensitivities

Allergic respiratory and skin reactions to proteins originating from shed dander and hair or contained in animal urine, saliva, serum, or tissue is a serious and common occupational health disease of labora-tory care staff and biomedical research personnel (12). Hypersensi-tivity reactions include nasal congestion and discharge, sneezing, itching of the eyes, and asthma. Dermatologic manifestations include wheal and flare reactions, urticaria, and eczema. In addition to animal products, aerosolized mold spores and proteins found in animal feed may also act as allergens (13). Recent studies indicate that dispersal of urinary protein from animal litter may be a possible cause of sensitization in laboratory personnel (14). A similar mechanism of sensitizing laboratory personnel to other allergens is plausible.

Persons with known allergies to a particular animal species should avoid contact with these animals and their products. Reduction of exposure to animal allergens is recommended for all personnel working with animals. Protective clothing including face masks or respirators should be worn especially during procedures such as the dumping of bedding, which causes the release of large amounts of potential animal allergens into the air. Increasing the ventilation exchange rates in animal-holding rooms, the use of ventilated caging systems, the housing of animals in exhaust hoods, and the utilization of filter tops on cages are all additional methods of reducing the amounts of animal allergen in the environment.

B. Pharmacologic and Anesthetic Agents Injurious to Man

The judicious use of pharmacologic agents for anesthesia and euthanasia in the laboratory animal facility poses minimal health and safety concerns. However, precautions need to be taken with several of the commonly used inhalational anesthetics because of their volatility and potential of occupational health disease from chronic low-dose exposure.

Inhalational Anesthetics

Ether is commonly used as an inhalational agent in small laboratory animals. Ether forms explosive peroxides when stored in metal containers and allowed contact with air. Even when treated to eliminate peroxides, ether is highly flammable and will continue to vaporize unless stored below -50°C (15). Basic laboratory safety should be instituted when using ether. It should be stored in either an approved flammable liquids cabinet, an explosion-proof refrigerator, or a fume hood. Ether should be purchased in the smallest practical size iron containers and should be used in a well-ventilated area, away from any flame or electrical equipment not specifically designed for use with flammable anesthetics. Carcasses should be adequately ventilated in an exhaust hood when animals are sacrificed by ether overdose before placing them in a refrigerator or incinerator.

Methoxyflurane, a nonflammable halogenated hydrocarbon, has been substituted for ether in many laboratory animal facilities because its anesthetic characteristics are similar to ether but it is not as highly explosive. Methoxyflurane has a low vapor pressure, which makes it safe for use in open drop systems commonly used for small laboratory animal anesthesia. Methoxyflurane is nephrotoxic and should be used within an exhaust hood or in a well-ventilated area (16).

Halothane is another nonflammable and nonexplosive volatile liquid used for general anesthesia in laboratory animals. Halothane has been implicated as an etiology of hepatic necrosis in humans following its repeated administration (17). In rare circumstances, hepatic dysfunction progresses to a fatal hepatitis. Halothane use should be restricted to use in a leak-proof anesthetic machine with low airflows and in a closed system when feasible. Anesthetic waste gases must be scavenged. This is done by placing gas traps over the pop-off valve and exhaust hose and evacuating them into the exhaust outlet of a non-recirculating ventilation system or by absorption in an activated charcoal filter. The safe use of halothane has not been established with respect to possible teratogenic effects upon the fetus during pregnancy (18). Pregnant women should avoid surgery areas where halothane is in use.

Nitrous oxide has powerful analgesic properties and is an excellent adjunct to other anesthetics. Nitrous oxide has been shown to interfere with the production of blood leukocytes and red blood cells by the bone marrow following prolonged administration (19). This has led to concern about long-term, low-dose exposure of operating room personnel to the gas. As with halothane and methoxyflurane, scavenging systems should be used to remove waste gases. Chloroform, a once popular inhalational anesthetic, should be avoided as it is a hepatotoxin and nephrotoxin (20).

Parenteral Anesthetics

Parenteral anesthetics used in laboratory animals usually do not pose safety hazards if properly used. As with all animal injections, adequate animal restraint is necessary. Accidental inoculation of a human with an animal anesthetic should be treated by appropriately trained medical personnel. Medical treatment is required because many of the formulations of the parenteral anesthetics are extremely alkaline. Left untreated, extravascular injection of these drugs will cause severe necrosis of the skin and underlying tissues.

Urethane (ethyl carbamate), a parenteral anesthetic used for non-survival experiments because of its long duration, has been shown to be teratogenic and carcinogenic in laboratory animals (21). Urethane should be handled as a liquid carcinogen. Protective clothing should be worn whenever there is potential for skin contact, and all procedures should be performed in an adequately ventilated fume hood.

Pesticides

Pesticides may be utilized in animal research facilities for insect and rodent control. All pesticides are toxic, but some pesticides have a

very specific spectrum of activity, while others do not. Ideally a pesticide is selected that kills the pest but exhibits no toxicity to research
animals or staff. Pesticides should be used only when necessary in an
animal facility, and compounds should be selected that will not alter
the animals experimental responsiveness. Manufacturer's recommendations and precautions for use should be followed when using a particular product. Nontoxic products such as silica gels or boric acid should
be used for insect control, and live animal traps should be used for
wild rodent capture.

C. Hazards Related to Husbandry Practices, Equipment, and Products Unique to the Maintenance of Animals

A number of safety hazards arise as part of routine husbandry procedures performed by the animal care staff. As discussed earlier, the
dumping of bedding and spray washing of animal cages and pens are
excellent mechanisms for creating aerosols of infectious or contaminated
materials excreted in the urine or feces. Bedding may also be contaminated with a variety of toxic materials used experimentally (21).
Laboratory animal diets and drinking water have been shown to be
contaminated with many natural and synthetic compounds, including
chlorinated hydrocarbons, organophosphates, heavy metals, poly-
chlorinated biphenyls and aflatoxins (22). It is important to recognize
that rodents and rabbits are coprophagic and recycle 10-80% of the
metabolites or residues of toxic substances in comparison to a human
ingesting the same substance (23). These recycled toxic substances
accumulate in higher levels and increase the exposure to the animals
and to the humans involved in animal care and research. Hardwood
chips and shavings are commonly used as contact beddings for rodents.
A high incidence of nasal tumors have been reported in workers exposed to fine particulates of hardwoods or to volatile products generated during processing (24). Use of protective clothing including
gloves and masks should be worn when changing bedding.

Many products used as part of routine animal care practices pose
safety risks to their users if not used properly. Disinfectants, detergents, and other cleaning solutions are usually supplied in concentrated
forms which may be caustic to the skin. Dilute working concentrations
should be provided for general use. Protective clothing including
gloves and eye goggles should be worn when handling concentrated
forms of these compounds. Attention should always be paid to the
manufacturer's instructions and/or material safety data sheets when
available. By definition, no disinfectant can be considered safe in
terms of toxicity, corrosiveness, etc. The term safety is a matter of
degree when selecting or utilizing a particular disinfectant. Phenolics
are excellent disinfectants but are extremely corrosive and require ex-
ceptional precautions in handling in comparison to iodophors and hypo-

chlorites. Care should be exercised when using quaternary ammonium compounds, which have been shown to be toxic to a number of species (25-27). Peracetic acid solution is used as a disinfectant, which is atomized and sprayed on equipment or in rooms. Peracetic acid must be handled with caution as it has been shown to be a tumor promoter and a cocarcinogen (28). One should recognize that there is always a possibility of an adverse reaction when handling any of these compounds.

Steam released from equipment used in cleaning and disinfection procedures is an important hazard. Autoclaves, cage-washing machines, and portable steam cleaning units should be handled by trained personnel. Routine maintenance and periodic inspection should be provided to ensure that equipment is in proper working order.

Wet floors are common in all areas due to mopping and disinfection procedures. Falls due to the accumulation of water on floors are a common cause of injury in the facility. The hazard is especially prevalent in cage-washing areas and animal rooms that are cleaned with running water. The utilization of skid-proof flooring and appropriate footwear can reduce this hazard.

Many of the materials used in the facility have the potential for lacerating or puncturing the skin. Skin trauma can potentially lead to local or systemic infections. Tetanus prophylaxis should be provided to all personnel. Animal cages should be inspected prior to disinfection for possible safety hazards. Broken and bent wires on cages should be repaired, and cracked or fractured plastic shoebox-type cages should be discarded to ensure animal and human safety.

The hypodermic needle and syringe may be the single most hazardous piece of equipment used in a laboratory animal facility. Autoinoculation and aerosolization of infectious materials with a needle and syringe are together the leading cause of laboratory-acquired infections (5,29). The use of needlelocking hypodermic syringes, the disposal of needle and syringe into an impenetrable storage container, and adequate animal restraint will reduce the likelihood of autoinoculation. Needles should not be sheared, bent, removed from the syringe, or placed in the sheath after use. Utilization of a disinfectant-moistened pledget around the stopper and needle when removing a syringe from a rubber-stoppered bottle, and swabbing the injection site before and after injection of an animal will reduce the likelihood of generating aerosols. All excess fluid and bubbles should be expelled from a syringe into a disinfectant soaked pledget.

D. Hazards Associated with Necropsy

Certain safety procedures should be followed within the necropsy suite to avoid injury or exposure of personnel to chemicals and biohazardous agents. Hazardous chemicals commonly encountered in necropsy include formalin, ethanol, glutaraldehyde, and xylene. These substances are utilized for preservation and/or processing of tissues. A variety of other fixatives are also frequently used such as Zenker's, Helly's, Cornoy's, and Bouin's solutions. They contain chemicals such as acetone, hydrochloric acid, picric acid, and potassium dichromate, often in conjunction with formalin or alcohol.

The postmortem examination of animals include the collection of body fluids by aspiration and dissection, and collection of tissues. This necessitates the use of hypodermic needles and syringe, scalpels, blades, scissors, bone saws, knives, and chisels. Care must be employed in the use of these instruments not only because of the potential damage from trauma to oneself or others, but because these instruments are mechanisms that can spread hazardous agents.

Hazardous agents may be transmitted via aerosols by inhalation, via ingestion from contaminated hands, pens, pencils, and other instruments, and via penetrating wounds from contaminated sharp instruments. To protect against this eventuality, personnel working or observing in the necropsy laboratory should wear laboratory coats, disposable face masks, and gloves. These items should remain in the necropsy suite or be disposed of, and hand washing should be performed prior to leaving the necropsy room. Other safety precautions that should be followed include moistening the carcass with a disinfectant solution and performing the necropsy on plastic-backed, disposable, absorbent paper. Instruments should be thoroughly disinfected after necropsy and blunt-pointed scissors should be substituted for scalpels and other sharp instruments when performing necropsy on high-risk animals.

Necropsy performed on animals harboring agents classified as Biosafety Level 2 or above, containing radionuclides, or perfused with glutaraldehyde should be performed in an appropriate safety cabinet.

E. Hazards as a Byproduct of Experimental Utilization

Experimental utilization of a variety of biohazardous compounds including carcinogens, mutagens, toxic chemicals, irritants, and radioisotopes in laboratory animals poses obvious hazards. Laboratory animals are also routinely used to determine if a chemical compound is carcinogenic, mutagenic, or teratogenic. The effect of a particular compound on the animal and its environment is dependent on a number of factors including the route by which the compound is administered and the disposition of the substance and its metabolites.

Oral administration of a compound, either by gavage or addition of the substance to the feed or water, is often used for substances that are likely to be ingested by humans, substances or drugs readily absorbed or requiring activation in the intestinal tract, or when the gastrointestinal system is being studied. Gavage is the preferred

method of the two and when performed correctly with adequate manual
or chemical restraint of the animal is unlikely to cause environmental
contamination during dosing. Vomiting and regurgitation are the
only likely mechanisms for contaminating the environment when dosing
by gavage. Animals should be placed in solid-bottom caging housed
on absorbent plastic-backed paper for a period of time after gavage.
This is unnecessary when working with mice and rats, which are unable to vomit. In comparison, administration of substance in the feed
or water is likely to lead to environmental contamination with the formation of contaminated aerosols and particulate matter (30). In an effort
to reduce the aerosol hazard, chemicals should be administered in wet
cakes or agar gels rather than in powdered or pelleted diets (31).
However, these diets may adhere to the animals' fur and can be transferred to caging and bedding. Excretion of the substance unchanged
or its metabolites in the feces is likely with oral administration.

Administration of hazardous chemicals that are gases or volatile
compounds should be administered by tracheal intubation or tracheo-
stomy. Administration should be confined to an appropriate fume hood
or other safety cabinet to prevent personnel exposure.

Parenteral inoculation of hazardous material is the safest method
of introducing these agents. It decreases the potential for personnel
and environmental exposure and allows accurate doses of the material
to be administered. However, the administration by parenteral routes
may circumvent required metabolic activation in the gut, making interpretation of data difficult. Local tissue irritation or tumor production
may further confound results (23,32).

Substances may be applied directly to the skin or fur of laboratory
animals. This methodology is preferred with particular classes of
substances. The required dose with topical administration tends to
be lower than with other routes of administration, however, the animals
may ingest the substance by licking the area of application. Aerosolization and spread of topically applied materials into the environment
has been demonstrated (33).

Testing the effects of a particular agent on the respiratory system
may pose the highest risks for personnel (7). Many substances require administration in inhalation chambers or by tracheal installation.
Aerosolization is the mechanism of material dispersal in an inhalation
chamber, and tracheal installation may induce coughing or sneezing,
distributing the agent into the environment. Only the animal's head
should be exposed in the inhalation chamber. Whole-body exposure
contaminates the animal's coat, which can lead to contamination of
other animals and personnel (34,53).

Personnel and environmental protection should be instituted in
consideration of the specific substance. One must consider the compound's activity, including its known effects and its disposition. Protection should be addressed at two levels: primary containment of

the hazardous material at the cage or animal level and secondary protection utilizing personnel protective devices, training, and operational
practices. Personnel protection can range from a laboratory coat,
gloves, and face mask to a one-piece positive pressure suit ventilated
by a life support system. Primary containment should begin at the
animal level. Animals can be housed in caging that confines the feed,
feces, urine, and bedding in their primary enclosure. Spun-molded
polyester or polycarbonate filter tops and the utilization of ventilated
racks are examples of caging systems applicable to rodent housing.
Larger animals can be housed in ventilated cages or in metabolic cages
within negative pressure cubicles or rooms with nonrecirculating ventilation systems that remove or inactivate the biohazards before releasing them into the atmosphere. These items will be covered in greater detail later in this chapter.

Laboratory animal facility administrators should establish safety
programs which include standard operational and emergency procedures
in respect to the biohazards in use. Entrances to all containment,
isolation, and other high-risk areas should be identified with warning
signs that include identification of the risks and materials present,
requirements for admission into the facility or area, and person(s) to
be contacted in cases of emergency. Cages housing animals exposed
or carrying a hazardous agent should be similarly identified. Personnel involved in biohazardous studies should be minimized. All personnel involved should be educated in appropriate safety procedures.

III. HAZARD CONTROL

A. Facility Design and Specifications

Controlling hazards within the laboratory animal facility must be addressed from several vantage points. The weakest link in the system
is the one that usually leads to personnel or animal injury. The facility itself must be conducive to hazard control. Facilities should be
easy to service and sanitize to reduce cross-contamination. Although
all animal facilities must be designed to house animals in a safe and
comfortable environment, there are many specifications that may be
altered, depending on a facility's primary functions.

Animal facilities must be separated from areas of human occupancy.
This separation may be accomplished by building, floor, wing, or room.
Traffic flow patterns have been integrated into facility design (36).
These patterns revolve around placement of cage sanitation areas and
movement of cages and equipment to and from it. Separation of "clean
and dirty" corridors provides superior sanitation and infection control.
Facilities for biohazard, chemical, and radioisotope containment should
be located close to cage sanitation areas so that caging does not need
to be moved an excessive distance within the facility.

Animal rooms are commonly subdivided into smaller cubicles. Cubicles are modules, usually about 28 square feet with glass doors. The cubicle concept is useful in that it gives the ability to provide multiple isolated animal-housing spaces. They also limit the potential for cross-contamination to a small area. Cubicles may also be equipped with HEPA filters and be pressure-balanced to house animals on biohazardous protocols.

Heating, Veniltation, and Air Conditioning

Heating, ventilation, and air conditioning (HVAC) are also important factors in disease and hazard control. Air should not be recirculated within the animal facility unless it is treated to remove particulates and toxic gases. Air should not be recirculated from animal-holding areas to personnel occupancy areas, and air from animal facilities should be exhausted away from intake stacks of air-handling units.

Air Pressure Balance. Airflow patterns can be established with air pressure balance. Air pressure control is an adjunct safety measure for preventing cross-contamination. Animal facilities located adjacent to human occupancy areas should be negative pressure with respect to these areas. Areas within the facility used for quarantine, isolation, cage sanitation, and hazard containment should be pressure-balanced so that they are negative with respect to adjoining areas of the animal facility. Inclined manometers, which measure the difference in pressures, should be installed in containment facilities that depend on air pressure differentials for control. The supply and exhaust fans servicing containment areas should be interlocked to provide shutdown of the supply fan in the event of exhaust fan failure. Facilities using clean-dirty traffic flow patterns will often pressure-balance corridors and rooms such that air flows from clean areas to dirty. For such directional airflow systems to remain effective, doors should be kept closed and traffic minimized. Air locks can be utilized as a safety precaution for entrance into containment areas.

Filtration Systems. Air can be passed through a course (roughing) filter followed by passage through high efficiency particulate air (HEPA) filters before exiting or entering an area. HEPA filters are capable of filtering out 99.9% of all particles greater than 0.3 μm in diameter. Studies utilizing infectious agents classified by the CDC as a Biosafety Level 3 or 4 must be housed in negative pressure rooms whose air has been passed through HEPA filters prior to release into the atmosphere. The system should be designed such that maintenance of equipment can be performed with little risk to personnel. Containment facilities may require air treatment, which can include passage through scrubbers, filters, absorption beds, incinerators, or a combination of these prior to release into the atmosphere.

Design Criteria

Facilities should be constructed of materials that are easily sanitizable and that reduce the likelihood of occupational injury. Floors should be monolithic, chemical- and stain resistant, skid-proof when wet, yet relatively smooth and easy to sanitize. They should be coved and carried 4-6 inches up adjoining walls. Epoxy resin flooring meets these requirements. Walls should be smooth and capable of withstanding scrubbing, cleaning, disinfecting, and impact from high pressure water. Concrete block coated with epoxy paint suits this purpose and is desirable, especially in areas housing large animal species. Ceilings, like walls, should be monolithic and easily sanitizable. Dropped ceilings are not ideal, as they are not easily sanitizable and may harbor vermin. Doors should be metal or metal-covered, coated with a material that is easily sanitized. Animal rooms should be provided with sinks to facilitate room cleaning and personal hygiene.

Research institutions may require specialized laboratory suites for housing and manipulation of animals utilized in studies with highly hazardous agents. Specialized equipment such as biological safety cabinets and autoclaves should be provided in these areas. A double-door autoclave, changing area with shower for personnel, and air locks may be required for passage of materials and personnel into and out of these areas.

B. Utilization of Specialized Equipment for Hazard Control

Caging Systems

Selection of the appropriate caging system is an important consideration in hazard control. Caging systems for rodents and other small laboratory animal species are usually portable, allowing cages and racks to be washed in equipment specifically designed to clean, sanitize, and sterilize caging systems. There are two commonly used caging systems for animals of this size. The shoebox cage is usually made of plastic, e.g., polycarbonate, polystyrene, or polyethylene. Shoebox cages have solid bottoms and sides. Contact bedding or an elevated rack must be provided. Contact bedding is provided to absorb excrement and provide a comfortable environment for the animal. Bedding should be selected that is dust-free and absorbent. The activity of animals on contact bedding can generate aerosols. Thus, its use may be inappropriate in particular studies. A raised floor may be used to keep the animals off the bedding and avoid their becoming contaminated with excreta. Polycarbonate shoebox cages can be autoclaved. Polystyrene cages are relatively inexpensive, making them ideal for studies in which caging must be disposable. Shoebox cages can be either suspended from or rested on portable racks. The other cage type, which can only be suspended from racks, has perforated or slotted bottoms and fronts with solid backs and sides. These cages are usually manufactured from stainless steel that allows for repeated autoclaving. Excrement falls through the slotted floor and collects on the shelf below. This can create aerosols, making this system inappropriate for studies employing hazardous agents.

Shoebox cages can also be equipped with filter tops that limit cross-contamination and disease spread and confine the feed, feces, urine, and bedding in the cage enclosure (38). Filter tops manufactured of molded spun polyester and polycarbonate frames with filter material are available for use. For the reasons outlined above, shoebox cages equipped with filter tops are preferred for containment in biohazardous studies.

Flexible or rigid-walled isolators may be ideal for holding small laboratory animal species requiring containment housing. These devices have been used extensively in gnotobiology during the past 30 years (37). Isolators are equipped with entry ports for passage of materials and incorporate rubber gloves into side panels that allow manipulation of animals or materials without direct contact. Air within the isolator can be pressure-balanced so that it is negative or positive with respect to the surrounding environment. Air entering and/or exiting the device can also be HEPA filtered. Transfer drums utilized for passage of materials into and out of the isolator can be autoclaved before materials contained within them are removed.

Caging systems utilized for larger animal species, e.g., cats, dogs, nonhuman primates, etc., can be either fixed or portable. Portable caging can be moved for cleaning and autoclaving, but may not be suitable for all animals because of size limitations. Fixed caging systems such as pens must be cleaned in situ, making the system more labor-intensive, impossible to autoclave, and thus not ideal for use in hazardous protocols. Cages should be constructed of stainless steel because it is strong, durable, and chemical- and heat-resistant. Galvanized steel and aluminum are not recommended because of the tendency of these materials to oxidize. Containment studies requiring excrement collection for specialized disposal can use pens equipped with floor racks that allow excrement to fall through and collect in pans below. Pans should be located directly beneath the rack to reduce the amount of urine splashing and the resulting aerosolization. Cages utilized for primate housing should have solid sides and backs in order to prevent the animals from reaching and traumatizing one another and be equipped with a movable back wall for restraint purposes. The use of transfer cages allow primates to be moved from cage to cage without handling.

Laminar Flow Equipment

Laminar flow equipment is commonly used in the laboratory animal facility. The development of specific pathogen-free laboratory animals

requires specialized caging to maintain their microbiological status.
Specific pathogen-free animals are free of selected infectious agents
that have been shown to alter research results or that may cause disease in animals and/or man (39). Laminar flow principles in conjunction with HEPA filtration systems have been incorporated into a variety
of portable ventilated caging systems. These systems have several
functions: They are designed to keep animals free of contaminants,
they can be used for containment, or they can provide both. The
airflow direction and filtration system placement will differ dependent
upon the equipments principle function.

Ventilated caging equipment utilized for maintaining animals free
of contaminants are designated positive flow. They bathe the animals
in HEPA filtered, positive pressure, laminar flow air. Units utilized
for containment are negative or reverse flow. They remove the air
surrounding the animals and pass it through the filtration system before returning it to the room. Air can be passed vertically or horizontally over the cages. Mass air displacement units (MADU) house
cages on shelves and direct air horizontally over the cages. MADUs
are available as positive or negative flow units. Portable laminar flow
clean rooms are also available. These enclosures come in a range of
sizes and can be used to house entire animal racks or even large animal pens. These units are vertical flow with either positive or negative flow capabilities. As discussed earlier, cubicles or entire room
HVAC systems can be equipped with HEPA filtration systems, laminar
flow air, and be pressure balanced to provide similar functions. There
are also ventilated racks available, which can filter both the intake
and exhaust air of individual cages.

The greatest risk for contamination of animals housed in positive
flow caging systems or exposure of personnel and the environment
to hazardous agents when animals are housed in reverse flow containment systems occurs when the animals are removed from the cage for
manipulation or cage servicing. A variety of specialized equipment
can be used to maintain containment under these circumstances. Animals requiring the highest degree of containment should be manipulated and changed in an appropriate biological safety cabinet. Cabinets can be purchased equipped with trash collection systems, allowing bedding and other materials to be disposed of within the cabinet.
Portable bedding disposal cabinets which draw aerosols away from personnel into HEPA filters can be used when dumping contaminated bedding that does not require the high degree of containment provided
by biological safety cabinets.

Caging and animals that are to be kept free of contamination can
be manipulated and serviced utilizing a laminar flow workbench. The
laminar flow workbench is a portable workbench which provides positive pressure, horizontal flow, HEPA filtered air from the back of the
unit that is exhausted along the front edge of the bench. The uni-

directional airflow prevents particulates form blowing into the user's face and cross-contaminating the animals when they are manipulated within the unit.

Cagewashers, Autoclaves, and Other Specialized Equipment

Specialized equipment is essential in the laboratory animal facility for cleaning, sanitation, and sterilization. Mechanical cagewashers are necessary to clean and sanitize caging systems. Hand-washing cage equipment is labor-intensive and requires the use of detergents and disinfectants which, if not adequately rinsed, may leave unwanted chemical contaminants on the cage. Mechanical cagewashers sanitize cages by exposing them to temperatures in excess of 82°C (180°F) for a period of time in order to destroy vegetative pathogenic organisms (4).

There are two basic types of cagewashers in use. In the batch washer, cages, cage racks, bottles, etc., are placed in a chamber and subjected to high pressure and volume, washing and rinsing. These machines come in a variety of sizes. Smaller machines may only be capable of cleaning a single rabbit cage at a time, while larger rack washers can wash an entire rack of cages. These machines usually have spray headers on all sides and have individually timed wash and rinse cycles. Batch washers can either supply fresh water for each cycle of operation or may hold either the wash or rinse water from the previous cycle in holding tanks for reuse in the next wash cycle (40). Washers that reuse rinse or wash water can contaminate cages washed in the machine that were never exposed to any contaminate, if the cages in the prior cycle were contaminated (41). Large rack washers can be equipped with two doors for pass through operation. This type of machine is ideal in facilities that have separated the "clean and dirty" activities of the cage wash facility. Equipment is brought into the dirty side of the cage wash area and is placed in the rack washer. After cleaning, cages are removed and brought out the opposite side of the washer into the clean side of the cage wash area.

The second type of cagewasher is the tunnel washer. In this type of machine, the equipment to be cleaned is transported on a continuously moving conveyor belt through the various sections of the machine for prerinse, detergent wash, rinse, and drying. Tunnel washers are ideal for cleaning shoebox cages, bottles, sipper tubes, and other small items. Tunnel washers usually recycle water except for the water used for the rinse cycles.

Cagewashers should be equipped with an exhaust fan that prevents the release of aerosols when used to service equipment from studies employing hazardous agents. They can also be equipped with dispensers that can automatically dispense detergent during the wash cycle. Detergent concentration is measured and its release determined by a conductivity probe, which determines electrical resistance based on a certain concentration of detergent in a given water sample. The conductivity probe must be protected from physical damage and be kept free of mineral deposits which often form (42). Failure of inadequate cage rinsing can result in serious problems. Alkali residues from detergents left on equipment may cause chemical burns or, if present on sipper tubes, may lead to dehydration because animals refuse to drink.

It is imperative that cage wash and/or rinse temperatures reach or exceed 82°C and adequate volumes and pressure of wash solutions reach the caging equipment. Temperatures should not exceed 91°C because cavitations may occur in water pumps, resulting in less effective spray patterns (42). Clogged filter screens and spray headers result in reduced pump pressure and reduced spray patterns. Filter screens and spray nozzles need to be monitored and cleaned as required. Temperature gauges located on the washer usually measure water tank temperature, not the temperature in the washing chamber. It has been shown that water sprayed into the washing chamber can undergo a loss of up to 10-15°C (43). In addition, the multitude of cage positions, spray deflections, shielded surfaces, and crevices all act to reduce the temperature and volume of spray water exposure (42). For these reasons, thermometers or heat-sensitive tape should be placed on equipment to be sanitized to verify that adequate temperatures are reached. In addition, microbiolgoical monitoring of cage equipment may be necessary (44). One proposed program evaluates wipe samples from equipment for bacterial growth. Cage equipment is considered adequately sanitized when samples are free of Gram-negative rods and Gram-positive cocci (45). Other programs employ the use of agar contact (RODAK) plates (46).

Facilities should be equipped with an autoclave for heat sterilization of materials. Heat sterilization is necessary if facilities are housing immunocompromised animals. Sterilized bedding, feed, cage equipment, and water should be provided to these animals in order to prevent opportunistic infections. Materials utilized in infectious disease studies should be sterilized before being disposed of or cleaned in conventional equipment, and surgical supplies require sterilization, before use. Sterilization monitoring should be performed. Recording the highest temperature and the length of time that temperature was kept is the single most important mechanism for assuring sterility (42). Biological indicators provide an excellent means of verifying product sterility. Indicators consist of ampules containing heat-resistant bacterial spores, which are placed within the material to be autoclaved or are distributed throughout the load. Ampules are incubated and monitored for bacterial growth after sterilization (47).

An ethylene oxide sterilizer is useful for sterilization of materials that cannot withstand heat and moisture. Ethylene oxide sterilization

kits are available. Materials are sterilized by placing them, along with
an ampule of ethylene oxide, in a sealed plastic bag. The ampule is
then broken, releasing the ethylene oxide gas. The bag is then stored
in a locking plastic case. Ethylene oxide is a carcinogen, so appropriate safety precautions should be instituted (21). Gloves should be
worn when handling the ampules or materials before they are aerated.
Sterilization should be performed within an appropriate fume hood.
Materials must be aerated according to manufacturers' instructions
before use and handling. Chemical-sensitive indicators should be used
to show that each package has been exposed to the gas. Biological
indicators similar to those available for steam sterilization are also available to monitor effectiveness. Mechanical ethylene oxide sterilizers
are available for large-volume operations. Ethylene oxide should not
be used to sterilize feed or bedding, as it reacts with moisture to form
toxic ethylene glycol and possibly other compounds (42).

Other commonly used equipment for sanitation includes high pressure sprayers and steam generators. These are useful for sanitizing
fixed caging systems and outdoor pens. Foggers, which are mechanical
devices that distribute a fine mist of disinfectant or similar material
into the environment, are practical for sanitizing rooms and other dif-
ficult-to-sanitize equipment. Both wet and dry vacuums are useful in
the facility but should be equipped with HEPA filters to avoid spreading dust and infectious agents. Equipment such as polishers and
scrubbers is also commonly utilized. One must be careful not to cross-
contaminate cubicles, rooms, or facilities with equipment. When possible, equipment should be committed to an individual area or be
thoroughly sanitized before transfer.

Protective Equipment

The appropriate use of protective clothing and equipment in conjunction with training and good personal hygiene habits can dramatically
reduce occupationally acquired disease among facility personnel. Protective clothing and equipment is utilized by animal care and research
staff to protect against infectious, toxic, corrosive, and physical hazards that may be present within the facility. The selection of appropriate equipment should be based upon the level of risk associated
with the activity performed. Certain protective clothing and equipment are safer, more practical, and provide greater comfort than
others. However, comfort should not take precedence over safety in
selection.

Protective Clothing. Clothing is a commonly used protective device in the laboratory animal facility. Both reusable and disposable
clothing is available from suppliers. Reusable clothing should be manufactured from closely woven fabric and be made of cotton and polyester

so it acts as a barrier and can withstand decontamination in an autoclave or gas sterilizer. Clothing that is gas-sterilized should be properly aerated before it is worn. Animal care personnel should be issued
laboratory clothing for use in the animal facility. Street clothing
should be removed upon entering the facility. Clothing can carry contaminants into work areas or the home. Clothing should have long
sleeves to protect the arms from exposure to chemicals, animal bites,
scratches, and other types of trauma. They should be tight fitting in
order to reduce the dispersal of bacteria normally found on the skin
(48). A one-piece coverall or pants and shirt are suitable for animal
care personnel. Color-coding clothing may aid in preventing employee
access to inappropriate areas. Plastic or rubber aprons can be worn
over the laboratory garment when working with hazardous chemicals,
steam or hot water in cage-sanitation areas.

Disposable clothing may be attractive because the cost and time
required in decontaminating and laundering reusable items may exceed
the cost of the disposable item. Disposable items selected should be
durable and prevent exposure of the skin to harmful agents. Synthetic polyethylene or cellulose fiber materials meet these requirements
and are popular materials used in the manufacturing of disposable garments. Disposable gowns or coveralls worn in conjunction with other
protective devices are ideal for use in containment, isolation, or quarantine facilities.

Animal care personnel should be provided with safety shoes that
are acid-resistant and have steel toes. Safety shoes reduce the likelihood of trauma from heavy equipment or burns from caustic detergents.
Boots should be able to withstand gas sterilization if contaminated and
be restricted for use in the animal facility or specific areas within the
facility.

Investigators working within the facility should be provided with
3/4-length laboratory coats for many of the same reasons listed above.
Cross-contamination can be a major problem, thus, lab coats should
be designated for use within specific areas. Disposable booties will
reduce cross-contamination between areas when they are worn and
removed before leaving specific areas.

Respiratory Protection. Respiratory protection equipment such
as face masks should be worn during activities that are likely to create
aerosols, to reduce exposure to allergens, or when hazardous agents
have been employed. Contagion or hospital masks are suitable for
protection in most instances. Masks composed principally of glass
fibers or a combination of glass fibers and other synthetic materials
are extremely efficient as respiratory protective devices for most activities within the animal facility (48). The peripheral seal between
the face and mask is extremely important for the mask to function
properly. Investigators and animal care personnel working with non-

human primates should wear a hospital mask and gloves. Primates,
especially Old World species, are extremely susceptible to tuberculosis.
In addition, monkeys may be carriers of a number of zoonotic agents.
Hospital masks will reduce the likelihood of aerosol exposure to infectious agents. Respiratory devices with greater protective capabilities
are available and should be selected for use according to risk. Full
or half facepiece masks with mechanical and/or chemical filters or air-
supplied respiratory devices or suits may be appropriate in particular
high-risk containment studies.

   Gloves. Gloves are important protective devices. Depending on
use, they should be impermeable to chemical and infectious agents,
strong, provide resistance to penetration from sharp objects, protect
against heat and cold, and provide dexterity. The type of glove selected depends upon its intended utilization. Gloves utilized to provide
a barrier to chemicals and other agents should overwrap the cuff and
lower sleeve of the laboratory apparel. A disposable plastic armshield
can be used for additional protection of the lower arm. Gloves for
animal care personnel should be selected that have sufficient strength
to maintain the gloves integrity during cage-sanitation procedures.
Primates requiring physical restraint should be handled cautiously
with restraint glvoes. Restraint gloves are usually constructed of
thick padded leather and are available with leather gauntlets which
protect the forearm. Steel mesh gloves are ideal for handling rodents
and other small species. The provide protection from bites and still
provide dexterity for manipulation.

   Protective Eyewear. Protective eyewear should be worn in addition to a mask when aerosol exposure of the eyes is likely, e.g., when
pressure washing cages and drop pans. Safety glasses with metal or
plastic frames, impact-resistant lenses, and side shields are sufficient
in most situations. However, goggles, face shields, or a combination
of both should be selected when working with cages or animals that
are contaminated with hazardous agents.
   Additional personnel safety equipment may need to be provided
in high-risk operations. Disposable items such as jumpsuits, masks,
gloves, booties, and caps can be utilized. Material utilized should be
decontaminated before disposal or reuse. Reusable laboratory apparel
should not be placed in conventional washing equipment before decontamination.

C. Personal Hygiene and Safety

The practice of good personal hygiene by animal care and research
personnel will reduce the possibility of occupational injury and cross-
contamination within the facility. Each facility should establish and

enforce policy in these areas. Hand washing is an important practice to be utilized within the facility. Personnel should wash their hands before leaving the animal holding rooms, laboratories, necropsy and surgical suites, and after removing protective gloves and garments. A suitable detergent or disinfectant should be provided.

Personnel with fresh or healing cuts and abrasions should wear gloves and should not be permitted to work with any infectious agents. The hands are an excellent mechanism by which contaminants are transferred to the mouth. Personnel should be trained not to touch their hands to the nose, eyes, or mouth. Eating, drinking, and smoking should be prohibited from the animal facility and laboratories except in areas provided and designated for such use.

Personnel working within high-risk areas of the facility such as the containment suite must be thoroughly briefed by a knowledgeable supervisor in the use of appropriate safety equipment and precautions and should be trained to respond to emergency situations such as accidental exposure or spillage of biohazardous materials. General principles that should be conveyed include avoiding the inhalation of hazardous aerosols by breath holding. Personnel should leave the immediate area as soon as possible so that contaminated clothing can be removed. Clothing is removed by folding contaminated areas inward, and contaminated areas of the body should be thoroughly washed with soap and water.

D. Housekeeping, Decontamination, and Waste Disposal

A program of adequate housekeeping should be established and monitored in the laboratory animal facility. A housekeeping program should aim to reduce the amount of physical clutter and contamination, thus providing an environment that will not be a source of injury or infection. The selection and use of appropriate cleaning equipment and supplies is crucial to a successful housekeeping program.

Housekeeping Procedures

Bedding should not be dumped form the cages or cage pans within the animal room. Cages and cage pans should be transported to the cage sanitation area for bedding removal and cleaning. Floors, walls, and ceilings should be cleaned and sanitized on a regular basis. Schedules for cleaning should be established and followed. Rooms or cubicles should be thoroughly cleaned before placing a new shipment of animals or a different species into an empty holding space.

Dry sweeping should be minimized as it is a proven mechanism for creating aerosols. Vacuum cleaners with HEPA filtered exhaust should be utilized to clean up debris and particulate matter (49).

Special precautions are required when changing filters and bags on vacuum cleaners so that personnel and animals are not exposed to contaminants. In situations when dry sweeping is the only practical method of cleaning, sweeping compounds are available which reduce aerosolization of dust. These compounds, which are available from janitorial supply houses, fall into three categories: (1) wax-based compounds for use on vinyl and wax flooring, (2) oil-based compounds for concrete floors, and (3) oil-based compounds with abrasives for use on heavily soiled flooring (48).

Wet mopping or vacuuming is a safe and effective cleaning practice to be used in the animal facility. As with the dry vacuum, wet vaccums should be HEPA filter-equipped and must be emptied carefully and routinely disinfected. The two-bucket method of wet mopping is a convenient and efficient method of sanitization (42). The key feature of this technique is that fresh cleaning solution is applied from one bucket while all used liquids are collected and wrung from the mop into the other. Dolly-mounted double-bucket units are commercially available. A new or sanitized mophead should be used daily and be designated for use in specific areas within the facility. Housekeeping equipment used in the containment suite should be sterilized or chemically decontaminated before disposal or washing in conventional equipment. A squeegee can be used to remove the disinfectant solution from the floors when drains have been provided. Careful application of the cleaning and disinfecting solution is required to avoid aerosolization of contaminants when using any of the wet cleaning techniques. Disinfectant should be left on the surface for 10-15 minutes before it is picked up or removed. Disinfectant solution can be poured into open floor drains to prevent escape of sewer gases into the facility. Mechanical foggers or atomizers can be used to produce a mist of disinfectant to assure adequate wetting of all surfaces.

Disinfection

The selection of appropriate cleaning and disinfecting agents for use in the facility should be based on sound judgment. Agent selection should be tailored to the needs of the program. Agents used in animal-holding areas should be broad spectrum because of the variety of microorganisms likely to be present. Supplementation or substitution with narrow-spectrum agents may be desirable in the containment, quarantine, or isolation suites where specific agents needing control may be identified. The ideal disinfectant is broad spectrum, stable, not adversely effected by salts, organic matter, pH, or the hardness of the water used as a diluent, and is not corrosive, toxic, or irritating.

Inorganic and organic mercurials were once commonly used disinfectants. These substances are now used primarily as fungicides. Their effectiveness is markedly reduced in the presence of organic

material. Quaternary ammonium compounds are commonly used as germicides and are effective against vegetative cells. They have good wetting properties and have considerable detergent activity. They are
not compatible with soaps and are inactivated by organic material. Surfaces must be thoroughly rinsed so no soap residue or debris is present. The halogens, which include chlorine, idoine, and bromine, have
been widely used as disinfectants. Chlorine is an excellent disinfectant
against all microorganisms, including bacterial spores. Organic materials reduce the effectiveness of chlorine disinfectants, requiring surfaces and materials to be cleaned before application. Sodium hypochlorite, which can be prepared from household bleach, is ideal for
use as a virucide and surface disinfectant. Precautions must be taken,
as it is corrosive and causes skin irritation. Tincture of iodine and
aqueous iodine preparations are commonly used as antiseptics in the
surgical facility. Iodophors, which contain iodine and a solubilizing
agent, are stable, relatively odor- and stain-free, and have the antimicrobial properties of iodine. Iodophors are commonly used for skin
disinfection of animal and man. The alcohols, usually prepared in 70-
80% aqueous solutions, have limited usefulness within the facility.
They are slow in their germicidal action and are not effective against
certain viruses and spores. The bactericidal action of isopropyl is
slightly greater than that of ethyl alcohol. Synthetic phenols, which
are similar to the original compound phenol but lack its undesirable
characteristics, are commonly used as disinfectants. These compounds
are effective germicides against vegetative bacterial cells and tubercle
bacilli. Synthetic phenols are compatible with detergents, making the
detergent-disinfectant combination ideal for general sanitization purposes. Skin contact should be avoided, as phenolics are extremely
irritating (47). Peracetic acid (PA) is sometimes used as a disinfectant
and is usually dispersed as a spray. The solution is sporicidal in the
vapor as well as the liquid phase and thus will kill suspended organisms in the air and on the surface that escape wetting (38). Gloves
as well as eye and respiratory protective devices should be worn when
working with this agent. Preferably, the material should be handled
with gloves in a suitable ventilated cabinet (38).

Waste Disposal

Careful consideration must be given to the final disposal of waste materials from the animal facility. Facility administrators should establish
policy in these areas. The final disposition of a particular material
depends on the type of waste, whether biologic, radioactive, or chemical (50). Suitable storage facilities must be provided for waste storage until processed for final disposition. Refrigerators or freezers
are required for biodegradable wastes such as animal carcasses. Waste
material should be stored separately from other storage areas. Waste
disposal should be performed in accordance with all applicable local,
state, and federal regulations.

Animal excrement and soiled bedding are by far the largest waste produced within the facility. Excrement may be flushed into the sewerage system. Bedding should be stored in sealed, leak-proof plastic bags and can be incinerated or transported to a landfill. Biologic wastes such as animal parts or carcasses should be refrigerated in sealed plastic bags until incinerated or transported to a sanitary landfill.

Hazardous solid or liquid biologic wastes, i.e., microbiologic-contaminated laboratory animals or excrement must be decontaminated by sterilization before conventional disposal or should be immediately incinerated. Materials should be placed in impervious biohazard bags before careful transport to the incinerator or autoclave. Autoclaved urine, feces, or animals may cause unpleasant odors in the surrounding area.

Chemical-contaminated solid wastes that contain noncombustible materials are generally incinerated at temperatures of 1800-1900°F, which is sufficient to break down the chemical into carbon dioxide, water, and oxides of nitrogen and sulfur (51). Chemical-contaminated wastes can also be buried in suitable sealed containers at appropriately designated disposal areas. Small quantities of chemically contaminated liquid waste can be disposed in the sewerage system, provided it is not prohibited by regulation. The state or federal Environmental Protection Agency should be consulted before disposal of any carcinogenic chemical or substance. Large quantities of chemical waste or flammable substances should be disposed of by properly trained individuals.

Radioactive substances or biologic wastes contaminated with radionuclides must be segregated and handled in accordance with regulations promulgated by the U.S. Nuclear Regulatory Commission (NRC). Isotopes with half-lives less than 65 days can be stored until a safe level for conventional disposal is reached (49). Materials can be sealed in suitable containers and buried in designated sites when contaminated with radionuclides having longer half-lives or when storage of radioactive materials is not practical. Materials contaminated with particular isotopes, i.e., ^{14}C and ^{3}H, can be incinerated and the ashes collected and appropriately disposed of if a suitable NRC license is obtained. All hazardous wastes, storage vessels, and storage areas should be identified with appropriate biohazard or radionuclide warning signs. Identification should include the date and the identity of the material present.

E. Animal Procurement, Quarantine, and Surveillance

Vendor Selection

Selection of a suitable vendor for animal procurement will reduce the likelihood of facility and personnel contamination with unwanted or

hazardous infectious agents. Reputable vendors should be selected
that produce high quality animals for research. Vendors that institute
surveillance programs to monitor the health status of the research animals they sell should be given preference. When available and economically feasible, investigators should select purpose-bred animals that
are free of specific pathogens. These animals are often monitored to
ensure their specific pathogen-free status. Health-monitoring reports
are available from vendors that perform them. They should be reviewed prior to vendor selection and animal purchase. As discussed
earlier, random source animals often harbor zoonotic agents. Suppliers of random source animals may not perform any diagnostic evaluation to determine the suitability of the animal for use in a research
setting. They will usually ship to customers after a short stabilization
period. In these instances, the animals should be monitored by the
receiving institution for specific agents that may be infectious for personnel or could interfere with research results. Suppliers who do
test will often cull or treat animals that have been identified as unsuitable.

Health Monitoring and Surveillance Programs

Some agents may be shed intermittently. The stress of shipment or
stress arising from the animal's experimental manipulation may cause
a previously undiagnosed carrier to shed an infectious agent into the
environment. Laboratory animal facilities acquiring animals from the
wild, from vendors who do not perform health monitoring, or other
sources in which the health status is questionable, should establish
and administer their own health-monitoring programs. Programs consist of sampling animals to determine their status in respect to a specific agent or condition. Individual animal testing is usually performed
with large animal species, or a representative sample of animals from
a colony is examined, as in the case of rodents (52). Evaluation may
include a physical examination, bacteriological culturing, serological
analysis for viruses and mycoplasma, endo- and ectoparasite examinations, and, in the case of rodents, necropsy examination with histo-
pathologic examination of select tissues (53).

Quarantine and Isolation

Quarantine procedures instituted for newly arrived animals will vary
dependent on the origin, species, and particular facility. A quarantine period allows animals to recover from the stress of shipment, ac-
climate to their new environment, reduces the possibility of transmitting active infections that may be incubating or are clinically inapparent and allow animal facility staff to determine the animals' health
status (54,55). Quarantine procedures can vary from a satisfactory
visual inspection, when animals such as rodents or rabbits are pur-

chased from commercial vendors with satisfactory health profiles, to isolation and testing for periods up to several months, as is customary with new shipments of nonhuman primates. Primates are usually subject to a variety of diagnostic tests, including chest radiographs and a series of tuberculin tests (56). Animals requiring quarantine should be physically isolated from established colonies of animals. This isolation can be provided by a room, use of specialized laminar flow equipment or cages, or, if necessary, separate facilities. If animal care personnel service both quarantine and established colonies, the established animals should be serviced first in order to reduce the possibility of cross-contamination. Personnel access to quarantine animals should be limited to those that are required to provide for the animals' husbandry and health needs.

Transplantable Tumors and Biologics

It is well recognized that many infectious agents, especially viruses, can be transmitted through passage of biologic materials in animals (57). This has been recognized in rodents where transplantable tumors were contaminated with agents such as lymphocytic choriomeningitis virus, a zoonotic virus (58). Tumor preparations, cell lines, or other biological materials may be contaminated with infectious agents, and inoculation of these materials into animals may contaminate colonies. Biological materials that have been passed in or originated from rodents should be tested prior to use in rodents within the laboratory animal facility. The testing procedure utilized for viruses is called the antibody production test (53). This test consists of inoculating a suitable host recipient with the suspect material and evaluating seroconversion. The host recipient should be free of the agents prior to use in the test.

Animal facilities should have access to veterinary professionals who specialize in laboratory animal medicine and facility management. Diagnostic support is necessary. Large animal care programs may require diagnostic laboratory support at the institution, whereas in smaller programs outside contract laboratories may suffice. Diagnostic support should allow for prompt and accurate disease diagnosis so that contagion can be minimized and animal isolation and treatment can be instituted expeditiously.

F. Risk Assessment

Assessment of the risks associated with activities performed within the laboratory animal facility must be addressed in order to provide appropriate safeguards and to minimize ill effects on personnel. Many of the hazards within the facility can be clearly identified, however, many are undefined, arising as a part of new scientific investigations. The process of risk assessment requires the knowledge and judgment to

recognize the risks and institute the necessary precautions. Risk
assessment should involve examination of the experimental protocol,
agent or materials utilized, equipment and facilities available, and training and qualifications of personnel. Medical surveillance programs
should be established to detect personnel who may be at great risk for
adverse effects if exposed to the agent or materials in consideration
and also to detect changes in the state of health of personnel involved
in the project.

An integral part of risk assessment involves appraisal of the hazards involved. Hazards can be real or potential. Real hazards are
those that can cause or have the capacity to cause short-term or long-
term disease. Other substances fall into the category of potential hazards because definitive evidence is unavailable, but related information
supports its hazardous character. The toxicity of many chemical substances have been clearly identified. Nevertheless, the toxic nature
of many stubstances remains undetermined because of the vast number
of new materials being synthesized through laboratory investigation
and manufacturing.

Although limited, there are materials available to aid in hazard
assessment. As discussed earlier, the Centers for Disease Control's
(CDC) manual, *Biosafety in Microbiological and Biomedical Laboratories*
(11), classifies infectious agents and places them into one of four biosafety levels. For chemical substances, information can be found in
the *Registry of Toxic Effects of Chemical Substances* (59), published
by the CDC and the National Institute for Occupational Safety and
Health (NIOSH), and the National Toxicology Program's (NTP) *Annual
Report on Carcinogens* (21), which contains lists and monographs on
known and suspected carcinogens. Many manufacturers have compiled
safety information on materials they produce in material safety data
sheets, and references such as the *Merck Index* (60) can be useful in
assessing the safety of a particular compound. The National Cancer
Institute's *Biological Safety Manual for Research Involving Oncogenic
Viruses* (48) and the National Institutes of Health's *Guidelines for Research Involving Recombinant DNA Molecules* provide useful information when working with oncogenic viruses and recombinant DNA. Assimilation of this material in conjunction with the use of appropriate
safety equipment, personnel training programs, and common sense
can greatly reduce the risk associated with use of a particular agent
or substance.

G. Occupational Health Programs

Occupational Health programs, which include medical surveillance for
personnel working with hazardous materials, should be mandatory for
personnel working in the laboratory animal facility or those having
extensive animal contact (4). Programs should be tailored to address

the specific risks that apply in an individual facility. The program should include preemployment evaluation in addition to a regularly prescribed evaluation after employment commences. Preemployment evaluation should allow for identification of individuals that may be at increased risk to hazards during employment. This can include musculo-skeletal disorders, e.g., neck and/or back pain, immunosuppressive therapy, immunodeficiency syndromes, cancer, hormonal imbalances, tuberculosis, allergies, and, in particular circumstances, pregnancy. A preemployment physical examination is performed to identify similar or additional risk factors which have impact on current or future employee health.

Immunization therapy should be applied as necessary. Personnel should have been immunized for tetanus within 10 years. Immunizations may also be given if a tetanus-prone injury occurs and more than five years has elapsed since prior immunization. Rabies prophylaxis should be provided to those person(s) who have contact with livestock, carnivores, nonhuman primates, and wildlife. Specialized prophylactic therapy may be instituted in specific circumstances. Personnel with exposure to chimpanzees or their biological materials should be tested and vaccinated for hepatitis B. Individuals working with newly imported chimps should receive human immune globulin for hepatitis A (61,62).

Appropriate diagnostic tests should be performed during the preemployment examination and subsequent evaluation. Laboratory tests may include a complete blood count, routine urinalysis, tuberculin test, hepatic enzyme analysis, fecal culture, and endoparasite examination. Positive findings should be treated or handled appropriately. Tuberculin reactors should have chest radiographs. Personnel working with nonhuman primates should be evaluated for tuberculosis annually, whereas this may not be necessary for other personnel. Persons having contact with pregnant sheep should have Q-fever antibody titers determined annually (63). A serum storage program should be given careful consideration. Preemployment and regular employment serum samples should be stored for future diagnostic purposes and epidemiological evaluations.

Medical records emanating from occupational health programs should be maintained long after the employee has terminated or the project has concluded. Adverse effects from hazardous agents may arise many years after exposure.

IV. CONCLUSION

It should be apparent that control of hazards within the animal facility involves the integration of a number of factors. Hazard recognition is an integral part of control. However, many hazards are not apparent

even to the trained eye of investigators, facility managers, and veteri-
narians. Systematic analysis of facility practices along with an adequate physical plant, use of appropriate safety equipment, selection
of high quality animal subjects, administration of quality assurance
programs, and personnel training programs all reduce the likelihood
of personal injury and environmental contamination. In contrast to
the containment suite where biohazards are clearly recognizable, the
remainder of the facility may pose dangers that are not readily apparent.
Development of facility policy and standard operating procedures that
address this possibility and that utilize available safety measures will
make the facility a safer workplace for all who interface with it.

REFERENCES

1. CFR (Code of Federal Regulations), Title 9; Subchapter A, Ani-
    mal Welfare. Washington, DC: Office of the Federal Register
    (1984a).
2. CFR (Code of Federal Regulations), Title 43; Good Laboratory
    Practice Regulations. Washington, DC: Office of the Federal
    Register (1978).
3. CFR (Code of Federal Regulations), Title 29; Part 1910, Occupa-
    tional Safety and Health Standards; Subpart G, Occupational
    Health and Environmental Control. Washington, DC: Office of
    the Federal Register (1984b).
4. *Guide for the Care and Use of Laboratory Animals.* (1985).
    DHEW Publications No. (NIH) 85-23, Bethesda, MD.
5. Pike, R. M. (1976). Laboratory-associated infections: Summary
    and analysis of 3921 cases. *Health Lab. Sci.*, *13*(2):105-114.
6. Phillips, G. B., and Jemski, J. V. (1963). Biological safety in
    the animal laboratory. *Lab. Anim. Care*, *13*(1):13-20.
7. Gerone, P. J. (1978). Hazards associated with infected labora-
    tory animals. *Laboratory Animal Housing.* National Academy of
    Sciences, Washington, DC, pp. 105-117.
8. Hatch, T., and Gross, P. (1964). *Pulmonary Deposition and
    Retention of Inhaled Aerosols.* Academic Press, New York.
9. Landahl, H. D. (1963). Particle removal from the respiratory
    system: Note on the removal of airborne particulates by the human
    respiratory tract with particular reference to the role of diffusion.
    Bull. Math. Biophysics 25:29.
10. Wedum, A. G. (1964). Laboratory safety in research with infec-
    tious aerosols. *Public Health Rep. 79*(7):619-633.
11. *Biosafety in Microbiological and Biomedical Laboratories.* (1984).
    HHS Publication No. (CDC) 84-8395, Bethesda, MD.
12. Lutsky, I. I., and Toshner, D. (1978). A review of allergic
    respiratory diseases in laboratory animal workers. *Lab. Anim.
    Sci. 28*(6):751-756.

13. Patterson, R. (1964). The problem of allergy to laboratory animals. *Lab. Anim. Care 14*:466-469.
14. Schumacher, M. J. (1980). Characterization of allergens from urine and pelts of laboratory mice. *Mol. Immun. 17*:1087-1095.
15. Steere, N. V. (1967). Control of peroxides in ethers, in *Handbook of Laboratory Safety*, Steere, N. V. (Ed.). Chemical Rubber Co., Cleveland, p. 190.
16. Taves, D. R., Fry, B. W., Freeman, R. B., and Gillies, A. J. (1970). Toxicity following methoxyflurane anaesthesia, II. Fluoride concentration in nephrotoxicity. *J.A.M.A. 214*:91.
17. Lind, R. C., Shively, J. W., et al. (1985). Halothane-associated hepatitis. *Comp. Path. Bull. 17*(10:3-4.
18. Ad Hoc Committee, American Society of Anesthesiologists (1974). Occupational disease among operating room personnel. *Anesthesiology 41*:321.
19. Marshall, B. E., and Wollman, H. (1980). General anesthetics, in *The Pharmacological Basis of Therapeutics*, Gilman, A. G., Goodman, L. S., and Gilman, A. (Eds.). MacMillan, New York, p. 289.
20. National Cancer Institute Research Progress Report. (1976). Chloroform: Tumor induction in laboratory animals. *Hew News* (June 10).
21. *Fourth Annual Report on Carcinogens* (1985). HHS Publications No. (NTD) 85-002, Bethesda, MD.
22. Newberne, P. M., and McConnell, R. G. (1970). Dietary nutrients and contaminants in laboratory animal experimentation. *J. Environ. Pathol. Toxicol. 4*:105-122.
23. Newberne, P. M., and Fox, J. G. (1978). Chemicals and toxins in the animal facility, *Laboratory Animal Housing*, National Academy of Sciences, Washington, DC.
24. Acheson, A. D., Cowdell, R. H., Hadfield, E., and Macbeth, R. G. (1968). Nasal cancer in woodworkers in the furniture industry. *Br. Med. J. 2*:587-596.
25. Serrano, L. J. (1972). Dermatitis and death in mice accidentally exposed to quaternary ammonium disinfectant. *J. Am. Vet. Med. Assoc. 161*:652-655.
26. Grier, R. L. (1967). Quaternary ammonium compound toxicosis in the dog. *J. Am. Vet. Med. Assoc. 150*:984-987.
27. Reuber, H. W., Rude, T. A., and Jorgenson, T. A. (1970). Safety evaluation of quaternary ammonium sanitizer for turkey drinking water. *Avian. Dis. 15*:211-218.
28. Bock, R. G., Meyers, H. K., and Fox, H. W. (1975). Cocarcinogenic activity of peroxy compounds. *J. Natl. Cancer Inst. 55*: 1359-1361.
29. Chapel, H. M., and August, P. J. (1976). Report of nine cases of accidental injury to man with Freund's complete adjuvant. *Clinical Exp. Immun. 24*:538-541.

30. Sansone, E. B., Losikiff, A. M., and Pedleton, R. A. (1977). Potential hazards from feeding test chemicals in carcinogen bioassay research. *Toxicol. Appl. Pharmacol. 39*:435-450.
31. Sansone, E. B., and Fox, J. G. (1977). Potential chemical contamination in animal feeding studies: Evaluation of wire and solid bottom caging systems and gelled feed. *Lab. Anim. Sci. 27*:457.
32. Grasso, P. (1970). Carcinogenicity testing and permitted lists. *Chem. Br. 6*:17-22.
33. Darlow, H. M., Simmons, D. J. C., and Roe, F. J. C. (1969). Hazards from experimental skin painting of carcinogens. *Arch. Environ. Health 18*:883-893.
34. Kruse, R. H., and Wedum, A. G. (1970). Cross-infection with eighteen pathogens among caged laboratory animals. *Lab. Anim. Care 20*(3):541-560.
35. Phillips, G. B., Broadwater, G. C., Reitman, M., and Alg, R. L. (1956). Cross-infection among brucella-infected guinea pigs. *J. Infect. Dis. 99*:56-59.
36. Henke, C. B. (1978). Design criteria for animal facilities, *Laboratory Animal Housing*. National Academy of Sciences, Washington, DC.
37. Kraft, L. M., Pardy, R. F., Pardy, D. A., and Zwichel, H. (1964). Practical control of diarrheal disease in a commercial mouse colony. *Lab. Anim. Care 14*:16-19.
38. Trexler, P. C. (1983). Gnotobiotics, in *The Mouse in Biomedical Research*, Vol. III, Foster, H. L., Small, J. D., and Fox, J. G. (Eds.). Academic Press, New York, pp. 1-16.
39. Parker, J. C. (1980). The possibilities and limitations of virus control in laboratory animals, in *7th ICLAS Symposium*, Spiegel, A., et al. (Eds.). Gustav Fischer Verlag, Stuttgart, pp. 161-172.
40. Ament, R. K. (1971). A study of automatic detergent dispensers. *Lab. Anim. Sci. 21*(6):927-931.
41. Fox, J. G., and Newberne, P. M. (1971). Environmental safety and chemical hazards in animal research, Part I. *Lab. Anim. 21*(6):927-931.
42. Small, J. D. (1983). Environmental and equipment monitoring, in *The Mouse in Biomedical Research*, Vol. III, Foster, H. L., Small, J. D., and Fox, J. G. (Eds.). Academic Press, New York, pp. 83-100.
43. Zwarun, A. A., and Weisbroth, S. H. (1979). Development of an 82.2°C (180°F) temperature indicator system for monitoring equipment washing and sanitizing programs. *Lab. Anim. Sci. 29*(3):395-399.
44. Favero, M. S., McDade, J. J., Robertsen, J. A., Hoffman, R. K., and Edwards, R. W. (1968). Microbiological sampling of surfaces. *J. Appl. Bacteriol. 31*:336-343.

45. Weisbroth, S. H., and Weisbroth, S. P. (1982). A proposed microbial evaluation program. *Lab. Animal 11*:25-26.

46. Hall, L. B., and Harnett, M. J. (1964). Measurement of the bacterial contamination on surfaces in hospitals. *Public Health Rep. 79*:1021-1024.

47. Favero, M. S., and Skaliy, P. (1981). Sterilization and disinfection, in *Diagnostic Procedures for Bacterial, Mycotic and Parasitic Infections, 6th ed.*, Balows, A., and Hausler, W. J. (Eds.). American Public Health Assoc., Washington, DC, pp. 93-104.

48. *Biological Safety Manual for Research Involving Oncogenic Viruses* (1976). DHEW Publication No. (NIH) 76-1165, Bethesda, MD.

49. Hessler, J. R., and Moreland, A. F. (1984). Design and management of animal facilities, in *Laboratory Animal Medicine,* Fox, J. G., Cohen, B. J., and Loew, F. M. (Eds.). Academic Press, Orlando, pp. 505-526.

50. Mackel, D. C., and Mallison, G. F. (1981). Waste disposal in microbiology laboratories, in *Diagnostic Procedures for Bacterial, Mycotic and Parasitic Infections, 6th ed.*, Balows, A., and Hanler, W. J. (Eds.). American Public Health Assoc., pp. 105-108.

51. *Chemical Carcinogen Hazards in Animal Research Facilities* (1979). Office of Biohazard Safety, National Cancer Institute.

52. Institute of Laboratory Animal Resources (1976). Long-term holding of laboratory rodents. *ILAR News 19*:L1-L25.

53. Small, J. D. (1984). Rodent and lagomorph health surveillance-quality assurance, in *Laboratory Animal Medicine,* Fox, J. G., Cohen, B. J., and Loew, F. M. (Eds.). Academic Press, Orlando, pp. 709-723.

54. Lang, C. M. (1983). Design and management of research facilities for mice, in *The Mouse in Biomedical Research,* Vol. III, Foster, H. L., Small, J. D., and Fox, J. G. (Eds.). Academic Press, New York, pp. 37-50.

55. Landi, M., Krieder, J. W., Lang, C. M., and Bullock, L. P. (1982). Effect of shipping on the immune fucntions of mice. *Am. J. Vet. Res. 43*:1654-1657.

56. Institute of Laboratory Animal Resources (1980). Laboratory animal management: nonhuman primates. *ILAR News 23*:P1-P44.

57. Collins, M. J., and Parker, J. C. (1972). Murine virus contaminants of leukemia viruses and transplantable tumors. *J. Natl. Cancer Inst. 49*:1137-1143.

58. Bhatt, P. N., Jacoby, R. O., and Barthold, S. W. (1986). Contamination of transplantable tumors with *Lymphocytic Choriomeningitis Virus. Lab. Anim. Sci. 36*:136-139.

59. *Registry of Toxic Effects of Chemical Substances 1981-1982* (1983). U.S. Department of Health and Human Services, Public Health Service, Centers for Disease Control, National Institute for Occupational Safety and Health, Cincinnati, OH.

60. Wendolz, M. (Ed.). (1976). *Merck Index*, 9th ed. Merck and Co., Rahway, NJ.
61. Rivera, J. C., Bayer, R. A., and Johnson, D. K. (1984). The National Institutes of Health Animal Handlers Medical Surveillance Program. *J. Occup. Med.* *26*(2):115-117.
62. Muchmore, E. (1975). Health program for people in close contact with laboratory primates. *Cancer Res. Saf. Monogr. Ser.* *2*:81-89.
63. Bernard, K. W., Parham, G. L., Winkler, W. G., and Helmick, C. G. (1982). Q fever control measures: Recommendations for research facilities using sheep. *Inf. Cont.* *3*(6):461-465.

II. Biosafety Principles and Practices

7

Biosafety: The Research/Diagnostic Laboratory Perspective

DANIEL F. LIBERMAN *Massachusetts Institute of Technology, Cambridge, Massachusetts*

LYNN HARDING *Harvard University, Cambridge, Massachusetts*

I. INTRODUCTION

Practical planning for safety is hampered by the fact that safety cannot be measured directly. The words "safe" and "safety" are ideal concepts which, while desirable, are unattainable in absolute terms. Practical planning for safety is performed by evaluating its opposite, namely, risk (1). Safety in a research laboratory is an exercise in recognizing what the risks are and then introducing procedures, practices, equipment, and facilities to control the identified risks or hazards or reduce them to acceptable levels.

The recent advances in genetic engineering, cell fusion, immobilized cells, and enzymes, etc., have provided a new dimension to applied microbiology. Technology is advancing so rapidly that it is not possible for safety specialists to anticipate each use of potentially hazardous biological or chemical systems and to monitor, appropriately, every operation that involves these materials. It has become essential for the researcher to have sufficient knowledge to identify the potential hazards associated with the work and to institute procedures and practices necessary to conduct the activity in as safe a manner as possible.

The purpose of this chapter is to emphasize the need for adequate safety precautions in the planning, initiation, and termination of activity involving biological systems that represent a real or potential hazard to the worker. There are two principal reasons for this emphasis: first, the potential threat of infection in the work environ-

ment has long been recognized and appreciated by the microbiologist;
and second, the number of nonmicrobiologists engaged in biotechno-
logical activities is increasing. Researchers may not be sufficiently
trained to take the necessary precautions to protect themselves, their
co-workers, and the environment. The safe handling of an infectious
agent requires an objective evaluation of the hazard potential associated with each aspect of work activity. Such an evaluation should
include consideration of the microorganism(s) involved; the procedures
and practices associated with their use; the adequacy of equipment
and facilities required to carry out the program in a safe and prudent
manner; a review of available information on related laboratory acquired illness; the training and experience of personnel; and the
appropriateness of vaccination or other medical surveillance requirements prior to program initiation (2).

II.   HISTORICAL PERSPECTIVE

A.   Review of Laboratory-Acquired Illness

Available Information

Considerable information has been compiled in recent years concerning
the accidental infection of laboratory workers with pathogenic microorganisms (1, 3-9). An analysis of available information indicates that
only 20% of these infections are due to recognized accidents and that
80% can be attributed to unknown or unrecognized causes (5,6).

Over the past 30 years a concerted effort has been made to determine the extent of laboratory-acquired infections. Studies by Gross
on erysipeloid among veterinary students (10), by Furcolow et al. on
histoplasmosis (11), by Huddleson and Menger (12) and Meyer and
Eddie (13) on brucellosis, and by Robbins and Rustigion on Q fever
(14) clearly demonstrate that bacterial, fungal, and rickettsial agents
are potentially hazardous to individuals within the laboratory, as well
as to those in surrounding areas.

The potential for laboratory-acquired viral infections was first
addressed in a systematic fashion by Sulkin and Pike (7). In 1951,
in collaboration with the laboratory branch of the American Public
Health Association, they initiated a surveillance program to obtain information on laboratory-acquired infections. This program included
the development, distribution, and analysis of a questionnaire that
was submitted to over 5,000 laboratories, a search of the published
literature, and personal communication (5,6,8,9,15,16).

As of 1978, this registry included 4,079 cases of laboratory-
acquired infections (6). Bacterial agents accounted for 1,704, viruses
for 1,179, rickettsia for 598, fungi for 354, chlamydia for 128, and
parasites for 116 of these illnesses. Thirty-eight different bacterial
species, 84 viruses, 16 parasites, 9 rickettsia, 9 fungi, and 3 chlamydia

were identified as causative agents. Certain agents accounted for a
higher number of illnesses than others. For example, brucellosis,
typhoid fever, tularemia, and tuberculosis accounted for 65% of the
1,704 recorded bacterial infections. Hepatitis, Venezuelan equine
encephalitis, and Kyasanur Forest Disease virus accounted for 41% of
the viral infections (17,18). Dermatomycosis and coccidiomycosis accounted for 72% of the fungal infections. Q fever (*Coxiella burnetii*),
psittacosis, and scrub typhus accounted for 80% of the rickettsial infections. (See references 5 and 6 for the specific genus/species listings.)

It is clear from the cases in the registry that the laboratory manipulation of infectious disease agents represents an occupational risk
that requires rigorous control to prevent illness and save lives.

Pike (6) and Grist (1) have stressed that our knowledge regarding the number of infections is far from complete because there has
been no requirement for reporting laboratory-acquired infections.
Sixty-one percent of reported cases have been published, and the
remainder have come form personal communications and questionnaire
responses (1,5). It is important to keep in mind that there is no way
to determine precisely how many laboratory-acquired infections have
occurred or whether all such reported infections were, indeed, laboratory-acquired. The latter comment applies particularly to chronic
diseases. For example, the relatively slow development of pulmonary
tuberculosis introduces some uncertainty when attempting to determine
if a given infection is laboratory-acquired (19,20).

With these limitations in mind, the analysis of illness that could be
traced to identified causes has provided information to determine the
probable sources of infection, the type of accidents or procedures
involved, the work activity that resulted in or was associated with the
laboratory illness, and, finally, the personnel at risk (5,8,9,16,21,22).

As indicated previously, identifiable accidents accounted for only
20% of the laboratory-acquired illnesses reviewed (5,6,23); the remainder were due to unknown or unrecognized causes and included a
group for whom all that was known was that the individuals affected
had worked with the responsible agent. An analysis of the outcome
of identified illnesses indicated that 70% experienced complete recovery,
26% suffered permanent disability, and 4% died (5,6). The case fatality
rate (e.g., the number of deaths due to a specific class of agents
divided by the number of illnesses caused by that class of agent) was
6.6% for bacterial-caused disease; 4.7% for viral; 4.2% for rickettsia;
1.4% for fungi; 7.8% for chlamydia; and 1.7% for parasites.

Causative Agents

Over the past 50 years there has been a definite change in the class
of eiologic agents associated with laboratory-acquired infections, e.g.,

65% of the brucellosis and 97% of the typhoid fever cases were reported before 1955, whereas almost half the hepatitis and lymphocytic chorio-meningitis infections have been reported since 1955 (1,6).

From the data presented in Table 1 on the changes that have occurred over the past 50 years in the agents responsible for labora-tory-acquired infections, it would seem that bacterial-associated illness has declined from 67% to 13%; viral illness has increased from 15% to 59%; and fungal illness has increased slightly during the three periods considered.

Before we can conclude that the apparent changes in bacterial, viral, and fungal illness, etc., are indicative of reliable trends, we must consider the source(s) for this information. As indicated previously, case reports came from questionnaire responses, published case reports from the medical literature, and scientific reports. Many of the early infections were reported as case studies because of their clinical significance rather than because they were laboratory-acquired. Since 1932 there have been 20 reports of laboratory infections from agents not previously known to infect man, e.g., *Herpes simiae* (24), *Coxiella burnetii* (25), Louping ill virus and Zika virus, *Plasmodium cynomolgi* (26), and swine vesicular virus (27). Few case reports of laboratory-acquired infections have been published during the last 10 years (17,28,29). This may be due to the concern that illness reports may reflect unfavorably on the laboratory in which they were acquired, or to a lack of interest in the clinical details of the illness.

Thus, the true incidence of laboratory-acquired illness and the precise distribution of causative agents is not known.

Population at Risk

When the reported episodes were classified according to the primary purpose of the work, it was found that research activity accounted

TABLE 1 Laboratory-Acquired Infections: Causative Agents and Changing Trends

Agent	Period reviewed		
	1925 to 1934	1945 to 1954	1965 to 1974
Bacteria	67%	40%	13%
Viruses	15	22	59
Rickettsia	6	22	3
Fungi	2	8	20
Other	10	8	6

for almost 59% of the infectious episodes; diagnostic work, 17%, biological production, 3.4%; teaching, 2.7%; and 21% could not be classified (5). Pike has cautioned that comparisons of this kind may be
misleading because they do not take into consideration the attack rate
(i.e., number stricken divided by the number exposed) of the infections in the various groups (6). A reliable calculation of the attack
rate would require accurate information on the number of technical
people in each activity classification during the period under study,
as well as more precise information on the number of laboratory-associated illnesses. Approximations have been made, and the estimated
attack rates suggest that the risk for researchers is seven to eight
times greater than for public health and hospital laboratory workers
(8). Phillips has estimated that the frequency rate of laboratory-acquired infection (using available United States and European data) results in an expected number of one to five infections per million working hours. If we assume that the average laboratory person works 8
hours per day for approximately 225 days (allowing for vacations, holi-
days, and weekends), then the probability of a laboratory worker acquiring a work-related infection is on the order of 10^{-2} to 10^{-3} (31).
He has also suggested that research personnel are more likely to handle
more hazardous agents, as well as new or rare agents, and thereby
increase the risk for this group of workers. Nevertheless, Grist reported that diagnostic laboratory workers who handle clinical blood
specimens are subjected to a higher infection risk than other workers
(28-30). Whether or not this is confirmed, the essential point is that
all personnel who handle infectious agents are at some risk, and laboratory-acquired infections are not confined to scientific personnel. Data
supporting this feature of laboratory-associated illness are presented
in Table 2. Although trained investigators, technical assistants, animal caretakers, and graduate students experienced over three quarters of the illnesses, the remainder occurred among clerical staff, dishwashers, janitors, and maintenance personnel. Wedum reported that
of the 369 illnesses evaluated at Fort Detrick, scientific personnel accounted for 82.3%; janitors, dishwashers, and maintenance personnel
accounted for 13.7%; and clerical personnel accounted for 3.7% of the
illnesses reported. These results are consistent with those reported
by Pike and Sulkin (15). This latter study reviewed 1,286 laboratory
infections that occurred in the United States between 1930 and 1950.
The results of these two studies are remarkably similar. In the Pike
and Sulkin report 78.1% of the infections occurred in trained personnel,
10.3% in janitors, etc., and 6.6% in clerical personnel. They also reported that 4.9% of the infections were experienced by students, whereas none occurred at Fort Detrick. Fort Detrick was a military facility,
where few, if any, students were present.

TABLE 2 Distribution of Infection According to Occupation

	Percentage of infections	
Personnel at risk	Study 1 (N 369)[a]	Study 2 (N 1286)[b]
Principal scientists, technicians	82.3	78.1
Animal caretakers, janitors, dishwashers, maintenance personnel	13.7	10.3
Clerical personnel	3.7	6.6
Students	0	4.9

[a]Adapted from Reference 9.
[b]Adapted form Reference 15.

Mode of Exposure

The most common mechanisms of exposure are direct inoculation by
needle and syringe, cuts or abrasions from contaminated items and
animal bites, inhalation of aerosols generated by accident or by work
practices and procedures, contact between mucous membranes and
contaminated material (hands or surfaces), and oral ingestion either
as a result of inhalation, mouth pipetting, or accident. Pike (5) has
indicated that approximately 18% of these episodes resulted from known
accidents that were due to either carelessness on the worker's part or
human error. Twenty-five percent of the accidents involved hypodermic needles and syringes, exposure by self-inoculation, or leaky
syringes. The other major categories were spills and sprays, 27%; injury with broken glass or other sharp objects, 16%; aspiration through
a pipette, 13%; and animal bites, scratches, and contact with ectoparasites, 13.5%.

    The precise source of exposure for the remaining cases (82%) was
difficult to determine. Although some could be ascribed to discarded
glassware, handling infected animals, and clinical specimens, all that
was known with any certainty was that the individual had worked with
the agent.

B. Role of Aerosols and Their Significance

Laboratory studies of potential sources of infection have focused on
hazards associated with routine microbiological techniques (9,32-35).
Considerable information has been accumulated to indicate that nearly

all routine bacteriological and virological procedures are capable of
producing aerosols. In their studies on aerosol dispersion, Dimmick
et al. (35) used the term "spray factor" for evaluating the hazard potential of various techniques. They used concentrated bacterial suspensions in standard laboratory procedures to generate aerosols. They
collected these airborne organisms in an Andersen cascade sampler
and determined the number of organisms released to the air. The spray
factor was then defined as the number of organisms released by the
technique divided by the initial concentration of organisms. For example, if the starting concentration was 10^8 organisms per milliliter,
and a lab procedure, e.g., blending, liberated 1.5×10^3, then the
spray factor was 1.5×10^{-5}. The larger the number of organisms released as an aerosol, the larger the spray factor. Spray factor then
could be used to characterize the biohazard potential associated with
laboratory techniques. Table 3 provides illustrative data on the
number of viable particles recovered within two feet of a work area.
These data are based on an extensive series of air sampling determinations (33,34).

The simple presence of organisms in the air is insufficient to cause
disease. To initiate a respiratory infection, infectious aerosols must
be deposited and retained within the respiratory tract. Studies have
shown that particles less than 5 μm in diameter are most effective in
establishing airborne infections in laboratory animals (36-39). Particles
of 1-2 μm deposit preferentially in the alveoli (39), whereas particles
larger than 3.5 μm are likely to deposit in the upper respiratory tract.
Particles in the 2.0-3.5 μm range deposit in both sections. The data
in Table 3 indicate that particles of respirable size are generated by
routine laboratory procedures.

Droplets less than 5 μm in size lose their liquid content rapidly
when airborne. These dried particles are called *droplet nuclei*, and
they can remain suspended in air for long periods and can be carried
about the room and building by both convective air currents (39) and
building ventilation systems. The information presented in Table 3
indicates that standard laboratory procedures generate aerosols that
are respirable and therefore potentially hazardous to the investigator
and to other personnel in the vicinity. The infectious dose for man
varies and is dependent on the type of agent, the route of exposure,
and the health status of the host in addition to dose. This interrela-
tionship is important because it helps to explain why not all laboratory
workers who handle infectious agents become ill. Viruses pose a greater hazard than the other agents because the infectious dose for man
is lower than that for other agents. Data to illustrate this are presented in Table 4.

With the current expansion of interest in diagnostic and research
virology, and in view of the aerosol production capability of standard
laboratory techniques as mentioned previously, a corresponding increase

TABLE 3 Concentration and Particle Size of Aerosols Created During
Representative Laboratory Techniques

Operation	Number of viable colonies[a]	Particle size[b] (μm)
Mixing culture with:		
pipette	6.0	3.5
vortex mixer	0.0	0.0
mixer overflow	9.4	4.8
Use of Waring blender:		
top on during operation	119.0	1.9
top removed after operation	1500.0	1.7
Use of sonicator	6.0	4.8
Lyophilized cultures:		
opened carefully	134.0	10.0
dropped and broken	4838.0	10.0

[a]Mean number of viable colonies per cubic foot of air sampled.
[b]Mean diameter of the particle.
Source: Adapted from References 33 and 34.

in laboratory-acquired viral illnesses would be expected among personnel who handle these agents. The large doses of bacterial agents required to produce disease via ingestion (Table 4) help to explain
why all microbiologists do not become clinically ill from their cultures.
In addition, the very low inhalation dose for certain agents (e.g., *F.
tularensis*, *C. burnetti*, measles virus, and coxsackie A21 virus) points
out the importance of preventing aerosol formation in the laboratory
environment and illustrates why the direct cause of the majority of
laboratory associated infections may go undetected.

III. BIOSAFETY CONTROL MEASURES

Based upon these various pieces of information, preventative measures
(containment) have been developed that provide some protection to at-
risk workers. Containment represents the integration of personnel
procedures and practices with laboratory design and engineering features to minimize the exposure of workers to hazardous or potentially
hazardous agents or substances (2,9,21,40-42). By using increasingly
stringent procedures and better designed facilities, work can be conducted with a higher degree of safety on agents of correspondingly

TABLE 4 Infectious Dose for Man

Disease or agent	Dose[a]	Route of inoculation
Scrub typhus	3	Intradermal
Q fever	10	Inhalation
Tularemia	10	Inhalation
Malaria	10	Intravenous
Syphilis	57	Intradermal
Typhoid fever	10^5	Ingestion
Cholera	10^8	Ingestion
Escherichia coli	10^8	Ingestion
Shigellosis	10^9	Ingestion
Measles	0.2[b]	Inhalation
Venezuelan encephalitis	1[c]	Subcutaneous
Polio virus 1	2[d]	Ingestion
Coxsackie A21	18	Inhalation
Influenza A2	790	Inhalation

[a]Dose in number of organisms.
[b]Median infectious dose in children.
[c]Guinea pig infective dose.
[d]Median infectious dose.
Source: Adapted from Reference 45.

increasing hazard or risk potential (3,41). The objective of incorporating containment measures into the design of a research program is to minimize the potential for exposure of investigative and support personnel, as well as minimizing the escape from the laboratory of experimental materials that may pose a health hazard to the surrounding community or cause some ecological perturbation.

A. Containment Practices and Procedures

It would be difficult to describe in detail all the specific practices and procedures that are employed to prevent or control laboratory-acquired illness or product contamination. These procedures vary and depend to a great extent on the agent, the type of experiment, the equipment,

facilities, the proficiency of personnel, etc. Fortunately, there are a
number of excellent reviews on this subject, and the reader is encouraged to consult them (3,40-43). Each laboratory that is involved with
or contemplates the use of hazardous or potentially hazardous agents
should develop protocols that meet the associated safety concerns.
There are, however, several basic practices and facility considerations
that are worth reviewing here.

Personnel Practices

The laboratory director should determine whether or not the use of a
particular agent requires special entry provisions. When entry restrictions are necessary, a hazard warning sign incorporating the universal
biohazard symbol should be posted on all access doors to the restricted
area. This sign should identify the infectious agent(s) currently in
used, list name and telephone number of the responsible supervisor,
and define specific requirements for entering the area.

As inadvertent ingestion is a realistic means of exposure to microorganisms, supervisors should prohibit eating, smoking, and drinking.
This prohibition should also be extended to chewing gum and the use
of throat lozenges, as well as application of cosmetics in the laboratory.
Nail biting is strongly discouraged, as this is another means of exposure to hand contamination. Clearly, mouth pipetting is not to be permitted. Mechanical pipetting devices are required. These devices
are more accurate than standard pipettes and eliminate aspiration/in-
gestion as a source of laboratory-acquired illness. It is important to
keep in mind that 13% of confirmed accidental infections were caused
by aspiration as a result of mouth pipetting. Hand washing provides
another means of preventing ingestion of organisms and their spread
to other surfaces. Hands should be washed after working with infectious microorganisms or cleaning up a spill.

Protective Clothing

A second group of protective practices includes the use of appropriate
gloves and laboratory clothing. Gloves function to prevent the worker's hands, fingers, and nails from being contaminated. This helps to
reduce the hazards associated with ingestion (hand-to-mouth transfer)
or the penetration of material through broken or unbroken skin. Although there are relatively few organisms that can penetrate inbroken
skin, current experiments also involve chemicals and radionuclides,
therefore, using gloves should be regarded as a minimum requirement.

The recent description of an accidental human vaccination (44)
with Vaccinia virus demonstrates the importance of proper procedures
and practices. While studying the immunogenicity of Vaccinia viral
recombinants, which expressed two proteins of a second virus (Vesicu-

lar Stomatitis Virus (VSV)), one of the workers accidently was exposed
to the recombinant virus and seroconverted to the proteins under study.

Analysis of this incident indicated that the exposure occurred when
some of the viral culture contaminated a cut on the worker's ring finger.
The worker promptly washed the affected area. Unfortunately, hand
washing was insufficient to prevent the subsequent infection. The
worker experienced a mild infection and developed antibodies to one of
the VSV proteins as well as to the Vaccinia virus. This laboratory-
acquired infection would have been prevented if the worker had worn
gloves. This illness was not reported in the scientific literature because of its laboratory origins. It was reported because it represented
the first human efficacy data for a Vaccinia vaccine delivery system.
But for those interested in both science and safety, this exposure
reinforces the need to understand and appreciate the importance of
personnel protection requirements.

Laboratory clothing designed to protect one's personal clothing
should be worn. The purpose of this clothing is to keep street clothing, forearms, hair, face, hands, etc., free of contamination. It is
important to realize that wearing these items, especially lab coats, to
the cafeteria, libraries, meetings, or to other buildings provides a
mechanism for spreading contamination to others as well as to oneself.
The use of laboratory clothing must be restricted to the laboratory
and not worn outside the laboratory area.

Housekeeping and Waste Management

Housekeeping. Well-defined housekeeping procedures and schedules are important in reducing the risks of working with biohazardous
material and in protecting the integrity of the work activity. Housekeeping should be done in a manner that reduces or prevents the generation of aerosols. While vacuuming with a system that exhausts
through a HEPA filter or the use of a two-bucket wet-mopping method
are routinely recommended (41), they can be a bit excessive for level
1 or level 2 laboratories. In these areas, the use of a wet-mopping
procedure with a disinfectant solution or the use of dry-mop floor-
care products which contain a disinfectant should be sufficient.

Clearly for areas where agents of higher hazard level are handled,
cleaning techniques such as the two-bucket mopping method and HEPA-
filtered vacuums can be and should be considered (41).

Work Surfaces. Work surfaces must be decontaminated daily and
immediately following spills. This will reduce the spread of contamination to one's person and at the same time reduce the potential for
contaminating one's own or someone else's experiment or process.

Waste. All biological waste and contaminated equipment should
be decontaminated or inactivated prior to disposal or reuse. This is

especially true if known pathogens are involved. Contaminated materials such as fermenters, culture flasks, glassware, laboratory equipment, etc., should be decontaminated before washing, reuse, or disposal. This again will help protect personnel not directly associated with the laboratory activity (e.g., glassware workers, janitors, repair personnel, animal care technical support, etc.). Such personnel should never be exposed to contaminated materials without first being informed and thoroughly trained.

Care should be exercised whenever hypodermic needles and syringes are used. Twenty-five percent of laboratory illnesses due to defined causes were attributed to syringe mishaps (5,43). Hypodermic needles and syringes should only be used for the parenteral injection or aspiration of fluids. Only needle-locking syringes or units where the needle is an integral part of the syringe should be used. Syringes and needles should be placed directly into a puncture-resistant container immediately after use; they should not be clipped or the needle resheathed. The contents of the puncture-resistant container should be chemically inactivated and/or autoclaved prior to disposal. A recent case reported (45) of a laboratory-acquired gonococcal infection describes an accident which could have been avoided if needle-locking syringes had been used. Contaminated broken glass also presents a hazard and must be decontaminated and disposed of properly. Once contaminated needles or other sharps are decontaminated, the physical hazard they present is still real and should be addressed accordingly.

B. Containment Equipment

The purpose of safety equipment is to provide physical containment to protect workers from exposure to aerosols of infectious agents. Properly maintained and used equipment provides a significant level of protection to the worker, but such equipment must always be tested to assure that it is functioning correctly. Safety devices that prevent the escape of aerosols into the laboratory environment are available (39,40,43). These include biological safety cabinets and a variety of air-tight enclosures designed to house various pieces of laboratory equipment such as sonicators and centrifuges. The biological safety cabinet is the principal device used to provide physical containment of infectious aerosols.

The three types of biological safety cabinets are described elsewhere in this volume. The personnel protection provided by Class I and Class II cabinets is dependent on the maintenance of an adequate inward airflow. These cabinets provide only partial containment and are not adequate for containment of high-risk infectious agents (sometimes designated Class 4 organisms). This is due to the chance for inadvertent escape of aerosols across the open cabinet front. This type of escape is prevented when a Class III cabinet (glove box) is

used. Class III cabinets are designed to completely isolate the worker
from the hazard, thus this cabinet provides the highest level of personnel and product protection available. When such a degree of containment is needed, most (if not all) procedures involving infectious agents
can be contained within them. Class III cabinet systems have been
designed as an interconnected network of individual cabinets which
contain various pieces of equipment such as incubators, refrigerators,
and centrifuges, etc. Double-door autoclaves and pass-through chemical dunk tanks also have been included in cabinet networks to allow
safe introduction and removal of supplies and equipment during the
course of the work (40,46).

The Class II cabinet is the most commonly used piece of containment
equipment in the biomedical research laboratory. This unit is used in
preference to Class I cabinets because the latter does not provide protection of the research located in the work area. The Class I cabinet,
like the chemical fume hood, draws unfiltered room air into the work
area. Protection of research materials from contamination is attained
in a Class II cabinet by the downflow of HEPA-filtered air over the
entire work space.

C. Facility Design/Program Requirements

The National Institutes of Health (NIH) and the Centers for Disease
Control (CDC) have developed a monograph which provides guidance
for the selection of appropriate biosafety precautions (3). This monograph provides specific information on laboratory hazards associated
with a variety of agents as well as recommendations on practical safeguards and facility considerations that can significantly reduce the
risk of laboratory-associated diseases.

Anyone who is contemplating research activity with infectious agents
should consult this monograph as well as two other guides published
by NIH (46,47) to determine the necessary safety procedures, practices, and facility requirements prior to initiating any program that
involves infectious or genetically engineered organisms.

In the event that the agent is not described in this document, the
reader should consult a second publication from CDC, entitled *The
Classification of Etiological Agents on the Basis of Hazard* (48). This
is a useful document that provides a more extensive listing of organisms and their hazard classification than the NIH/CDC monograph (3).

For reference purposes, four biosafety levels are described in the
NIH/CDC publication (3). Each corresponds to the containment re-
quirements for the four classes of microorganisms 1, 2, 3, 4 in the
CDC publication (48). The agents of minimal hazard are in Class or
Biosafety Level 1, and dangerous and exotic microorganisms are in
Class or Biosafety Level 4. Each biosafety level specifies work prac-
tices, containment equipment and facility design appropriate for the

different hazard classes of microorganisms. The facility design requirements for each level and the type of work suitable for the level are outlined below. Features such as ease of cleaning, impervious bench surfaces, sturdy furnishings, handwashing sinks, and window screens where windows can be opened are incorporated into even the lowest level of containment.

Biosafety Level 1 Containment

At this level of containment the laboratory provides an environment for work with low hazard agents that can be controlled by standard laboratory practice. Viable microorganisms not known to cause disease in healthy adult human beings are handled at this level. Work activity is routinely conducted on an open bench. No special design features are required, although laboratory space should be separated from general offices, food service and patient areas.
   This laboratory is suitable for experiments involving:

1. Recombinant DNA requiring BL1 (P1) containment (46).
2. Microorganisms of minimal or no biohazard potential under ordinary handling conditions. Such agents are designated Class I (48) by the CDC or Biosafety Level 1 in the NIH/CDC document (3).
3. Nonrecombinant cell or tissue culture studies which do not involve infectious plant or animal virus (49).

   Some microorganisms in this class include *Escherichia coli* K12, *Saccharomyces cerevisiae*, and *Bacillus subtilis*.

Biosafety Level 2 Containment

This type of laboratory is identical in construction to the level 1 facility. While work that does not produce significant aerosols can be conducted on an open bench, biological safety cabinets are frequently present, and if contamination control is important (e.g., keeping cultures clean) they should be used.
   This laboratory is suitable for experiments involving:

1. Recombinant DNA activity requiring BL2 (P2) physical containment (46).
2. Microorganisms of low biohazard potential such as those in Class 2 (48) or Biosafety Level 2 (3,48).
3. Nonrecombinant cell and/or tissue culture systems that require this level of containment (49).
4. Oncogenic viral systems classified as low risk (40).
5. Certain suspected carcinogens and other toxic chemicals (42).

The microorganisms handled in a BL2 laboratory include many of the indigenous infectious agents that produce disease in man (*Staphylo-*

coccus, Streptococcus, measles, polio, enteric pathogens, etc.). Most academic, diagnostic, or industrial laboratories operate under levels 1 and 2.

Biosafety Level 3 Containment

At BL3, one encounters a facility that includes special engineering design features and containment equipment. These facilities are usually separated from the general traffic flow by controlled access corridors, air locks, locker rooms, or other double-door entries. Biosafety cabinets are used for all technical manipulations that involve unsealed cultures. The surfaces of all walls, floors, and ceilings are constructed to be sealed. This means that all penetrations are chalked, collared, or sealed to prevent leaks.

The ventilation system in this facility is designed to exhaust more air than is supplied, resulting in a directional airflow from outer clean corridors into the laboratory, where the higher potential for contamination exists. The air is usually discharged to the outdoors and not recirculated to other parts of the building without appropriate treatment.

This laboratory is suitable for experiments involving:

1. Recombinant DNA molecules requiring physical containment at the BL3 (P3) level (46).
2. Microorganisms of moderate biohazards potential such as those in Class 3 (48) or Biosafety Level 3 (3).
3. Oncogenic viruses that have human cells in their host range (40,46).
4. Nonrecombinant cell tissue culture experiments that involve large volumes or high concentrations of virus-infected cells (where the virus is infectious for man or require level 3 containment) (49).
5. Nonrestricted carcinogens and other toxic chemicals (42).

Biosafety Level 3 organisms differ from those at level 2 because there is a real potential for infection due to aerosol exposure. Organisms such as *Mycobacterium tuberculosis*, St. Louis encephalitis virus, and *Coxiella burnetii* belong in this category.

Biosafety Level 4 Facility

These are extremely sophisticated facilities in terms of design, which provide a very high level of containment for research involving biological agents that present a life-threatening potential to the worker or may initiate a serious epidemic disease.

The distinguishing characteristic is the use of barriers to prevent the escape of hazardous material to the environment, as well as additional barriers to protect laboratory personnel.

Barriers that serve to isolate the laboratory area from the immediate area include:

1. Monolithic walls, floors, and ceilings in which all penetrations such as air ducts, electrical conduits, and utility pipes are sealed to ensure the physical isolation of the laboratory area.
2. Air locks through which supplies and materials can be brought safely into the facility.
3. Contiguous clothing change rooms and showers through which personnel enter and exit the facility.
4. Double-door autoclaves to sterilize wastes and other materials prior to removal from the facility.
5. Biowaste treatment systems to sterilize liquid wastes.
6. Separate ventilation systems that control air pressures and airflow directions within the facility.
7. Treatment systems to decontaminate exhaust air before discharge into the atmosphere.

A description of specific operational procedures and laboratory practices recommended for use in maximum containment facilities can be found in the monograph entitled *Safety and Operational Manual, High Containment Research Facility* (50). These facilities are usually operated by very well-trained workers who work under rigorous supervision.

This laboratory is suitable for experiments involving:

1. Recombinant DNA molecules requiring physical containment at the BL-4 (P-4) level (46).
2. Microorganisms of high biohazard potential such as those classified as Class or Biosafety Level 4 (or higher) (3,48).
3. High-risk oncogenic viruses (40).
4. Carcinogens on the restricted list (42).

Practical Considerations

As indicated above, these are extremely sophisticated facilities that can provide the highest level of containment for working with agents that have life-threatening potential. Any more detailed descriptions of these facilities and associated research activities are beyond the scope of this seciton. (See References 49 and 50 for specific information.)

While four distinct biosafety levels are described, there are instances where a combination of procedures from two different levels provide a safer work environment. An absence of information and experience regarding new microorganisms often stimulates laboratory personnel to select work practices that offer increased protection. Laboratory manipulations of the human immunodeficiency virus (HIV) have prompted biosafety committees and safety personnel to take a hard look at containment practices. Based upon observations that this virus does not have a documented aerosol route of exposure, the need to require the additional containment of a level 3 facility (as described in Reference 3) does not seem critical at this time, and a level 2 facility is currently routinely used. Level 3 procedures work practices (3) in conjunction with the appropriate safety equipment, safety centrifuge cups, biosafety cabinets, etc., do afford a greater margin of safety for personnel. The resultant BL2 plus "hybrid" containment provides a greater measure of safety for work with agents for which there is limited safety information available, e.g., HIV. As more information on the virus becomes available, then the practices can be altered accordingly.

IV. EDUCATION AND TRAINING

All personnel who work with hazardous organisms or substances must receive sufficient information and training to enable them to work as safely as possible.

This instruction should include a review of the following areas: a thorough review of operations and procedures with emphasis on material transfer and other possible sources or exposure; adverse health effects of specific work practices detailing which are acceptable as well as the ones that are unacceptable; engineering controls (hazard control ventilation and acceptable perimeters); containment equipment (contained centrifuges, safety cabinets, etc.) in use or being considered for use; proper disposal of contaminated waste; decontamination of surfaces; and specific emergency procedures to be followed in the event that there is an accident or spill. Each member of the work team must be familiar with the biology of the system or process. It is vital that they understand that human factors are associated with accidents, such as tiredness, haste, and inattentiveness.

The risks associated with work involving infectious agents are minimized when appropriate attention is given to all biological safety practices. Each person involved in such work must accept full responsibility for biological safety to protect himself, his colleagues, and the general public.

Although primary emphasis is placed on preventing laboratory-acquired infections in humans, there are a number of infectious agents that affect animals or plants. Release of such agents to animal colonies and the environment must be prevented with equal vigor. Foot and mouth disease of cattle is an example of an animal disease that is so contagious and so costly economically that research on it is confined to a single laboratory located on an island several miles from the mainland.

Ectromelia (mousepox) and lymphocytic choriomeningitis (LCM) are less dramatic but more important examples of agents that require careful procedures to prevent spread of infection to animal colonies. Ectro-

melia can render an animal research program useless, and LCM is extremely hazardous to animal handlers.

V. CONCLUSION

The potential threat of infection in the laboratory has long been recognized by the medical microbiological community as an ever-present occupational hazard. Published reports of the occurrence of laboratory-
acquired infections have served as reminders that the potential can
easily become an actuality. They have pointed out the need for unre-
mitting adherence to appropriate precautionary measures. Only in
recent years have we come to appreciate the magnitude of the problem
of laboratory-acquired infections and the many factors, both human
and environmental, that may be involved (1,6).
It is crucial that management provide laboratory facilities that are
commensurate with the requirement for the work to be conducted in a
safe manner.
As indicated previously, considerable information has been accumulated that clearly indicates that nearly all routine laboratory procedures are capable of producing aerosols. Therefore, any operation
that generates a significant aerosol or involves a human, animal, or
plant pathogen, or some other agent that could disrupt the environment
if inadvertently released, should be contained within safety equipment
or facilities. These containment systems must be subjected to periodic
inspection and certification to assure proper function (41).
In the laboratory, the procedures used should be appropriate for
the highest level of risk to which personnel, the experiment, and the
environment will be subjected. Such an approach will avoid multiple
practices and the constant retraining of personnel.
Physical containment is dependent upon the safety awareness and
the techniques of the investigative staff, the availability and proper
use of safety equipment, and the design of the laboratory or facility.
It should be recognized that physical controls alone cannot create
a facility that is safe. Containment is achieved through the combination of equipment, engineering features, and *the scrupulous adherence
to good laboratory or facility practices*. Poor practices can override
the protection provided by equipment and facility design and place all
personnel in jeopardy. Poor practices relate to more than simple poor
technique or bad technique (e.g. sonication on an openbench). Poor
practice also includes rushing through activities without thinking, being
easily distracted or constantly distracting one's colleagues, working
alone at night, or even worse working alone at night and feeling ill or
tired. The effect of these practices should not be underestimated; too
many of the exposures to infectious materials occur under these conditions.

REFERENCES

1. Grist, N. R. (1982). *Yale J. of Biol. and Med.* 55:213.
2. United States Public Health Service. (1978). For laboratory safety at the Centers for Disease Control. Atlanta, GA. CDC 7.68118.
3. Barkley, E., and Richardson, J. (Eds.) (1984). Biosafety in microbiological and biomedical laboratories. Centers for Disease Control. Atlanta, GA, and National Institutes of Health, Bethesda, MD. HHS CDC 848395.
4. Liberman, D. F. (1979). Occupational hazards. *Pub. Health Lab.* 37:118.
5. Pike, R. M. (1979). *Health Lab.* 37:118.
6. Pike, R. M. (1978). *Arch. Pathol. Lab. Med.* 102:303.
7. Sulkin, S. E., and Pike, R. M. (1946). *N. Eng. J. Med.* 241:205.
8. Sulkin, S. E., and Pike, R. M. (1951). *JAMA* 41:1740-1745.
9. Wedum, A. G. (1969). *Pub. Health Rep. U.S.* 79:619.
10. Gross, H. T. (1940). *J. Kansas Med. Soc.* 41:329.
11. Furcolow, M. L., Guntheroth, W. G., and Willis, M. J. (1952). *Lab. Clinic. Med.* 40:182.
12. Huddleson, F., and Menger, M. (1940). *Am. J. Public Health* 30:944.
13. Meyer, K. F., and Eddie, B. (1941). *J. Infect. Dis.* 68:24.
14. Robbins, F. C., and Rustigion, R. (1946). *Am. J. Hyg.* 44:64.
15. Pike, R. M., and Sulkin, S. E. (1952). *Sci. Monthly.* 75:222.
16. Pike, R. M., and Sulkin, S. E. (1951). *Am. J. Public Health* 147:769-781.
17. Hanson, P., Sulkin, S. E., Buescher, E. L., McD. Hammon, W., McKinney, R. W., and Work, T. H. (1967). *Science 158:* 1283.
18. Harrington, J. M., and Shannon, H. S. (1977). *Brit. Med. J.* 1:626.
19. Hellman, A., Oxman, M. N., and Pollack, R. (Eds.) (1973). Cold Spring Harbor Laboratory, New York.
20. Sulkin, S. E. (1961). *Bacteriol Rev.* 25:203.
21. Sulkin, S. E., Hanel, E., Phillips, G. B., and Miller, O. T. (1956). *Am. J. Public Health* 46:1102.
22. Wedum, A. G., Barkley, W. E., and Hellman, A. (1972). *J. Amer. Med. Assoc.* 161:157.
23. Sulkin, S. E., Pike, R. M., and Schulze, M. L. (1965). *Am. J. Public Health* 55:190.
24. Sabin, A. B., and Wright, A. M. (1934). *J. Exp. Med.* 59:115.
25. Dyer, R. E. (1938). *Public Health Rep. (U.S.)* 53:2277.
26. Most, H. (1934). *Am. J. Trop. Med. Hyg.* 22:157.
27. Brown, F., Talbot, P., and Burrows, R. (1973). *Nature 245:* 315.

28. Grist, N. R. (1973). *J. Clin. Pathol. 26*:388.
29. Grist, N. R. (1973). *Proc. R. Soc. Med. 66*:255.
30. Grist, N. R. (1975). *J. Clin. Pathol. 28*:255.
31. Phillips, G. B. (1969). *Amer. Ind. Hyg. Assoc. J. 30*:170.
32. Anderson, R. E., Stein, L., Mopes, M. L., et al. (1952). *J. Bacteriol. 65*:473.
33. Kenny, M. T., and Sabel, F. L. (1968). *Appl. Microbiol. 16*:146.
34. Reitman, M., and Wedum, A. G. (1956). *Pub. Health Rep. (U.S.) 71*:659.
35. Dimmick, R. L., Vogl, W. C., and Chatigny, M. A. (1973). In A. Hellman, M. N. Oxman, R. Pollack (Eds.), *Biohazards in Biological Research*. Cold Spring Harbory Laboratory, p. 246.
36. Brown, J. H., Cook, K. M., Ney, F. G., and Hatch, T. (1950). *Am. J. Public Health 40*:450.
37. Druett, H. A., Henderson, D. W., Packman, L., and Peacock, S. (1953). *J. Hyg. 51*:359.
38. Goodlow, R. S., and Leonard, E. A. (1961). *Bacteriol. Rev. 29*:182.
39. Hatch, T. F. (1961). *Bacteriol. Rev. 25*:237.
40. U.S. Dept. of Health and Human Services. (1979). Biological Safety Manual for Research Involving Oncogenic Viruses. Public Health Service. National Cancer Institute. Bethesda, MD. (NIH).
41. Laboratory Safety Monograph. (1979). U.S. Dept. of Health, Education and Welfare, Public Health Service. National Institutes of Health. Bethesda, MD.
42. Prudent Practices for Handling Hazardous Chemicals in Laboratories. (1981). National Research Council. National Academy Press. Washington, D.C.
43. Richardson, J. N., and Huffaker, R. H. (1980). Biological safety in the clinical laboratory. In Manual for Clinical Microbiology. American Society for Microbiology, Washington, D.C.
44. Jones, L., Ristow, S., Yilma, T., and Moss, T. (1986). *Nature 319*:543.
45. Briuns, S. C., and Tight, R. R. (1979). *JAMA 41*:274.
46. Guidelines for recombinant DNA research activity. (1986). *Fed. Register 51*:16958.
47. National Institutes of Health Biohazard Safety Guidelines. (1974). U.S. Dept. of Health, Education and Welfare, Public Health Service. National Institutes of Health, Bethesda, MD.
48. Classification of Etiologic Agents on the Basis of Hazard. (1976). U.S. Dept. of Health, Education and Welfare, Public Health Service. Centers for Disease Control, Atlanta, Georgia.
49. Liberman, D. (1980). *Public Health Lab. 38*:199-207.
50. Safety and Operations Manual. 1977). High Containment Research Facility. U.S. Dept. of Health, Education, and Welfare. Public Health Service. National Cancer Institute. Bethesda, MD.

8

Hospital Epidemiology and Infection Control: The Management of Biohazards in Health Care Facilities

DIANE O. FLEMING* *The Johns Hopkins University School of Hygiene and Public Health, Baltimore, Maryland*

I. INTRODUCTION

Biohazards of concern in a health care facility are those that can cause disease in patients, personnel, and visitors. Situations that may result in exposures to such materials include: failure to use effective barriers and appropriate procedures; inappropriate handling and disposal of specimens and wastes from patients with communicable diseases; improper cleaning, sterilization and storage of instruments, kits, and packs; failure to protect the hospital environment by routine maintenance and cleaning especially during construction and renovation of facilities. Biohazards in health care facilities are usually managed under the auspices of a hospital epidemiology or infection control department. Hospital epidemiology is a discipline in which epidemiologic methods are applied to identify situations in which hospital-axquired (nosocomial) infections may be prevented or controlled. An active, organized infection control program is the foundation for control of biohazards in the health care setting.

II. HISTORICAL PERSPECTIVE

There are good historical reviews of attempts at control of communicable diseases, a recent one being LaForce, 1986 (1). Although hospitalized patients with smallpox, leprosy, and plague were segregated by early physicians to prevent spread to other patients, the fever hospitals in England were the first to establish routine isolation practices for infec-

---

Current affiliation: Sterling Drug, Great Valley, Pennsylvania

tion control. Hospital epidemiologists attribute the emergence of their specialized field of medicine to Semmelweis, whose careful study identified the hands of physicians in the transmission of puerperal fever. Procedures that helped to reduce infection include antisepsis, as introduced into the operating room by Lister, and the use of gloves by Halstead.

III. INFECTION CONTROL AS AN ORGANIZED PROGRAM

Organized infection control programs were developed in the United States following the outbreaks of hospital staphylococcal infections in the 1950s and 1960s (1). The Centers for Disease Control collected data from hospitals and emphasized active surveillance and control programs in the mid to late 1960s. In 1958 the American Hospital Association (AHA) and in 1969 the Joint Commission on the Accreditation of Hospitals (JCAH) endorsed infection control programs and nosocomial infection surveillance in hospitals. From 1974-1985 the Centers for Disease Control (CDC) carried out the SENIC study (Study on the Efficacy of Nosocomial Infection Control), which, in the final analysis, verified the efficacy of surveillance and control programs in reducing the risk of nosocomial infections (2a). The CDC recommendation of at least one infection control practitioner for every 250 hospital beds became the norm during this period and was one of the methods that SENIC documented to be effective in preventing infections. Other necessary elements of the infection control program identified from SENIC included the presence of a trained hospital epidemiologist and an intensive infection surveillance and control program within the hospital.

In addition, the SENIC study indicated that hospital infection control programs should promote the surveillance of surgical wound infections and provide the results of surgeon-specific attack rates to each surgeon. A recent follow-up study of the hospitals that participated in SENIC found that the benefits that could be achieved with an infection control program, such as a reduction in nosocomial infections, had fallen short of the maximum (30%) to be expected (2b).

IV. THE SCOPE OF INFECTION CONTROL PROGRAMS

A. Factors that Influence Infection Control

Accreditation

Infection control is required in inpatient facilities in order to obtain accreditation from the Joint Commission for the Accreditation of Health Care Organizations (JCAHO). Requirements for accreditation of ambulatory care and long-term care facilities such as nursing homes and addiction treatment centers have extended the practice of infection control to new areas. As in many other fields of medical care, there is more art than science to infection control, due to the lack of hard scientific data on specific procedures as they relate to the prevention of infection. The SENIC study added credible data for infection control programs but not for specific infection control practices.

DRGs and PPSs

The impact that Diagnostic Related Groups (DRG) and prospective payment systems (PPS) have had on inpatient hospitals in the past few years has been to decrease the hospital stay for individual patients. Each diagnosis is alloted a specific number of hospital days for care, thus, the payback from a third-party payer is defined in advance. Without a fee-for-service system, the adjustment in payment allowed for nosocomial infections is not likely to cover the actual costs. Early discharge and outpatient care of patients is thought to reduce the risk of nosocomial infection. In response to the needs of their patients and as a result of the decreased utilization of inpatient beds, ambulatory care centers are being constructed within hospitals or as free-standing clinics. Specialty clinics with same-day surgery, fertility clinics, birthing centers, health maintenance organizations (HMO), and other similar outpatient facilities are performing procedures that used to require hospitalization (3).

Ambulatory Care Facilities

Infection control policies and procedures that were written for inpatient facilities may not apply directly to ambulatory settings. Institutional policies and infection control support may not even be available to the ambulatory care practioner. Problems that can result in exposure to biohazards arise from failure to use patient care procedures which could reduce the risk of infection; inappropriate sterilization of instruments used for invasive techniques; lack of knowledge or consideration of the appropriate method for disposal of infectious waste, including needles, and other hazardous waste; failure to provide employee health protection such as hepatitis B vaccine to those with daily blood contact; and lack of preemployment testing for rubella, measles, chickenpox, etc., prior to contact with susceptible patients. Even the identification of patients with active acute or chronic infectious diseases is made more difficult in an ambulatory care setting due to the lack of a kardex file or the lack of identifiers on the patient's chart.

As the scope of the infection control program expands to meet the demands of modern health care practices, research efforts must continue to provide data to support the practices being recommended. Studies should take into consideration the problems of same-day care as well as those relating to the typical inpatient hospitalization. Results of research studies and review of current literature should be followed by the implementation of appropriate practices and the discontinuation of unnecessary or ineffective ones. Daschner (4) was able to obtain significant cost reductions by discontinuing ineffective practices in an infection control program. He also found that less costly practices could be used without compromising the level of protection against infection.

V. ACTIVITIES OF AN INFECTION CONTROL PROGRAM

The following list of activities carried out in an organized infection control program was compiled by the Certification Board of Infection Control:

Collection and analysis of infection control data (surveillance)
Planning, implementation, and evaluation of infection prevention and
        control measures
Education of individuals about infection risk, prevention, and control
Development and revision of infection control policies and procedures
Investigation of suspected outbreaks of infection
Provision of consultation on infection risk assessment, prevention,
        and control strategies

A. Management Activities

Modern infection control differs from its earliest roots, e.g., the use of aseptic technique in patient care practice. Management activities have been introduced (2b) and infection control practioners now attempt to convince hospital staff to care for patients in ways that will prevent or reduce the chance of infection. To accomplish this they must have surveillance data on the activities in question and be able to use that data to intervene and alter the risks of infection. Intervention may mean stressing the use of aseptic technique in in-service programs, putting a new isolation technique into practice, or even convincing the hospital purchasing department to buy a product that has been demonstrated to be associated with reduced rates or infection, i.e., a different type of catheter. The Centers for Disease Control has responded to these "middle management activities" by adding a management course to their training sessions for infection control practioners.

B. Epidemiologic Investigations

Infection control demands the ultimate in problem-solving capabilities in the evaluation of epidemics of infection. An epidemic situation is

ldentified as the occurrence of disease at greater than a generally accepted threshold. Clusters of infections in patients under their care
may be recognized by nurses on the floors, while clusters of similar
organisms isolated from patient specimens during a short period of
time may be noticed by a technician in the clinical laboratory. Line
listings of patients help the infection control practitioner to put these
clusters into perspective. In small hospitals, an outbreak or epidemic
may be identified by the ICP in consultation with the Chairman of the
infection control committee or the State Health Department. There are
certain arbitrary thresholds for concern in nosocomial infection, and
an epidemiologic investigation is usually undertaken when these thresholds are exceeded (5). They include, but are not necessarily limited
to:

A recurrent (second) outbreak of nursery staphylococcal infection
Two or more isolations of a strain of bacteria with a worrisome new
    antimicrobial susceptibility pattern
Two or more isolations, within a 30-day period, of a new strain that
    has never or rarely been seen in your laboratory
Significant increase in the incidence of any infection
Two or more cases of nosocomial disease in consecutive 30-day periods
Sustained high incidence of infections in the following categories:
    a clean wound infection rate of over 3-4%
    urinary tract infection rate of more than 20% in catherized patients
    primary bacteremia rate greater than 4-5 per 1,000 patients with
        IV cannulae

    An epidemiologic investigation may also be needed to evaluate ex-
cessive numbers of endemic (community-acquired) disease; evaluate
effects of policies and procedures; identify errors in laboratory results;
identify problems with surveillance methods; and conduct quality assurance studies.
    Once the initial case in an epidemic is reviewed, the seriousness
of the problem will determine if a basic or a major investigation is required. A basic investigation involves the collection of the data and
any necessary specimens, the institution of control measures, and the
identification of affected patients. Case characteristics of time, place,
and person are reviewed and the case definition is refined. Further
decisions are based on the information obtained from the basic investigation and the effectiveness of the control measures.
    If a major investigation is needed, a consultation may be requested
from the Health Department or the Centers for Disease Control. A
tentative hypothesis must be formulated, tested, and then followed by
the evaluation of any control measures which have been initiated (6).
Fortunately, most health care facilities will only require basic investigations of periodic problems with exposures to chicken pox, measles,

and tuberculosis or clusters of diarrhea or pulmonary diseases. Causative agents need to be identified. Surveillance data will be needed to help decide if nosocomial transmission has occurred and if it could have been prevented. Early intervention is required to prevent secondary cases.

C. Priorities of an Infection Control Program

The main thrust of an infection control program is still infection surveillance and control and not the relatively few epidemiologic investigations needed. Surveillance may be conducted on all hospitalized patients or as a limited surveillance of those areas with patients at high risk for infection. These would include all intensive care, neonatal, and oncology units (6). Surveillance activities are usually documented by making a line listing of patients with potential nosocomial infection, and then determining if the infection can be related to the hospitalization and care. The identification of hospital-acquired infections requires a definition for each possible type of infection: bacteremia, pneumonia, wound infections, urinary tract infections, etc. The final line listing will include both nonpreventable and preventable nosocomial infections. Methods for controlling the preventable infections are determined from published data and guidelines. Observations of patient care and housekeeping practices that may impact on the nosocomial infections are a part of the practice of infection control. The presence of the infection control practitioner has been shown to result in compliance by the health care provider with infection control practices, such as hand washing. This effect has been demonstrated with clinical mentors (7) and might be expected to occur with other role models such as medical and nursing school instructors.

VI. KNOWLEDGE AND SKILLS REQUIRED IN INFECTION CONTROL PRACTICE

The practice of hospital epidemiology and infection control is a specialized field of medicine and nursing. It requires a working knowledge of the etiology of infectious diseases, their transmission, prevention, and treatment. Other elements of an infection control program include a knowledge of microbiology, epidemiology, statistics, patient care, and employee health practices.

Management and communication skills are also essential (8). Although hospital epidemiology may be practiced by anyone with training in infection control, ideally, a hospital epidemiologist should be in charge of the program. The majority of infection control practitioners have a degree in nursing, and approximately 75% of the members of the Association of Practitioners in Infection Control are nurses (APIC

annual report, 1986, personal communication). Although relatively
few microbiologists or sanitarians are infection control practitioners,
they usually join a department that already has at least one nurse practioner. Their skills and knowledge complement each other in the overall program.

The hospital epidemiologist is usually a physician with specialized
training in infectious diseases, who can perform patient care functions
in addition to directing the infection control program. The Society of
Hospital Epidemiologists of America (SHEA) promotes the concept that
the hospital epidemiologist be a physician as well as an epidemiologist.
Recent reports indicate that very few hospital epidemiologists have
received the training needed and that there has been a decline in the
number of hospital epidemiologists in the United States (2b). Several
training programs for hospital epidemiologists have been organized
under the auspices of SHEA. Certification is available through the
American Academy of Epidemiology.

Infection control training may be obtained through coursework at
the Centers for Disease Control, state health departments, the annual
APIC educational conference, seminars or training sessions sponsored
by SHEA, or through one of several Schools of Public Health in this
country.

VII. JCAHO STANDARDS FOR INFECTION
CONTROL PROGRAMS

The Joint Committee on the Accreditation of Health Care Organizations
(JCAHO) has published standards for infection control programs and
has defined the basic elements of the program in their interpretive
guidelines. They give responsibility to the infection control committee
to monitor the infection control program and recommend corrective
action and also stipulate that the pertinent findings of the ICC be made
part of the hospital's continuing education program and the orientation
program for new employees. Hospitals that seek accreditation through
the JCAHO must comply with guidelines, which require that infection
control policies and procedures be written by and for specific depart-
ments/services and areas throughout the hospital. The JCAHO also
requires that these guidelines be "available for all personnel involved
with procedures that are commonly used in patient care and known
to be associated with nosocomial infection potential" (9). Future requirements for the licensure or accreditation of organized ambulatory
care facilities will undoubtedly involve the formulation of policies and
procedures appropriately tailored to infection control and employee
protection in those settings.

Infection control manuals containing the policies and procedures approved by the ICC are disseminated throughout the hospital. It is important that the information they contain becomes a part of the curriculum of health care practioners and administrators. The behavioral modification that must be made by present-day health care providers to enable them to routinely wash hands between patients and to conform to the isolation precautions for specific diseases is a major change. Such changes require motivation, practice, and the presence of a role model. Infection control committees need to concern themselves with methods for enhancing motivation within their facilities. Accreditation by the JCAHO is of itself the only motivation a hospital usually needs to provide appropriate policies and procedures for the control of infections. The ICC needs to identify motivational techniques that can cause behavioral modification in health care personnel and result in the actual use of these infection control procedures.

VIII. GOAL OF AN INFECTION CONTROL PROGRAM

The basic goal of an infection control program is to minimize the risk of nosocomial infection for any given level of resource allocation. The money is not always available to provide adequate staff for essential educational efforts as well as surveillance activities. There are situations in which the staffing may be adequate, but the lines of communication and responsibility are ineffective. To achieve the basic goal, the program must have an adequate staff, administrative support, and peer status.

A. Making an Infection Control Program Work:
Policy Development and Implementation

In order to develop a workable infection control program, the policy-making body of the hospital must interact with the infection control committee (ICC). The ICC reveiws data on infections, approves departmental and institutional policies on infection control, and makes recommendations to the hospital administration. These recommendations may be made into policy by the hospital board to provide funding or enforcement. Hospital administration is responsible for dissemination of information about hospital policies and procedures which have been approved. The implementation of these policies and procedures reamins the responsibility of the department or functional unit. There is no strict enforcement mechanism in an infection control program. The infection control practitioner and the hospital epidemiologist usually act in an advisory capacity. According to the requirements of the JCAHO, they must be given the authority to institute appropriate control measures, such as initiating or terminating isolation precautions or cohorting patients when necessary.

B. Making the Infection Control Program Work: Reporting
   Responsibilities and Essential Interactions in the
   Modern Health Care Facility

Infection control programs impact on every area of the hospital. Although the early programs attempted surveillance of the entire hospital,
the data collected was not actually used to prevent further infections.
The surveillance was done as a nursing responsibility, and results
were merely reported to infection control committees. At the present
time, hospitals are tending to do limited surveillance in areas of high
risk, and reports to ICC include information on attempts to intervene
and prevent infection when surveillance data indicate control measures
are appropriate. Infection control practitioners interact in the practice
of their specialty with personnel from employee health, pharmacy, the
operating room, product standardization, the laboratory, safety, and
others as dictated by the problems that arise. The multidisciplinary
infection control committee usually includes representatives from areas
that interact frequently with infection control. While there seem to
be as many different infection control programs as there are types
of hospitals, the "best" program is the one that works well for a particular institution. The program may be run from an independent functional unit, with a reporting mechanism directly to administration. Such
a program might be found in a major medical center and would employ
several infection control practitioners. An infection control program
in a small hospital may consist of one nurse practitioner who reports
to a nursing service administrator. In some hospitals and long-term
care facilities, the infection control function may be placed under
quality assurance as a cost-containment mechanism, responding to decreased utilization of beds and the constraints imposed by the DRG
compensation schemes. The reduction of manpower or the elimination
of infection control as a separate department may have an effect on
the practice of this specialty in the future. However, as long as the
data obtained from active surveillance and infection control research
is used to intervene, educate, and motivate in an attempt to control
infections, the program should be effective.

C. Making the Infection Control Program Work: Setting
   Standards Tailored to the Health Care Facility

The CDC's Hospital Infections Program has developed sets of "Guidelines for the Prevention and Control of Nosocomial Infections," which
have been sent to all acute-care hospitals in the United States. They
include guidelines for hand washing and environmental control, surgical wound infections, catheter-associated urinary tract infections,
nosocomial pneumonias, intravascular infections, infection control in

hospital personnel, and isolation precautions. They have been published in the infection-control literature (10,11), and by the Government Printing Office. The CDE guidelines are essential general references for any infection control program and should be available in every
health care facility for use in the development of policies and procedures related to infection control. They should be tailored and prioritized to the needs and experience of each facility.

The Centers for Disease Control "Guidelines for Isolation Precautions in Hospitals" (11) gave hospitals a choice between two general
programs for isolation of patients with infectious diseases: disease-
specific or category-specific isolation. Infection control practitioners
and members of the infection control committee select or modify the
guidelines as appropriate for their facility. For example, a patient
with jaundice would be identified at admission as a possible case of
hepatitis. Disease-specific isolation would place the patient under
"hepatitis precautions." The category-specific isolation would be under
"blood and body fluid precautions," which could also be used for patients with acquired immune deficience syndrome (AIDS), malaria, or
other diseases that can be transmitted by blood or body fluids. In
1987, as a result of the transmission of the human immunodeficiency
virus (HIV) from patient blood to health care workers, "universal precautions" were recommended by the Centers for Disease Control. In
"universal precautions" all patients and patient specimens are handled
according to "blood and body fluid" precautions. In August, 1987, the
CDC suggested that the category specific for "blood and body fluid"
be eliminated since it was now to be used for all patients. Some hospitals changed to barrier precautions that leave only "strict" and "res-
priatory" as true isolation categories. The CDC clarified their intent
in June, 1988 indicating the need for the other isolation and precaution
categories (12).

D. Making the Infection Control Program Work:
Education of Health Care Personnel

Continuing education is also an essential part of the infection control
program. The relatively simple measures that physicians and other
health care workers can use to prevent hospital-acquired infections
must be emphasized in this program. Awareness of infection control
practices must be included with other aspects of clinical care taught
in medical and nursing schools. Dr. James M. Hughes, Former Director
of the CDC's Hospital Infection Program, prioritized those infection
control measures as follows:

Observe proper hand-washing technique
Be aware of types of infection/colonization a patient may have and the
    potential need for isolation precautions

Be aware of and adhere to hospital infection control policy
Train others to adhere to isolation precautions and aseptic technique

IX. CONCLUSIONS

We have reached the age of the prospective payment system (PPS),
which has brought radical changes in the provision of health care in
the United States. The payment for diagnosis related groups (DRG)
has added a monetary incentive to the ethical considerations for minimizing complications of hospitalizations, such as nosocomial infections.
The potential for monetary loss may provide the impetus and motivation
needed to make hospital infection control programs more effective (13).
Implementation of infection control policies and procedures remains a
problem. Hand washing, appropriate initiation of isolation precautions,
and the careful use and disposal of hypodermic equipment and other
sharps could provide a substantial reduction in nosocomial infections.
Given current technology and current knowledge, it is estimated that
approximately one-third of nosocomial infections are preventable. Specifically, about 22% of pneumonias and about 35% of surgical wounds
and bacteremias could be prevented by the use of appropriate infection
control practices (2b). This fact was pointed out to the entire nation
in the proclamation establishing Infection Control Week: Those who
provide health care to patients have the responsibility to use the infection control procedures that will reduce this unacceptable level of morbidity and health care costs. The quality of health care that we all
strive to attain is that defined by Williamson (14) as "achievable benefits." We do not strive for the impossible, but we have data to show
that these particular benefits to patients and health care facilities can
be achieved.

REFERENCES

1. LaForce, F. M. (1986). In *Prevention and Control of Nosocomial
    Infections*, R. Wenzel (Ed.). The Williams & Wilkins Co., Balti-
    more, pp. 1-12.
2a. Haley, R. W., Culver, R. H., White, J. W., Morgan, W. M.,
    Emori, T. G., Munn, V. P., and Hooten, T. M. (1985). *Amer.
    J. Epidemiol. 121*:182-205.
2b. Haley, R. W., Morgan, W. M., and Culver, D. H. (1985). *Amer.
    J. Inf. Control 13*:97-108.
3. Macklin, R. (1985). *Infection Control 6*:375-376.
4. Daschner, F. D. (1984). *Infection Control 5*:32-35.
5. Dixon, R. E. (1985). In *Hospital Infections*, J. S. Bennett and
    P. S. Brachman (Eds.). Little Brown and Co., Boston, pp. 73-93.

6. Wenzel, R. P. (1986). In *Prevention and Control of Nosocomial Infections*, R. P. Wenzel (Ed.). The Williams & Wilkins Co., Baltimore, pp. 94-109.
7. Larson, E., and Larson, E. (1983). *Amer. J. Inf. Control 11*: 146.
8. Crow, S. (1984). *Infection Control 5*: 38-40.
9. Joint Commission Accreditation of Hospitals (1986). AMH/86 Manual for Hospitals. Joint Commission on Accreditation of Hospitals, Chicago.
10. Garner, J., Favero, M. S. (1986). *Infection Control 7*: 231-243.
11. Garner, J. S., Simmons, B. P. (1983). *Infection Control 4*: 245-325.
12. *Morbidity and Mortality Weekly Report* (1988) *37*: 377-388.
13. Farber, B. (1984). *Infection Control 5*: 425-426.
14. Williamson, J. W. 1978). In *Assessing and Improving Health Care Outcomes*. Ballinger Publishing Company, Cambridge, pp. 51-69.

9

New Frontiers in Biosafety:
The Industrial Perspective

JOSEPH VAN HOUTEN   *Schering Corporation, Bloomfield,*
New Jersey

I. INTRODUCTION

With the widespread application of recombinant DNA techniques, industrial microbiology has entered a new era, which has been dubbed The
Age of Biotechnology. In the pharmaceutical industry, products previously unavailable due to their scarcity in nature will soon be widely
distributed for consumption by the general public. Examples include
human insulin, growth hormone, and interferon, which are already
being marketed, and other human lymphokines, which await FDA approval as anticancer and antiviral therapies. Gene splicing will improve
the safety of blood-derived products such as human serum albumin
and antihemophilic factor. Current methods require the isolation of
these materials from whole blood, which has unfortunately resulted in
sporadic incidents of product contamination with bloodborne viruses
such as HIV, the causative agent of Acquired Immune Deficiency Syndrome (AIDS) (1). Biosynthesis of these compounds by recombinant
organisms will eliminate the risk of this type of infection for those receiving such therapy. In the area of vaccine development, one can
expect improvement in the safety of existing preparations and the
emergence of new therapies, especially for parasitic diseases such as
malaria and schistosomiasis. For viral agents such as hepatitis B,
herpes, and influenza, the surface proteins that trigger the production of antibodies have been identified along with the genes which code
for them (2). Using recombinant DNA techniques, it should prove
possible to develop a vaccine that contains only a piece of the infectious agent. Such subunit vaccines would eliminate the risk of infec-

tion for the patient, since at no time during the production process would an intact agent be used.

Exploiting the efficiency with which microorganisms convert organic substrates into products for human consumption is by no means a new venture for the pharmaceutical industry. For nearly 40 years, bacteria and fungi have been cultivated in large volumes to produce such notable antibiotics as penicillin, erythromycin, and tetracycline. Perhaps less well known, but of significance nonetheless, is the ability of these agents to produce organic and amino acids, enzymes, and vitamins. In contributing to the control of infectious diseases and the overall improvement in the public health, this industry has made vaccines containing live and attenuated bacteria and viruses commercially available. Diseases such as whooping cough (3) and viral hepatitis (4) are controllable, and smallpox has been eradicated worldwide (5) due to efforts in this area.

Whether the product is the metabolite of a microorganism or the microorganism itself, pharmaceutical processes have required the manipulation of large volumes of viable biological agents. Smaller quantities are usually handled by research scientists and those involved in technical support, such as antibacterial and antiviral assays, carcinogenicity testing, sterility testing, and environmental monitoring. Corporate biosafety programs have been developed to prevent employee infections and environmental contamination by controlling the hazard potential associated with these operations. With rare exceptions, this goal has been successfully achieved.

The use of recombinant products is the next logical step in the evolution of industrial microbiology. Drawing on the experiences of the last 40 years, the industry is well prepared to respond to those biosafety issues that may arise. As we enter the "age of biotechnology," younger companies can draw on this expertise to assure the safety of their employees and protection of the environment.

II. ADMINISTRATIVE MECHANISM FOR BIOHAZARD CONTROL

The management of biological hazards in an industrial setting must be considered as part of an integrated approach to the control of all hazardous operations. Toxic levels of chemicals and radiation, extremes of heat and cold, electrical shock, and pinch points in mechanical equipment demand equivalent consideration with the potential for human infection. To this end, the biosafety specialist must work in consort with the radiation safety officer, industrial hygienist, and safety professional to assure the protection of company employees.

For a corporate safety program to be effective, it must be based on a clear, concise statement of company policy and assignment of responsibilities to management, employees and safety personnel. An example of such a policy statement would read:

Safety Policy

The Company is committed to providing its employees with a safe and healthful work environment. In recognition of this responsibility and to comply with the Occupational Safety and Health Act of 1970, the company will establish safe work procedures which are to be enforced by management, followed by all employees and supported by a corporate safety department.

Specific responsibilities of the groups mentioned would include:

A. Management

1. To provide safe standard operating procedures for all activities conducted under his/her supervision.
2. To incorporate safety and health concerns into the design of all laboratory experiments and production operations.
3. To design and construct facilities that are safe and contribute to a healthy work environment.
4. To provide appropriate protective clothing and equipment when necessary.
5. To provide safety orientation to new employees and periodic retraining of veteran workers.
6. To correct identified hazards.
7. To assure the complete transfer of all hazard information as a project moves from research through development and into manufacturing.
8. To periodically review the work of employees and conditions of the workplace to identify unsafe work practices and facilities.

B. Employees

1. To follow safety rules and procedures established for his/her operating unit.
2. To relay to management new hazard information that becomes available.
3. To make management aware of hazards in the work area.

C. Safety Staff

1. To develop safety programs and policies aimed at reducing injuries and protecting company assets.
2. To serve as a resource to management and employees.

3. To represent the company during inspections by regulatory agencies.
4. To establish an audit system to assure that operating units comply with government regulations and internal guidelines.

To assure its effectiveness, the corporate safety policy and assignment of safety responsibilities must be endorsed and actively supported by senior management and incorporated into the company policy manual.

With an all-encompassing safety policy in place, statements establishing specific programs can be promulgated by the safety director and incorporated into a safety manual. For biological safety, the statement would define the scope of application, overall purpose of the program, and specific responsibilities of management and employees. For example:

Biological Safety Policy

Application

All divisions where biological agents* are used.

Purpose

To define the policy regarding the safe handling of biological agents.

Policy and General Information

Management has the following responsibilities with respect to biological safety:

1. To provide all employees with a safe working environment. 2.To protect its employees, the community and the environment
    from potential hazards which may arise in the handling of biological agents.

Operating unit managers will establish a biological safety program with procedures designed to ensure employee safety and health and to contain biological agents within the premises whenever they are used.

---

* Biological agents include: A. bacteria, viruses, fungi, and parasitic agents, B. infected or potentially infected human or animal cells, C. recombinant DNA molecules and organisms and viruses containing recombinant DNA molecules, D. infectious nucleic acids.

Development and implementation of biohazard control to prevent
laboratory/facility-acquired infections and the potential spread
of contamination will be part of every activity in which biohazardous materials are used. Primary responsibility for biohazard control resides with line management and for this each operating unit
will be held accountable. All employees are required by this policy
to follow the procedures and operating instructions provided to
them along with the general safety rules of each operating unit.

An effective network of safety committees is critical to the successful
communication of safety information. Director-level committees, manager-
level committees, and supervisor/worker-level committees should be
established to encourage safety and health awareness at all levels within the organization. With respect to biological safety, the company
should establish an institutional biosafety committee (IBC) and define
its role, responsibilities, and membership in a written charter, which
is reprinted in the company safety manual.

The IBC should be based within the safety organization and chaired
by the safety manager for the site where work with biological agents
is performed. Among its members should be:

A representative from the senior technical staff for each operating
    unit using biological agents.
A representative from the senior technical staff for a nonmicrobiology
    operating unit.
A representative from the junior technical staff from a microbiology
    area. This individual should handle biological agents on a daily
    basis.
The biological safety officer.

Since many of the issues dealt with by the IBC have medical, legal,
or facility aspects, the company medical director and representatives
from the law department and facility management should be considered
ex officio members.

Most companies engaged in recombinant DNA research voluntarily
comply with the *Guidelines for Research Involving Recombinant DNA
Molecules* issued by the National Institutes of Health. These guidelines
recommend that two individuals who are not associated with the institution be included on the IBC to represent the views of the general public. One of these external members should have sufficient technical
background to understand the science associated with the proposed
work (e.g., a college professor). The second external member should
have some technical background but should be qualified to address
the concerns of the general public. A local health officer is a good
example of this type of external member.

The IBC should be chartered to perform the following on behalf of the institution:

Recommend biosafety policy.

Provide advise and direction regarding facility design and operating practices for all work with biological agents.

If recombinant DNA research is conducted, review and approve experiments included in Section III of the NIH *Guidelines for Research Involving Recombinant DNA Molecules*. This includes review of facility design and operating practices for compliance with Appendix G (laboratory scale) or K (large scale).

The IBC should meet as frequently as necessary to fulfill the responsibilities of this charter. For major projects, an IBC subcommittee consisting of three or four members, including the chairman and biological safety officer, may meet with the principal investigator to assist him in preparing his proposal and to conduct a preliminary review. When completed, the proposal is presented to the entire IBC at a meeting (for a facility design or major procedure approval) or by mail ballot (for minor protocol or equipment changes).

The extent to which decisions of the IBC are implemented and the overall success of the biological safety program often hinges on the effectiveness of one person—the biological safety officer. This individual must be able to both recommend and monitor programs which assure compliance with internal policies and government guidelines/recommendations. A typical job description would include the following responsibilities:

1. Development of policies and procedures for the safe handling of infectious biological agents.
2. Provide advice and support to the IBC, management, and operating units in matters of biological safety.
3. Conduct surveys and formal audits to assure compliance with internal requirements and government regulations/guidelines. Report findings to the IBC and management.
4. Assist in the design of facilities that are to be used for infectious biological agents.
5. Develop emergency response plans for dealing with incidents involving infectious biological agents.
6. Investigate incidents that involve infectious biological agents. Develop recommendations for corrective action and report findings to the IBC and management.
7. Develop and conduct training programs for employees who use infectious biological agents.
8. Keep current with regulations affecting biological safety and advise operating units.

In summary, the administration of a good corporate biological safety program begins with management's support for an all-encompassing occupational safety program. Subsequently, within the safety department there is created a program that specifically addresses the hazards of research with biological agents. An institutional biosafety committee is organized to recommend policy in this area and an individual is assigned the duties of biological safety officer. With this mechanism in place, control of biohazards in the industrial environment can be readily accomplished.

III. CONTROLLING BIOHAZARDS IN THE INDUSTRIAL SETTING

A. Risk Assessment

The development of a pharmaceutical product (a metabolite or the organism itself) from discovery to marketing requires handling of increasing volumes of viable biological agents. The risk of employee infection is a function of a number of factors: pathogenicity (ability to cause disease), virulence (ease with which disease is caused), invasiveness (likeliness to invade local tissues), and infectivity (likeliness to proliferate at site of invasion) of the agent and the volume being manipulated. Assessment of the hazardous nature of biological agents can be found in several government publications (6-8) as well as in other chapters of this book. Etiologic agents are categorized as Class 1 through 4, in increasing order of hazard, and oncogenic viruses are classified as low, medium, and high risk. While Appendix B of the NIH *Guidelines for Research Involving Recombinant DNA Molecules* is helpful in determining the pathogenicity of biological agents, it is not an all-inclusive list, and one should be careful not to conclude that an agent belongs in Class 1 just because it is not found in a higher category. For example, *Candida albicans* does not appear in this Appendix. However, under certain conditions, it can colonize human mucous membranes. Thus, it should be included with the Class 2 fungi. Likewise, HIV, the AIDS virus, is not listed. The CDC recommends that small quantities be handled at Biosafety Level 2 and production volumes at Biosafety Level 3 (9). The accuracy of an assessment of an organism's pathogenicity may require the use of literature searches, discussion among informed microbiologists [e.g., the biological safety officer, principal investigator, or other biosafety professionals in government (CDC, NIH, USDA) and academia], or personal experience.

The research scientist who examines a new organism or its product in hope of finding a novel pharmaceutical entity will begin his quest with a single colony growing on agar in a petri dish. Isolates of interest will be cultivated in submerged culture, first in the test tube and shaker flask, and then in 10 liters of medium contained in a fermenter.

Work of this nature, involving 10 liters of culture or less, has historically been regarded as small scale.

If the organism or product shows promise, it is delivered to a development group for scale-up. Fermenters and fermentation volumes
increase in modest increments until a process that can be run repeatedly at the production level is developed. Scaled-up uses of organisms
introduced throughout development and in production are, for the
most part, greater than 10 liters in volume and are classified as large
scale.

As a pharmaceutical product makes its way from research through
development and into production, service groups are mobilized to provide technical assistance to the project. Antimicrobial and antiviral
assays are performed to determine the concentration of product in the
fermentation medium. In vivo studies requiring induction of infections
in animals are necessary to establish the drug's efficacy. Finally, if
the product is formulated into a sterile preparation, the absence of
contamination with biological agents must be certified. All of these
operations require the handling of biological agents on a laboratory
scale.

B. Physical Containment at the Large Scale

The small-scale use of biological agents as part of research or technical support operations differs very slightly, if at all, from similar activities conducted in an academic environment. Physical containment
for laboratory uses of biological agents in academia has been thoroughly
addressed by Liberman and Harding elsewhere in this volume. This
section focuses on physical containment for large-scale (>10 liters of
culture) uses of biological agents, with special emphasis on designing
a facility to handle large volumes of recombinant organisms.

Appendix K of the NIH *Guidelines for Research Involving Recombinant DNA Molecules* specifies physical containment requirements for
large-scale uses of recombinant organisms. It delineates three levels
of large-scale containment—BL1-LS, BL2-LS, and BL3-LS—which are
to be used whenever organisms that require BL1 through BL3 containment at the laboratory level are cultivated in volumes exceeding 10
liters.

Most recombinant organisms currently in use throughout the biotechnology industry are exempt from the NIH *Guidelines* at the laboratory scale and require BL1-LS containment for pilot plant and manufacturing operations. The BL1-LS requirements of Appendix K can be
summarized as follows:

Cultures must be handled in a closed system that is designed to reduce
the potential for escape of viable organisms.

Equipment that cannot adequately contain viable organisms must be
enclosed in equipment that reduces the potential for escape.
Culture fluids must be inactivated using a validated inactivation procedure prior to removal from a closed system.
Sampling, addition of materials to a closed system, and transfer between closed systems must be done in a manner that minimizes the
production of aerosols and contamination of work surfaces.
Exhaust gases must be filtered or incinerated to prevent release of
recombinant organisms into the environment.
All equipment that has been used with recombinant organisms must be
sterilized using a validated sterilization procedure before any work
such as a repair or preventative maintenance is performed.
The emergency response plan must include procedures for handling
large-scale spills of recombinant organisms.

Among the means of complying with each of these requirements are
the following:

Handling of Cultures in Closed System

At the heart of any large-scale operation is the fermenter. Typically,
this vessel is expected to support the growth of one organism while
excluding all other biological agents. By designing for product protection, the engineer usually delivers a vessel which is in itself a closed
system capable of containing the organism.

Fermenters capable of BL1-LS containment are commercially available from several suppliers. Features that contribute to the containment aspects of the equipment and that should be evaluated for adequacy are:

Construction Materials. Highly polished stainless steel is an excellent material for constructing fermenters. It is readily sterilizable
with steam and will resist corrosion by many chemical inactivating agents.
The use of all-glass fermenters should be avoided due to the possibility
of shattering. The fermenter should be an ASME-coded pressure vessel
so that sterilization conditions are attainable.

Installation of Utilities. It is desirable to have a fermenter that
can be steam-sterilized in place. This eliminates the possibility of
spilling recombinant organisms while transferring a fermenter from
the production area to an autoclave.

Other services, such as compressed air, water, electricity, carbon
dioxide, and nitrogen may be required, depending on the work being
performed.

Agitation Drive Mechanism. Most fermenters are equipped with
top- or bottom-driven agitators that are used to maintain a homogenous

mixture of organisms in the medium. The agitator shaft should be
equipped with a double mechanical seal which is designed so that any
failure will result in containment of the recombinant organism. One
mechanism that is employed requires a positive steam pressure on the
exterior of the seal. If a leak develops, steam flows into the seal and
inactivates the organism. In an alternative mechanism, the seal assembly is surrounded by a collection system which directs any liquid that
leaks out to an inactivation tank where the organism is killed prior to
its release from the facility (Figure 9-1).

Drain Valve Protection. The drain valve should be protected so
that any liquid that leaks out is collected and inactivated. A system
similar to that described for the agitator shaft in which liquid that

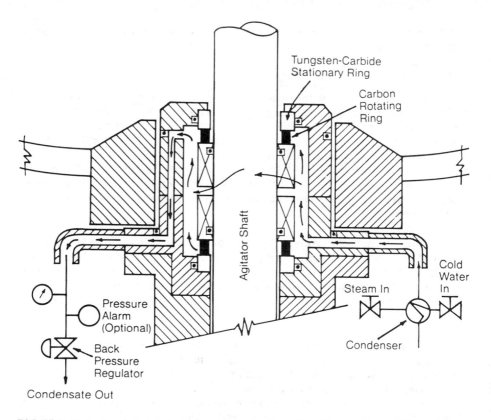

FIGURE 9-1 Contained double mechanical seal: Steam passes through
the seal and surrounds the agitator shaft. If a leak occurs, organisms
are rapidly inactivated and are directed to a dedicated inactivation
tank. (Courtesy of New Brunswick Scientific, used with permission.)

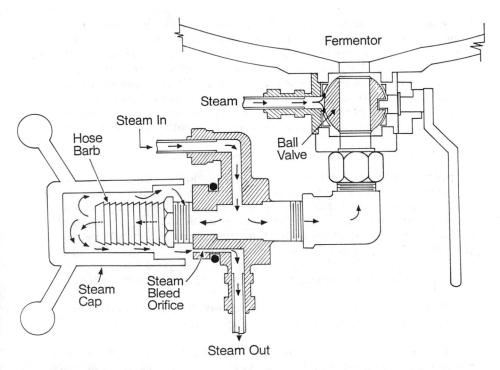

FIGURE 9-2 Contained drain valve: During fermentation, steam continuously bathes the valve and associated plumbing. Any organism
that may escape from the fermenter is quickly inactivated. Condensate
is directed to a special tank for collection and further treatment prior
to disposal. (Courtesy of New Brunswick Scientific. Used with permission.)

leaks out is directed to an inactivation tank for treatment is one possibility (Figure 9-2). A second method, albeit more crude, is to allow
drips to collect in a pail that contains a disinfectant.

Sealing of Vessel Penetrations. All fermenters are equipped with
several penetrations for sampling and measuring parameters such as
dissolved oxygen, pH, and foam level. Each of these penetrations
must be equipped with an O-ring seal to prevent the escape of aerosol
during fermentation, even if the tank is operating at positive pressure.

Containing Equipment that Leaks Viable Aerosol

Several pieces of equipment such as continuous flow centrifuges are
incapable of absolute containment. When it becomes necessary to use

these to process large volumes of viable organisms, they must be enclosed in containment equipment. Since product protection is often
not a factor in these operations, the most effective solution is enclosure
within a Class I biological safety cabinet. In cases where the agent
is highly hazardous or product protection is desired, use of a Class
III biological safety cabinet is indicated. This will completely seal the
operation so that personnel will not be exposed to hazardous aerosol
and the product will not become contaminated with environmental agents.

Inactivation Prior to Removal from Closed System

To protect employees and prevent contamination of the environment,
all process streams are treated to eliminate the biohazard before large
volumes are handled outside of a closed system. To accomplish this,
a variety of chemical and physical means are available. Selection of
the appropriate method depends on many factors including whether
the desired product is intracellular, extracellular, or the cell itself
and the effect the inactivating agent has on the yield and stability
of the final product. Among those which have been tried include:

Heat. This is useful for quickly inactivating vegetative forms
of the organism but may result in damage to the final product.

Chemical Disinfectants. Halogenated compounds, phenolics, and
quaternary ammonium salts have all been demonstrated to be effective
in killing biological agents. However, many are bacteriostatic in nature,
resulting in incomplete inactivation and many adversely affect the product. Frequently, it is observed that the concentration required to
kill the organism also destroys the product.

Solvents and Acids. These are useful as inactivating agents because they severely disrupt the integrity of the cell membrane. However, the flammability, corrosivity, and toxicity hazards are often
considerable, posing a significant risk of injury to the employee and
of destruction of company assets.

Mechanical Disruption. Use of equipment similar in principal to
that of the French pressure cell ruptures the organism by exposing
it to extreme differences in pressure. Usually several passages are
required to assure complete inactivation. In addition, the heat generated during the process may destroy the product.

Once selected, the inactivation method must be validated before it
can be introduced into widespread use. A minimum of three consecutive batches in which complete inactivation of the organism is demonstrated should be required to establish the efficacy of the procedure.
Validation is ordinarily conducted along the following lines:

At the conclusion of fermentation, the chemical agent is added, the
batch is heated, or the cells are subjected to mechanical disruption.
Strict adherence to physical containment is maintained during this
time.

Samples are withdrawn and analyzed for the presence of viable recombinant organisms. If bacteriostatic agents are used, samples must
be washed thoroughly to completely remove the chemical. Failure
to do so often leads to false negative results.

Once it has been determined that there are no viable cells in the fermentation medium, the batch can be released for further processing.

After a process has been validated, the requirement to hold a batch
pending results of testing for viable cells may be eliminated. However,
all wastes from downstream processing should be stored so that further
inactivation can be performed should results prove positive.

Sampling, Addition of Materials, and
Transfer Between Closed Systems

To minimize the creation of aerosols, special sampling devices, addition
vessels, and transfer lines have been designed and constructed.

   Sampling Assembly. Figure 9-3 depicts a sampling assembly whose
design is simple yet effective. It consists of an inlet line from the
fermenter with a valve, a vent line with an absolute filter for air dis-

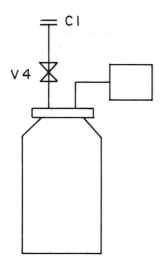

FIGURE 9-3 Contained sampling assembly: A simple yet effective design to contain biological aerosol while sampling the fermentation.

placement, a cap with an O-ring seal, and a bottle. All parts are steam-
sterilizable so that the unit is reusable. The sampling procedure begins
with steaming out the line. The appropriate valve is directed toward
the sampling device and an aliquot is collected. Steam is once again
passed through the sample line to inactivate any viable organisms that
remain there. The connection between the fermenter and sampling
device is broken, the connections are wiped with a disinfectant-soaked
cloth and capped. Finally, the contained specimen is transported to
the laboratory for analysis.

   Addition Vessels. In addition to the installation of contained fermenters, vessels designed for delivering acid or base for pH control,
nutrients and anti-foam under closed conditions are also installed.
This equipment is ordinarily sealed using the same arrangement associated with the fermenter. The connections between these vessels
and the fermenter are welded pipe, which cannot easily break or become disconnected.
   A special addition vessel is the inoculation assembly used for delivering viable cells to the seed fermenter. Pictured in Figure 9-4 is
one that has been developed using a side-arm Erlenmeyer flask, an
absolute filter for vacuum relief, and tubing with quick-disconnect
fittings. The procedure begins with the device being filled with viable
inoculum in a biological safety cabinet. It is then connected to the
fermenter via the quick-disconnect fitting. A clamp between the flask
and the fermenter is removed and inoculum flows by gravity into the
fermenter. When finished, the clamp is reinstalled and the device is
left in place. Upon completion of the fermentation, the fermenter is
sterilized. Disinfectant is poured into the flask and the clamp is once
again removed to allow disinfectant to sterilize the inoculum fitting.
The connection between the device and fermenter is broken and the
fermenter is cleaned.

   Transfer Lines. Connections between the fermenter, pumps, and
associated tanks that may come in contact with viable organisms should
be welded pipe and steam-sterilizable in place. The use of temporary
lines composed of rubber hose are to be avoided for several reasons.
First, any piece of temporary equipment creates a potential tripping,
access, or egress hazard. Second, small cracks in rubber hose may
harbor organisms and protect them from inactivation by steam or chemical disinfectants. Finally, the rubber may be incompatible with process
chemicals that could result in leaks of hazardous or potentially hazardous materials.

Exhaust Air Treatment

To provide for employee and environmental protection, all exhaust air
streams must be treated to prevent the release of recombinant organ-

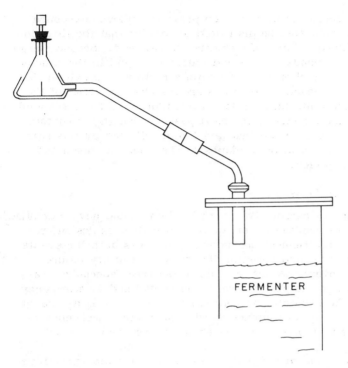

FIGURE 9-4 Contained inoculation assembly: Viable inoculum flows from the side-arm Erlenmeyer flask into the fermenter. Vacuum relief is through an absolute filter.

isms. Two methods commonly used are thermal oxidation (incineration) and filtration. Regardless of which is chosen, the overall system should include a primary means of treatment and emergency back-up which is designed to begin operation immediately upon failure of the primary system. To accomplish this, various combinations of incineration and filtration have been considered (e.g., two incinerators or two banks of filters installed in series or a filtration apparatus followed by incineration).

If incineration is chosen, there are two drawbacks that should be considered. First, operating two incinerators continuously throughout a fermentation may be a costly proposition, especially if they are electric. Second, the hazardous properties of chemicals that are added to the tank at the conclusion of fermentation need to be evaluated. If any are flammable, a hot incinerator could be a source of ignition and potentially cause serious injury to employees and damage to the facility.

Filtration also has drawbacks which need to be factored into the decision-making process. First, the housing that contains the filter

elements must be steam-sterilizable. To prevent employee exposure,
administrative controls must be instituted to require that the filtration
assembly be sterilized before the elements are removed. Second, large
concentrations of suspended solids and entrained liquid in the exhaust
air will cause the filter elements to foul quite rapidly. To extend their
life, coalescing and cyclone-type filters are installed upstream of the
HEPA filter. These units function to remove liquid and large suspended particles from the air stream before it passes through the containment filter. It should be noted that any liquid collected by this equipment must be inactivated prior to disposal. The methods discussed
previously can be applied.

Emergency Response Plans

From an overall safety perspective, each location should have a detailed
emergency response plan to deal effectively with fire, explosion, release of toxic material, medical emergencies, and less probable events
such as bomb threats, environmental disasters, and utility failures.
When any of these events occurs in a biohazard area, specialized response is required on the part of facility personnel and/or emergency
response teams. Accordingly, procedures for biohazard areas should
be developed to be used in conjunction with each site's emergency response plan. The following situations should be covered:

   Spills of Viable Biological Agents. Spills of viable biological agents
can be small or large, confined (e.g., within a biological safety cabinet)
or unconfined and liquid or aerosol in nature. Cleanup must be accomplished in a manner that considers both the safety of personnel and
limiting the release of viable biological agents from the facility. Only
those individuals who fully appreciate the potential hazards associated
with biological agents are to perform these operations. In most cases,
facility microbiologists will clean up their own spills.
   The following are recommended procedures:
   1. *Confined spill within a biological safety cabinet.* A spill that
is confined to the interior of a biological safety cabinet should present
no hazard to personnel in the area. Chemical disinfection procedures
should be initiated at once while the cabinet ventilation system con-
tinues to operate to prevent escape of contaminants from the cabinet.

The individual affecting cleanup will wear latex gloves, lab coat, and
    safety glasses.
Flood the work tray with a suitable disinfectant. Table 1 lists a
    variety of disinfectants and their applications. If the nature of
    the spill is unknown, use 5% Wescodyne or 5% hypochlorite (Clorox),
    since both are effective against vegetative bacteria, endospores,
    and viruses.

TABLE 1   Summary of Practical Disinfectants

	Quaternary ammonium compounds	Phenolic compounds	Chlorine compounds	Iodophor	Ethyl alcohol	Isopropyl alcohol	Formaldehyde	Glutar-aldehyde
Inactivates								
Vegetative bacteria	+	+	+	+	+	+	+	+
Lipoviruses	+	+	+	+	+	+	+	+
Nonlipid viruses	-	a	+	+	a	a	+	+
Bacterial spores	-	-	+	+	-	-	+	+
Treatment requirements								
Use dilution	0.1-2.0%	1.0-5.0%	500 ppm[b]	25-1600 ppm[b]	70-85%	70-85%	0.2-8.0%	2%
Contact time, minutes								
Lipovirus	10	10	10	10	10	10	10	10
Broad spectrum	NE	NE	30	30	NE	NE	30	30
Important characteristics								
Effective shelf life >1 week[c]	+	+	-	+	+	+	+	+
Corrosive	-	+	+	+	-	-	-	-
Flammable	-	-	-	-	+	+	-	-
Explosion potential	-	-	-	-	-	-	-	-
Inactivated by organic matter	+	-	+	+	-	-	-	-
Skin irritant	+	+	+	+	-	+	+	+
Eye irritant	+	+	+	+	+	+	+	+
Respiratory irritant	-	-	+	-	-	-	-	-
Toxic[d]	+	+	+	+	+	+	+	+

TABLE 1 (continued)

	Quaternary ammonium compounds	Phenolic compounds	Chlorine compounds	Iodophor	Ethyl alcohol	Isopropyl alcohol	Formaldehyde	Glutar-aldehyde
Applicability								
Waste liquids	-	-	+	-	-	-	-	-
Dirty glassware	+	+	+	+	+	+	+	+
Equipment, surface decontamination	+	+	+	+	+	+	+	+
Proprietary products[e]	CDQ End-Bac Hi-Tor Mikro-Quat	Hil-Phene Matar Mikro-Bac O-Syl	Chloramine T Clorox Purex	Hy-Sine Ioprep Mikroklene Wescodyne			Sterac	Cidex

+ = Yes; - = No; NE = Not effective.

[a]Variable results depending on virus.

[b]Available halogen.

[c]Protected from light and air.

[d]By skin or mouth or both. Refer to manufacturer's literature or Merck Index.

[e]Space limitations preclude listing all products available. Individual listings (or omissions) do not imply endorsement (or rejection) of any product by the National Institutes of Health or the U.S. Environmental Protection Agency.

Source: Adapted from *Laboratory Safety Monograph. A Supplement to the NIH Guidelines for Recombinant DNA Research.* pp. 104-105. National Institutes of Health, Office of Research Safety, National Cancer Institute, and the Special Committee of Safety and Health Experts, Bethesda, Maryland. January 1979. This table appeared in *Draft Manual for Infectious Waste Management,* pp. 4-35 and 4-36. United States Environmental Protection Agency, Office of Solid Waste and Emergency Response, Washington, D.C., September, 1982.

The disinfectant should be applied in a manner that minimizes aerosolization of the spill while assuring adequate contact with it.
Allow the disinfectant a minimum of 20 minutes contact time.
Soak up the spill with a disposable cloth and discard the cloth into an
autoclavable bag.
Wipe the walls, work surface, and any equipment in the cabinet with a
disinfectant-soaked cloth.
If the spill reached the exhaust grilles, flood the catch basin with disinfectant. After sufficient contact time (minimum 20 minutes),
drain the disinfectant into an autoclavable container and wipe the
exhaust grilles and catch basin with a disinfectant-soaked cloth.
Wipe the outside of the autoclavable container and bag with a disinfectant-soaked cloth.
When cleanup is finished, place all solid materials that came in contact
with the viable agent into an autoclavable bag. This includes
gloves, wiping cloth, lab coat, and any clothing that became contaminated. Autoclave this material and disinfectant at 121°C for
an appropriate amount of time. The length of a decontamination
cycle should be determined in advance and validated for a particular load size and distribution. If this has not been done, allow a
minimum of one hour exposure time at 121°C to decontaminate the
materials used for spill cleanup.
After contaminated gloves and clothing have been removed, wash arms,
hands, and face with a germicidal soap.

2. *Unconfined spill in the open facility.* An unconfined spill can
be small or large scale with the associated hazard being a function of
the volume of the spill, the pathogenicity of the agent, and its concentration within the spilled material. To deal effectively with small-scale
spills of biological agents, each facility should have a spill cart located
in an accessible area. The following items should be stored on this
cart:

Protective Clothing/Equipment: disposable lab coats, disposable latex
gloves, disposable safety glasses, autoclavable boots, half-face
respirator with HEPA filter cartridges.
Disinfectant: 5% Wescodyne, 5% Clorox.
Cleanup Supplies: Autoclavable forceps, autoclavable squeegee, autoclavable dust pan, autoclavable biohazard bags, disposable wipes,
spill pillows.

In the event of an unconfined spill in the open facility:

Warn others of the spill and leave the area immediately. Hold breath
to prevent inhalation of potentially biohazardous aerosol.

Remove contaminated clothing, folding contaminated areas inward, and
    discard into a autoclavable bag.
Wash all potentially contaminated body areas as well as the arms, face,
    and hands with germicidal soap. Shower if necessary.
Don a disposable lab coat, latex gloves, and safety glasses. If condi-
    tions warrant, also wear autoclavable boots and a half-face respira-
    tor fitted with HEPA filter cartridges.
Enter the spill area with the spill cart.
Isolate the floor drains with spill pillows to prevent release of viable
    biological agents to the sewer.
Encircle the spill with 5% Wescodyne, 5% Clorox, or other appropriate
    disinfectant (see Table 1). Apply in a manner that minimizes
    aerosolization of the spill while assuring adequate contact with it.
Allow the disinfectant a minimum of 20 minutes contact time.
Pick up broken glass and other sharp objects with the forceps, dust
    pan, and/or squeegee and place into a leak-proof autoclavable
    container.
Using disposable wipes or spill pillows, mop up the liquid and discard
    into an autocalvable bag.
Wipe the outside of the autoclavable container and bag with a disinfec-
    tant-soaked cloth.
When cleanup is finished, place all solid materials that came in contact
    with the viable agent into an autoclavable bag. Decontaminate
    this material and any contaminated clothing at 121°C for an appro-
    priate amount of time. The length of a decontamination cycle should
    be determined in advance and validated for a particular load size
    and distribution. If this has not been done, allow a minimum of
    one hour exposure at 121°C to decontaminate these materials.
Wash arms, hands, and face with germicidal soap.

     3. Large-scale spills of recombinant organisms. While the poten-
tial for catastrophic failure of a large fermenter seems extremely small,
the facility must be designed to completely contain a spill involving
the contents of the largest vessel. One method that has been used
is the enclosure of an area immediately beneath the fermenter with
a concrete dike. Should a massive leak occur, fermentation medium
containing recombinant organisms will collect in this area and can be
inactivated using steam or chemicals. It is not advisable that treatment
be performed in situ, since this would require that a swimming pool
of recombinant organisms be collected prior to addition of the inactivating agent. Rather, material in the diked area should be directed
to a sump, from which it can be pumped either to another fermenter
of similar size or through an in-line sterilizer, which injects steam
into the material to affect kill. Regardless of the method, inactivated
material must be collected and examined for the presence of recombinant organisms prior to disposal.

Employees who supervise this operation must wear disposable clothing, latex gloves, autoclavable boots, and a self-contained breathing apparatus.

Once the large-scale spill has been handled, puddles within the diked area should be decontaminated and cleaned up as described previously.

Fire and Explosions. To apprise the fire brigade or local fire department of the potential biological hazard associated with a fire/explosion in a microbiology area, use of the following labeling system is helpful:

The doors to each area in which biological agents are used should be labeled with the highest level of physical containment required for work in this area. For example, an area which uses Class 2 agents should be labeled with a biohazard symbol and "BL-2," the abbreviation for Biosafety Level 2.

For areas labeled "BL-3," "BL-4," BL2-LS," and BL3-LS," the following statement should be added: Emergency Personnel Do Not Enter Until Biological Safety Officer Is Notified.

Each piece of equipment used for storing, processing, or handling viable biological agents should be labeled with the biohazard symbol.

In the event of a fire or explosion, implement the following in addition to those steps outlined in the site emergency response plan:

1. *Facility Personnel*: (a) place all biohazardous material in an incubator, refrigerator or freezer; (b) turn off all gas burners.
2. *Fire Brigade/Local Fire Department*: (a) Check the sign on the door before entering any biohazard area. If the area is labeled "BL-1," "BL-2," or "BL1-LS," the fire hazard far outweighs the biological hazard. For areas labeled "BL-3," "BL-4," "BL2-LS," and "BL3-LS," the biological safety officer must be notified before anyone enters the area. (b) For "BL-1," "BL-2," and BL1-LS" areas, don a self-contained breathing apparatus, enter the area, and begin extinguishing the fire immediately. If possible, do not disturb any equipment labeled with the biohazard symbol. (c) Establish a perimeter and restrict the passage of persons and materials in or out so as to limit the spread of contamination. (d) When the fire has been extinguished, remain in the area and do not take equipment from the scene until cleared by the biological safety officer.
3. *Biological Safety Officer*: (a) Respond to all fire calls in biohazard areas. (b) Provide biosafety control during and after the fire. (c) Confer with the area supervisor to determine the extent of biological contamination of personnel, equipment and the area,

if any. (d) Recommend decontamination procedures for personnel, equipment, and the area. (e) Release personnel, equipment, and the area following decontamination.

Medical Emergencies. The main sources of accidental exposure to biological agents are ingestion, inoculation, and inhalation. Each of these can be controlled through prudent practices. However, accidents may occur.

Any accidental exposure to a biological agent must be reported to the employee's supervisor, employee health services, and the biological safety officer. It is important to note that if the exposure is associated with a spill, contaminated clothing must be removed before the individual leaves the accident area.

Environmental Emergencies.

1. *Floods.* If as a result of heavy rain or hurricane the potential for flooding of a facility exists, area personnel should take the following measures to prevent release of biological agents:

   Inactivate all cultures that could possibly enter flood waters.
   Relocate all stock cultures to areas that are not at risk of flooding.

2. *Tornadoes.* When the threat of a tornado exists, facility personnel will perform the following:

   Inactivate all cultures in areas at risk.
   Relocate all stock cultures to basement storage areas.

Utility Emergencies. In most cases, containment of biological agents is not compromised during utility failures provided that facility personnel respond in an appropriate manner. For example, if while decontaminating a fermenter there is a failure in steam or electricity, the operator should not disturb the fermenter. Any surviving biological agents will remain contained during this time. When service is restored, a full decontamination cycle will be run before the fermenter is opened.

One exception is an electrical failure while using a biological safety cabinet. If this occurs, the following should be done:

1. Discontinue work with biological agents immediately.
2. Seal all cultures securely.
3. Decontaminate the work area with a suitable disinfectant (see Table 1).

C. Validation

No plant or process design is complete without validation. Programs should be established to assure that the containment features and process safeguards designed to prevent the release of biological agents function properly.

Prior to the use of any recombinant organism, the following should be performed:

1. Biological safety cabinets should be certified.
2. The fermenter should be tested for leaks under steam pressure.
3. Each vessel line and piece of equipment that is designed for steam sterilization should be tested to assure that the inactivation temperature is achievable.
4. The method of culture inactivation should be validated using the parent strain.
5. Exhaust filter integrity should be tested using an aerosol generator.
6. The emergency containment system for catastrophic spills, e.g., the dike and in-line sterilizer, should be validated using water.

Once the plant goes on stream, the following should be implemented:

1. Biological safety cabinets should be tested annually and each time the HEPA filter is changed.
2. Exhaust filter integrity should be checked after several batches or whenever a filter element is changed.
3. Cell inactivation should be verified in duplicate prior to breaking containment.
4. Inactivation conditions should be recorded for each fermenter following transfer of an inactivated culture.
5. Samples of the inactivated broth should be collected and tested for viability.
6. Temperature, pressure and other recording devices on the fermenters should be calibrated on a quarterly basis.

IV. THE BIOSAFETY AUDIT

Assessment of compliance with corporate biosafety guidelines should be an ongoing process, with the biological safety officer conducting frequent informal surveys of areas using biological agents. A more formal review, the biosafety audit, should be performed at a regular frequency, perhaps annually or every 18 months.

A. Preaudit Assessment

The preaudit assessment is a review of documents relating to the area being audited. Items which should be read and fully understood include:

1. All submissions to the IBC.
2. Reports of incidents involving biological agents including spills and facility-acquired infections.
3. Guidelines issued by CDC, NCI, and NIH.

B. The Audit

The biosafety audit begins with an opening conference during which the purpose of the audit is explained to senior management. Meetings are then held with department supervision to discuss the details of their biosafety program. The following outline has proved useful in assuring an organized approach to a thorough review:

1. *Biohazard Classification*–What types of etiological agents or oncogenic viruses are being used?
2. *Scope of Work*–Is the work small scale in vitro ($\leq$10 liters of culture), large scale in vitro (>10 liters of culture), or in vivo animal studies?
3. *Facility Design*–Based on classification and scope of work, are facilities adequate to contain the organism?
4. *Facility Practices*–Are the appropriate biosafety practices used for the type and quantity of organisms being handled?
5. *Material Handling and Storage*–Are all organisms stored in a safe manner and shipped in accordance with DOT regulations, where applicable?
6. *Biohazard Waste Disposal*–Is all contaminated material treated via autoclaving, incineration, or other validated means prior to disposal?
7. *Equipment Certification*–Is the operation of equipment used for containing or inactivating organisms (e.g., autoclave, biological safety cabinet) certified on a regular basis?
8. *Employee Training*–What kind of training is provided to new employees, veteran employees who handle new agents, and support personnel to assure that each has sufficient knowledge of the biohazards in the area?
9. *Emergency Response Plan*–Does a plan exist, has it been documented, and is it reviewed periodically with all responsible individuals?
10. *Monitoring*–Is environmental monitoring required to assure that containment has not been breached?

11. *Medical Surveillance*—Has a program been developed and are all
    employees participating?

After meeting with management, the biological safety officer should
perform a physical inspection of the facility. Every attempt should be
made to interview several employees to ascertain their awareness of
the elements of the biological safety program.

The audit concludes with a closing conference at which the biological safety officer discusses his findings with senior management.
Recommendations that will appear in the final report are discussed
at this time.

C. The Report

The audit report is issued to the principal investigator with copies
to his and your management. It is composed of three main sections:

1. Summary—A one-paragrpah statement that embodies the salient
   points of the audit.
2. Recommendations—These should be clear and concise. It is useful
   to assign a unique number to each to facilitate easy follow-up.
   For example, BS-86-01, BS-86-02, BS-86-03 would be the first
   three recommendations for a biosafety audit conducted during 1986.
3. Findings and Discussion—This is the most lengthy portion of the
   document. In it, strengths as well as weaknesses are discussed.
   Supporting details for each of the recommendations should be included.

A final version of the report should be issued to the principal investigator within one month after the completion of the audit. In turn,
the principal investigator should provide you with a written response
to your recommendations within one month of receipt of the audit report.

V. CONCLUSION

The age of biotechnology can be thought to have been conceived in
the 1960s and born during the 1970s. Throughout the present decade,
it is approaching adolescence as products move out of the laboratory
and into manufacturing facilities. The potential benefits to be derived
from this new science are enormous and extend beyond the pharmaceutical industry into agriculture, environmental protection, and mineral
processing. As has been the case with other technologies, there are
risks associated with the large-scale use of biological agents. However,
these risks are controllable through strong corporate biosafety programs. This chapter has been presented to assist the reader in estab-

lishing a program of administrative and engineering controls aimed at preventing occupational injuries and environmental pollution resulting from the handling of biological agents in an industrial setting.

REFERENCES

1. *Morbidity and Mortality Weekly Report (MMWR).* 35(43):669 (1986).
2. *Commercial Biotechnology: An International Analysis.* U.S. Congress, Office of Technology Assessment, OTA-BA-218, Washington, D.C., January, 1984, p. 136.
3. *MMWR. 34*(27):405 (1985).
4. *MMWR. 34*(22):313 (1985).
5. *MMWR. 34*(23):341 (1985).
6. *Biosafety in Microbiology and Biomedical Laboratories.* CDC-NIH, 1984. Publication No. (CDC) 84-8395.
7. *Guidelines for Research Involving Recombinant DNA Molecules.* NIH, 51 *Federal Register* 16958 (1986).
8. *Biological Safety Manual for Research Involving Oncogenic Viruses.* NCI, 1976. Publication No. (NIH) 76-1165.
9. *MMWR. 35*(4):540 (1986).

10

Personal Protection and Hygiene

RALPH W. KUEHNE   *U.S. Army Medical Research Institute of Infectious Diseases, Fort Detrick, Maryland*

I. INTRODUCTION

The laboratory accident responsible for the redirection of my career goals happened one October morning in 1972. I was assisting an investigator who directed me to perform hematocrits in a Class III cabinet on blood from a rhesus monkey infected with Machupo virus. While pushing one of the glass tubes into the sealing wax, the obviously defective tube broke, penetrated the cabinet gloves, and punctured my finger. Emergency response involved "milking" the finger under 5% Lysol, followed by placement in medical isolation for 18 days and receiving five units of immune plasma. Fortunately, clinical infection did not result. During my confinement, I had plenty of time to reconsider the accident and the stupidity of using that accident-causing procedure. I came to the realization that very little attention was really being paid to the prevention of laboratory accidents in our (and probably other) infectious disease laboratories. Certainly great pioneering efforts were made and much valuable information resulted form the Fort Detrick Safety Office under Dr. Arnold G. Wedum and others prior to 1972 (1-33), but either (a) the practical translation from the data on hazard concepts to the bench level techniques had not been made, or (b) the information was not getting to the individuals who really needed it. Moreover, it seemed that much work still needed to be done, particularly as innovative and unusual techniques were being devised to keep up with evolving technological developments. Several months after my accident I became a safety manager and and I have learned much since. I now realize how easily my own mishap could have been

prevented by questioning and evaluating the potential dangers associated with the procedure I was using.

As a result of my subsequent years in the field of biohazard containment and safety, I have come to some general conclusions. Among
these are:

1. It is necessary to know the characteristics of agents prior to ini-
    tiating the proposed work in order to design and assign techniques
    that are effective in preventing their unique hazards, e.g., aero-
    sol infections, ocular infections, etc.
2. It is important to talk with laboratory personnel in advance to in-
    still a knowledge of what you are trying to accomplish and a re-
    spect (but not a *fear*) for the materials/agents they are to work
    with.
3. Most accidents can be prevented with either more forethought or
    more concentration at the moment of occurrence.
4. Some prohibitory rules are necessary, e.g., glass must be elimi-
    nated during high hazard manipulations whenever possible, no
    mouth pipetting is to be permitted, etc.
5. Rigid, mandatory rules should be kept to a minimum. Old estab-
    lished rules *should* be questioned. For example, is the practice
    of chewing gum in the laboratory or the practice of drinking coffee
    in an area outside of, but adjacent to, an infectious disease labor-
    atory really hazardous? (If done sensibly, probably not, but this
    will be discussed further in this chapter.) One should also be
    frank and admit that some prescribed procedures may be "overkill,"
    e.g., showering after working with some agents. Such procedures
    may be necessary even if the possibility of resultant infection is
    very remote for a variety of important reasons—health, political,
    legal, etc. The prevention of laboratory-acquired infections should
    not be a matter of dogma, but rather an application of common
    sense and a knowledge of agent characteristics, a principle that
    should be applied to overall containment requirements. It is recog-
    nized, however, that when one is responsible for a large workforce
    of varied disciplines, it is often necessary to impose hard and fast
    rules to provide guidelines for those who cannot make informed
    decisions.
6. In our laboratories, most of the significant potential personnel
    exposures to moderate or high hazard agents involved hand punc-
    tures or lacerations.
7. Proper protective devices and a knowledge of when, where, and
    how to use them should be readily available. Certainly if they
    had been used they would have prevented most, if not all, acci-
    dents at our institute with infectious agents. Proper use of pro-
    tective clothing and equipment in combination with engineering
    control measures and good laboratory techniques, can provide

adequate barriers against exposure to most hazardous materials. This chapter will examine some of the benefits and shortcomings of these devices.

In order to create a happier, healthier, more productive and at least a better informed workforce, I will discuss personal protection and hygiene in biomedical research in a practical rather than a despotic way.

II. RISK ASSESSMENT

Biological safety begins with risk assessment. The philosophy behind an effective program is based on an early estimation of risk and then applying appropriate containment and protective measures. Estimation of risk is based on the number of past laboratory infections (when such information is known), the natural morbidity and mortality rates, the human infectious dose, the efficacy of immunization and/or treatment (if either of these are available), the extent to which infected animals transmit the disease (e.g., whether the agent is excreted in the urine or the feces), the stability of the agent, and whether the experiment could result in the infectious agent being inoculated accidentally into the person conducting the experiment. Fortunately, several groups have already characterized most parasitic, fungal, bacterial, rickettsial, and viral agents in terms of the biosafety level of containment recommended for their laboratory use (34,35).

At any rate, once assessed, appropriate containment and techniques can then be selected and applied to interrupt the agent transmission and resultant infection of the laboratory worker. For example, if workers can be immunized against a particular agent, work can be performed at a lower level of primary containment than if they were unimmunized, while still maintaining the same level of secondary containment. The policies and procedures to be followed in achieving control should then be stated in writing and should include those that are considered the minimum for normal operations supplemented by operational procedures for each operation in which work of a special nature or on a particular substance is in progress. While an abundance of hard and fast rules is something to be avoided, adherence to whatever practices are developed for a particular study must be rigid and absolute. Therefore, written SOPs and practices are an important adjunct to safe practices, as each study or experiment can incorporate unique techniques developed in advance. The importance of preliminary planning, i.e., the evaluation and assessment of the potential risks, cannot be overemphasized, particularly since new technology generates new risks and greater ignorance. One interesting method to use to aid in the development of these operational procedures is to conduct a mock-up dress rehearsal.

By performing the required manipulations with noninfectious materials
prior to using infectious agents, you can determine which items of
protective equipment or which techniques are required at each step to
reduce the hazard potential. This rehearsal should be attended by
someone with biological safety knowledge and experience who can recognize and correct defects. It may be just a matter of eliminating or
substituting a dangerous step or item of equipment, such as a sharp
point or fragile glass.

III. RESPONSIBILITIES

The delineation of responsibilities for safety must be clear and documented beofre working with infectious microorganisms. The laboratory
director, principal scientist, department head, etc., is normally responsible for all of the safety aspects of the activity under their direction. Included in this is responsibility for the maintenance of adequate
precautions for protecting the surrounding community from substances
that employees are using under their supervision. Responsibility for
safe practices extends down through all supervisory levels, but the
final responsibility rests with the first-line supervisor. Each supervisor should be responsible for: (a) preventing accidents and infections during the course of work under his supervision and for accident
investigation, corrective action, and disciplinary measures, to the
extent that he is responsible for any other part of his job or assignment; (b) training persons under his supervision in the elements of
safety and in safe work habits; (c) conducting on-the-job refresher
safety for personnel under his jurisdiction; and (d) preparing special
standard operating procedures. Once the proper protective clothing
and equipment are selected, it is the task of the supervisor to provide
training in their use and to ensure that employees use and maintain
them properly. All organizations using infectious microorganisms should
have a safety committee and safety officer. The safety committee should
include representatives from management and all research units and
should meet at least quarterly to review accidents and discuss problems and other items of overall interest. The biohazard safety officer
should have a strong background in both microbiology and engineering
and is responsible for (a) formulating and interpreting safety policies
and procedures; (b) assisting supervisors in formulating special SOPs;
(c) incorporating adequate safe practices and operating procedures
in manuals, directives, and other instructions; (d) conducting surveys and inspections; (e) maintaining accident records and aiding in
the investigation of any occupational illness, ensuring that such incidence are reported promptly, the cause is properly determined, and
proper steps are taken to prevent a recurrence; (f) providing general
safety orientation to newly employed personnel; (g) ensuring that all

materials removed from infectious areas are properly decontaminated; (h) ensuring that all biological materials shipped from the laboratory are processed in accordance with procedures; and (i) ensuring that the following special procedures are carried out: periodic checks of building ventilation, measurement of ventilated cabinet leakage and airflows, ultraviolet light intensity measurements, filter efficacy testing, and decontamination of areas and equipment.

IV. ADMISSION TO INFECTIOUS AREAS

Infectious areas should be posted as such, and unauthorized personnel should not enter any area beyond this sign. It is recommended that an up-to-date list of all infectious microorganisms currently in use in the area or, preferably, a list of immunizations required for entry, be posted. Persons should be authorized to enter the area only if their presence is required on the basis of program or support need. It is the responsibility of each area supervisor to ensure that visitors and service or contract personnel not regularly assigned to the area are adequately protected from exposure to infectious substances. "Adequately protected" in interpreted to mean: (a) before admittance to an infectious area, every visitor will establish his need to enter and his immunization status will be determined; and (b) vaccination or baseline skin tests, x-ray or serological examination, as appropriate, will have been completed; (c) if (b) above is not complete and admission is necessary, then: (1) the visitor will use a respirator or face mask approved by the safety officer (a standard hospital or surgeon's gauze mask is not acceptable); or (2) infectious work will be suspended, decontamination performed, and the area declared nonhazardous; and (d) when within a hazardous area, the visitor will be accompanied by a member of the staff and exposure will be kept to a minimum, regardless of whether vaccination has been completed. Such persons shall be advised of the potential hazards and shall comply with all established entry and exit procedures.

Personnel who are to work in moderate to high containment (Biosafety Levels [BL] 3 and 4) laboratories should be screened to determine if they have a medical diagnosis associated with a risk of sudden loss of consciousness or other serious impairment which would result in inability to leave the laboratory unassisted. Applicable diagnoses include, but are not limited to: (a) neurological disorders, such as seizures; (b) cardiac disorders, such as ischemic heart disease; (c) endocrine disorders, such as diabetes mellitus; (d) psychiatric disorders, such as phychosis; (e) pulmonary disorders, such as emphysema with abnormal blood gases; (f) liver disorders, such as cirrhosis or chronic active hepatitis; or (g) use of prescription drugs which impair mental and motor performance. A physician should then make a judgment based on all relevant information as to whether a given diagnosis is reason for an individual's exclusion form laboratories.

Persons who work in BL-3 and BL-4 areas should also be screened on an annual basis to determine whether the possibility exists of a condition which would exclude them from work. Workers in these areas who develop signs and symptoms with or without evaluation by private physicians, which might indicate a condition where there is a risk of sudden loss of consciousness or other serious impairment, should be evaluated by medical personnel. When medications are prescribed, personnel should inquire whether these affect mental and motor performance before entering a high containment area. In addition, persons who have had recent surgical procedures, such as tooth extractions, or injuries involving significant alteration to the normal integrity of the skin, should also report for medical evaluation before entering infectious areas. Women laboratory workers should notify their supervisors as soon as pregnancy is suspected.

V. ENTRY AND EXIT PROCEDURES FOR INFECTIOUS AREAS

Entry and exit procedures for infectious areas are dependent on the hazard levels of the microorganisms under study. The hazard level should be permanently posted and necessary entry and exit procedures enforced.

The CDC/NIH guidelines on laboratory biosafety (35) recommends the wearing of supplemental laboratory clothing or a complete change of clothing before entering any infectious laboratory. Lab coats, gowns, smocks, or uniforms are recommended or required for BL-1 or BL-2 areas, respectively. When entering BL-3 laboratories, lab clothing that protects street clothes is specified, such as long-sleeved, solid front, wrap-around gowns, scrub suits, or coveralls. (Front-button lab coats are regarded as unsuitable.) For BL-4 areas, a complete clothing change is required, including shoes. A number of agents are studied in most BL-3 areas in our institute. Therefore, we require a change of clothes when entering all BL-3 laboratory areas. For example, some agents are potentially very infectious for unimmunized personnel, while others are nonindigenous pathogens of domestic livestock. Therefore, a complete clothes change on entry and a total body shower upon exiting appear to be prudent practices to enhance the margin of safety. When a clothes change and shower is required, the option should be offered to either cover the hair with a cap or wash the hair during the exit shower. We also recommend a complete change of shoes or the use of shoe covers when entering, or the use of a disinfectant soaked towel when exiting, even BL-2 laboratories or infected animal rooms.

The washing of hands after handling viable materials and animals and before exiting is the only requirement when leaving a BL-1 laboratory specified in the CDC/NIH guidelines. Before leaving a BL-2 laboratory for a nonlaboratory area, in addition to washing the hands, the lab coat, gown, etc., must be removed and left in the laboratory or covered with a clean coat not used in the lab. Laboratory clothing worn in a BL-3 area is not to be worn outside the area and is decontaminated before being laundered. In the event of laboratory contamination of street clothing, it must be decontaminated before leaving the area. When a complete clothes change and shower is required, upon exiting, all laboratory clothing should be placed in a laundry bag or stored in lockers before entering the shower area, and all personnel must shower using, preferably, a germicidal soap before leaving the area. Shoes are left within the area. The wearing of jewelry in containment areas is discouraged. However, if an exit shower is required and jewelry is worn within the area, it must be washed off with disinfectant or worn during the shower, and if eyeglasses are worn, they should be washed thoroughly.

VI. LABORATORY CLOTHING

A. Body

As indicated above, each biosafety level requires a specific type of protective laboratory clothing, and a wide variety of each is commercially available. Included are lab coats, smocks, gowns, total body suits, coveralls or jump suits, aprons, and two-piece scrub suits. Most of these items come in reusable or disposable models and a variety of materials including cotton, Dacron, nylon, polyester, olefin, rayon, vinyl, modacrylic, PVC, or rubber, or tradenames such as Tyvek (plain, polyethylene-coated, or Saranex-laminated), SEF, Kaycel, Chemklos II, MarGard, Nomex, Sontara FR, Safeguard, Fire-Stop, Micro-clean, Duraguard, and Disposagard. Many are designed for specific hazards, including abrasion, nuclear, chemical, and heat, and feature antistatic and flame-, caustic-, oil-, and acid-resistant materials. The selection of the optimum configuration and material depends on the regulatory guidelines, types of operations to be performed, potential hazards, availability of laundry facilities, type of decontamination available, environment, and personal preference (36).

For BL-3 areas or BL-4 areas that are not designed specifically for the wearing of a one-piece plastic positive-pressure suit, we prefer a cotton or cotton/polyester blend, two-piece scrub suit with a long-sleeved shirt which buttons down the front for rapid removal in case of a spill and which features integral ribbed wristlets. In addition, we stipulate that the scrub suits must be a unique color and cannot be worn anywhere outside of the moderate to high containment areas.

These latter restrictions are imposed to assure that a worker has at least changed his/her laboratory clothing when leaving the area, even though he/she may have violated the exit shower requirement.

When operations that generate infectious aerosols are performed outside of a biological safety cabinet, I recommend the additional wearing of a long-sleeved solid front wrap-around gown to minimize contamination of the basic laboratory outfit. Examples of such procedures would be inoculating animals with infectious materials, bleeding viremic animals, and otherwise handling or caring for animals who may be shedding hazardous viruses in the urine or feces. When exiting, the gown is removed and left in the room for reuse or steam sterilization.

Whenever there is a significant possibility of chemical splashes, radioisotopes, contamination, etc., heavy-duty rubber aprons or other specialized items of apparel can also be worn.

B. Head

The wearing of a head covering is normally not necessary in biological laboratories except in containment areas where a complete clothing change is required. As previously stated in Section V, in situations where a total body shower is mandatory upon exiting, the option usually offered is to require either washing the hair during the shower or the wearing of a head covering during the time spent in the containment area. Several styles are available: a simple cap, various hood styles, or a bouffant style for longer hair, and come in washable or disposable models in a variety of fabrics including cotton, cotton/polyester blend, polyolefin, or Tyvek.

C. Feet

Industrial safety shoes should be worn in any area where there is a significant risk of dropping heavy objects on the foot. When used in containment areas, these shoes, like all others, should be left within the area or decontaminated before removal. A large selection of types and configurations can be found in safety equipment supply catalogs. For general biological use, comfortable shoes, such as tennis shoes or nurses shoes, are used extensively. Sandals without socks are not recommended. The changing of street shoes when entering BL-3 areas is strongly advised and should be considered also for BL-2 areas, especially infected animal rooms. Shoecovers can be used in BL-2 or BL-3 areas when the requirement for a complete change of clothes is not necessary. Such covers are available in vinyl, polyethylene, Saranex, or Tyvek, and are usually considered disposable items. In animal rooms and other areas where the wearer may encounter the splashing of large amounts of water from the hosing of cages, racks, floors, etc., the wearing of butyl rubber, neoprene, or polyvinyl chloride (PVC) boots is strongly advised.

VII. PERSONAL PROTECTIVE EQUIPMENT

In addition to the wearing of the basic laboratory clothing in biological containment areas, as described above, it may be necessary to utilize specialized protective equipment designed to protect other bodily areas from injury or contamination. An understanding of the proper usage and limitations of these items is essential for maximum effectiveness.

A. Gloves

Gloves are used in the laboratory for protection against a wide variety of hazards, including heat, cold, acids, solvents, caustics, toxins, infectious microorganisms, radioisotopes, cuts, and animal bites. Unfortunately, there is no ideal glove which will protect against all of these. Therefore, the selection of proper gloves cannot be stressed too emphatically when hazardous tasks are to be performed. While it is tempting to use whatever is at hand (no pun intended), gloves are made of such a variety of materials (rubber, neoprene, neoprene/latex, Viton®, fluoroelastomer, nitrile, polyethylene, polyvinyl chloride [PVC], polyvinyl alcohol) and for such special uses that the user should seek advice or information on which glove is best for the task to be performed. For example, chemical solvents such as xylene, toluene, benzene, perchloroethylene, dichlorethane, and carbon tetrachloride normally degrade rubber, neoprene, and PVC (37) so that it is important to avoid these materials in favor of gloves made of polyvinyl alcohol or Buna-N. Because a detailed discussion of gloves for handling chemicals is beyond the scope of this chapter, the reader is advised to consult provided references (37-40) for further information.

For protection of hands, wrists, and forearms against steam or for handling hot objects, insulating gloves or mittens made of Zetex® aluminoborosilicate fibers or Kevlar® aramid fibers have replaced the traditional asbestos gloves. For handling very cold materials, Zetex® or insulated latex or neoprene gloves are available, and for liquid nitrogen, loose-fitting leather gloves are preferred.

The use of protective gloves in laboratories where microbiological hazards may be present often creates the situation where protection is sacrificed in the name of dexterity. Unfortunately, most gloves (latex, rubber, vinyl) that offer the good tactile sensitivity and dexterity required for many procedures offer very little, if any, protection against needle sticks, sharps, and animal bites. This has resulted in countless laboratory infections or potential infections due to the unconscious false sense of security associated with the wearing of surgical gloves. A review of the accident history of our institute shows that 7 out of the 10 significant potential exposures to high hazard infectious microorganisms (Biosafety Level 4), as well as the majority of laboratory accidents using agents of a lower hazard level, resulted from sharp

items (commonly hypodermic needles) and animal bites, usually involving the hands. Some of these accidents involved workers who were
wearing ventilated suits with integral .018" latex gloves or working at
Class III biological safety cabinets which incorporate arm-length .015"
neoprene gloves. It is important to acknowledge and remember that
gloves are the weakest component of the personal protection equipment
spectrum.

If the possibility of self-inoculation exists, a nonabsorbent glove
should always be selected to prevent contamination of the hands whenever infectious material is used. I recommend wearing the heaviest or
thickest glove possible without sacrificing the sensitivity or dexterity
required to accomplish the procedure or task. For example, we have
substituted .030" neoprene gloves for the .015" gloves on some areas
of our Class III cabinets where animals are routinely handled to reduce
penetration from bites. Also, workers in ventilated suits wear an additional pair of leather gloves when performing certain tasks and handling
animals, whenever possible.

B. Respiratory Protection

With the possible exception of a BL-4 containment area, which is designed for the wearing of ventilated suits, respiratory proection equipment should be worn only when engineering (ventilation, containment
of the agent) or administrative (substitution of the agent with one that
is less hazardous) controls are not achievable.

There are two general types of respiratory protection equipment:
tight-fitting and loose-fitting. Both types can be either atmosphere-
supplying (supplied air from pressurized tanks or air compressors) or
air-purifying (particle-removing filters). Atmosphere-supplying equipment can be operated at constant flow, demand flow, or pressure-
demand. Air-purifying respirators are usually breathing demand, but
are also available as constant flow or pressure-demand items which
incorporate a battery-powered fan and motor assembly.

Tight-fitting respirators include half- or full-face, both of which
require quantitative fit testing. According to 29 CFR 1910.134 (e)(5)(1)
(41), "Respirators shall not be worn when conditions prevent a good
face seal. Such conditions may be a growth of beard, sideburns, or
temple pieces on glasses." The American National Standards Institute
(42) recommends that a person who has hair (moustache, beard, stub-
ble, sideburns, low hairline, bangs) which passes between the face
and the sealing surface of the facepiece of the respirator, shall not
be permitted to wear such a respirator.

Single-use, paper "dust" masks are not classified as true respirators, are difficult to fit properly, and are used primarily for the prevention of fibrosis and pneumoconiosis (43). They do not offer adequate respiratory protection in areas where infectious aerosols (infected

animal rooms) may be present. They may be worn in "clean" animal rooms to reduce possible irritation from allergens, such as fur and dander, and to help keep hands away from the mouth. In addition, they are acceptable for use during necropsies and other procedures on noninfectious animals to help maintain a sterile surgical field. When used for these purposes, the appropriate size must always be worn to assure an optimal fit for maximum effectiveness.

Loose-fitting respirators include hoods, helmets, blouses, and full suits. With loose-fitting equipment, proper fit is not important, so the wearing of beards is not prohibited. While full suits are usually recommended for use in BL-4 areas, other configurations may be acceptable depending on the circumstances and operations to be performed, e.g., where there is absence of splashing of infectious blood or urine, no hosing of infected animal cages, or where no other infectious aerosols are present that may contaminate clothing. The substitution of any configuration other than a full suit or the use of any powered air-purified respiratory protection equipment must be compatible with both the work requirements and exit requirements. The requirement for a disinfectant shower upon exiting the BL-4 area may negate some of the alternative options.

Continuous or pressure-demand flow air-line or powered air-purifying helmets, hoods, or suits offer greater protection than demand equipment and are preferred for BL-4 use. Demand units or tight-fitting mechanical filter respirators with HEPA filters might conceivably be acceptable in BL-4 areas for emergency escape or rescue, if and only if, a good fit is assured. Demand or mechanical respirators are normally unacceptable in high hazard areas for routine use because of the possibility of leakage during inhalation. Positive pressure or pressure-demand SCBA provide a higher degree of protection than demand or closed-circuit equipment, but because of limited air supply (5-60 minutes), these were recommended until recently only for escape or rescue from high hazard areas. With the current availability of constant-flow air-purifying hoods which operate for 8-hours on rechargeable battery packs, this recommendation should be reconsidered. These devices have application in high hazard areas for emergency exit or emergency assistance, in high hazard field situations, or in atmospheres containing high concentrations of irritating chemicals such as formaldehyde gas after an area decontamination. They have the further advantage of readily fitting individuals with beards or who wear glasses.

Whenever supplied-air equipment is utilized, the respiratory protection program should include a back-up provision in the event of compressor failure. I recommend either an auxiliary compressor or bottled breathing air. If either of these is not feasible or is unavailable, then a combination pressure-demand breathing apparatus should be worn which permits the wearer to escape from the dangerous atmosphere in case the primary air is interrupted. Such equipment usually incorporates an approved rated 5- to 10-minute escape device.

When circumstances require the use of a respirator in BL-2 or BL-3 areas, e.g., in rooms housing animals where the possibility exists of infectious aerosols caused by inoculation, excretions, bleeding, agitation of contaminated bedding, etc., half-face or full-face respirators with HEPA filters are perfectly acceptable *provided a good fit is attained*. Full-face masks (or half-face masks plus eye protection) should be worn in atmospheres contaminated with aerosols of certain infectious viruses or other biological agents or toxins, which have the potential for being infectious or toxic by the ocular, as well as the respiratory route, and in other specific instances, such as when handling monkeys (especially male monkeys) infected with an infectious or toxic agent which is excreted in the urine.

Federal regulations (41) also state that "the wearing of contact lenses in contaminated atmospheres with a respirator should not be allowed." The American Standards Institute (42) interprets this to mean any respirator equipped with a full-facepiece, helmet, hood, or suit. The use of contact lenses in a laminar flow environment, such as a ventilated suit, may aggravate the dehydration of the tear layer upon which the lens rides, giving rise to subsequent corneal abrasions and other related phenomena (44). In addition, foreign bodies or contaminants that penetrate the respirator may get into the eye and cause severe discomfort, compelling the wearer to remove the respirator (45). Lastly, full facepieces can pull at the side of the eye and pop out the lens (46).

All employees required to use respirators should be enrolled in the facility Occupational Health Respirator Program prior to receiving a respirator or ventilated hood/suit. All suit wearers should also be enrolled in the Hearing Conservation Program, as the noise level in these suits sometimes exceeds 85 dB.

C. Eye/Face Protection

It has been my experience that most workers request safety glasses for the wrong reasons. When questioned as to why they need them, employees who work in laboratories will invariably answer, "Because I work with chemicals." After it has been explained that safety glasses are intended for impact protection and not chemical splash protection, they withdraw their request in favor of approved acid or chemical splash goggles or face shields. While the wearing of any type of eye protection, even ordinary spectacles, offers better splash protection than wearing nothing at all, goggles or shields that are designed for this purpose offer maximum eye protection. Such protective devices are inexpensive, and therefore should be made readily accessible in all laboratories where eye hazardous chemicals are used. Safety glasses are intended to provide impact protection against flying objects in shops

or maintenance areas, glassware handling areas, etc. We have also
found that monkeys in our research or animal holding areas will often
reach out and grab the glasses of animal caretakers and throw them
on the floor. Since we do not consider it fair for the employee to pay
for replacement of ordinary spectacles broken under these circumstances,
we issue safety glasses.

Eye protection in the biological laboratory is important for many
reasons. Concentrated acids, alkalis, or other corrosive or irritating
chemicals are often used routinely. In addition, concentrated disinfectants, including phenolics and quaternary ammonium compounds, can
cause severe damage or blindness if splashed in the eye. Infection
can occur through the conjunctiva if certain pathogenic microorganisms
are splattered into the eye. According to Krey et al., the mature retina
is vulnerable to virus invasion and susceptible to infection. Virus
infection can lead to retinal dysplasias and chronic retinal degeneration
(47). At least one virus, herpesvirus, has been shown to propagate
in the brain following intraocular inoculation (48). Full-face respirators or half-face respirators plus splash goggles are recommended when
operations with specific microorganisms or toxins may result in the
generation of respirable aerosols or droplets which may enter the eyes.

The position that contact lenses do not provide adequate protection
against eye damage following exposure to chemicals, fumes, or foreign
bodies is not universally accepted. There are reports (49-51) that
support the idea that contact lenses present an increased hazard to
the wearer, particularly in the event of a chemical eye splash. In
contrast, however, others (52,53) claim that *more* protection is provided when hard or soft contact lenses are worn than when absent,
including one controlled study with rabbits utilizing hard lenses in one
eye and exposed to an acid, a base, and a solvent (52). The authors
postulate that the lens may act as a barrier to an irritant, suggesting
that an irritant will cause lid spasm to occur, causing the lens to
tighten against the cornea, thereby effectively sealing off the area
under the lens. This theory is supported by actual cases described
by Rengstorff and Black (53). On the other hand, Rose (54) cites
the observation that rigid lenses are soluble in or swollen by many
organic solvents and that aqueous chemical solutions are readily soluble in the water phase on contact lenses, while Ennis and Arons (55)
argue that lenses will increase the concentration in contact with the
eye of chemicals that become trapped underneath the lens and prevent
thorough irrigation of the eye, thus accentuating corneal damage.

In a biological laboratory the situation may be a little more complicated. In the event of an accident of any kind or even the repositioning of a lens, sufficient decontamination of the hands must be done
prior to the removal or said repositioning of contact lenses. In the
hurry to remove the lens following a chemical splash, this hygiene
step may be neglected. In addition, as stated earlier in Section VII.B,

contact lenses should never be worn when wearing a respirator equipped with a full-facepiece, helmet, hood, or suit.

Contact lenses were never intended to be substitutes for properly designed safety eyewear in hazardous situations. Wearers of contact lenses, as well as wearers of ordinary spectacles, should use the same approved protective devices, such as splash goggles or safety shields, worn by other workers.

D. Ear Protection

As mentioned previously, the use of protective hearing devices are recommended when wearing ventilated suits. Such devices include plugs, muffs, and helmets, and may be also required in areas of the laboratory where significant noise is generated by certain items of equipment, either on a continuous or intermittent basis.

VIII. PERSONAL PROTECTION

A. Removal of Potentially Contaminated Gloves and Clothing

After handling potentially infectious or toxic substances, protective gloves should be removed in such a way as to avoid skin contamination. One technique is to insert the gloved left index finger between the right-hand glove and the wristlet cuff of the laboratory gown or scrub suit top. The right-hand glove is then peeled down, turning inside out, until almost off. The top of the right index finger is then inserted between left glove and clothing cuff. The glove is peeled off and then both gloves are discarded. Hands are then washed thoroughly.

The same caution should be exercised in removing contaminated clothing. With gloves on, pull off booties, touching only the outside. Then remove the cap, touching only the outside. Untie the gown in back and pull it forward over shoulders. Grab the right shoulder of the gown and pull it down to the wrist. Pull the left shoulder of the gown over the left hand, removing the glove in the process. With the bare left hand inside the right sleeve, pull the remaining gown over the right hand, removing the glove and folding the contaminated outside of gown inward. Place the gown and gloves in receptacle to be autoclaved and wash hands.

B. Hand Washing

The washing of hands when exiting biologically hazardous areas is generally considered the most important procedure in preventing laboratory infections (56). Although hand washing with an antiseptic agent is theoretically desirable, hand washing with soap, water, and mechanical friction is sufficient to remove most transiently acquired organisms.

C. Eating, Drinking, and Smoking

Although some controversy exists on this general issue, there is, I believe, general agreement that food, candy, and beverages should not be taken into infectious laboratories and smoking should never be permitted in any area in which work with infectious substances is in progress. In some containment facilities, smoking, drinking, and eating is permissible if gloves have been removed and hands washed, if no overt contamination has occurred, and only if restricted to an area away from the laboratory room or animal room. In this instance, food and drinks are stored only in a refrigerator in that area and designated for that purpose alone. Opinions and institutional practices differ on whether to permit eating, smoking, and drinking in research facilities. Some place the entire containment area off limits, while others designate a specific room or office. Some limit these practices to smoking and drinking coffee, if and only if disposable cups, individually wrapped sugar packets, and powdered cream substitutes are used. In such cases, it is required that employees who have been working with toxic or infectious or radioactive substances remove their gloves and wash hands thoroughly before smoking. Opinions also differ on whether to allow employees to chew gum in containment areas. Some facilities permit this practice, providing employees keep their fingers away from their mouths, while others flatly prohibit it.

IX. CONCLUSION

Experience in biohazard research in the past 40 or so years has shown that if personnel use approved techniques and use/wear proper protective equipment in a well-designed facility, it is possible to work safely with agents of the highest hazard level. Experience has shown that proper use of proper equipment can prevent most accidents from occurring and reduce the severity of those that do. Management should make the necessary protective equipment and clothing for eyes, face, ears, head, body, and extremities, and respiratory devices available to workers involved in biomedical activities. The equipment must be maintained in a sanitary and reliable condition, and the workers must be trained in its use, importance, and limitations.

APPENDIX

TRADEMARK	*OWNED BY*
Chemklos II	J. P. Stevens and Co., Inc.
Dacron	E. I. du Pont de Nemours and Co., Inc.

TRADEMARK	OWNED BY
Disposaguard	Kimberly-Clark Corp.
Duraguard	Kimberly-Clark Corp.
Fire Stop	Cotton, Inc.
Kaycel	Kappler Disposables
Kevlar	E. I. du Pont de Nemours and Co., Inc.
Mar Gard	Mar Mac Manufacturing Co., Inc.
Micro-Clean	American Convertors
Nomex	E. I. du Pont de Nemours and Co., Inc.
Safeguard	Kimberly-Clark Corp.
Saranex	Down Chemical Co.
SEF	Monsanto
Sontara FR	E. I. du Pont de Nemours and Co., Inc.
Tyvex	E. I. du Pont de Nemours and Co., Inc.
Viton	E. I. du Pont de Nemours and Co., Inc.
Zetex	Newtex Industries

REFERENCES

1. Anderson, R. E., Stein, L., Moss, M. L., and Gross, N. H. (1952). *J. Bact.* 64:473.
2. Anderson, R. J. (1950). *Pub. Health Repts.* 65:463.
3. Anonymous. (1950). *New England J. Med.* 242:106.
4. Anonymous. (1956). *Lancet* 2:880.
5. Barbeito, M. S., and Taylor, L. A. (1968). *Appl. Microbiol.* 16:1225.
6. Chatigny, M. A. (1961). In *Advances in Applied MIcrobiology*, W. W. Umbreit (Ed.), Vol. 3. Academic Press, New York, p. 131.
7. Darlow, H. M. (1969). In *Methods in MIcrobiology*, J. R. Norris and D. W. Robbins (Eds.), Vol. 1. Academic Press, New York, p. 169.
8. Fish, C. H., and Spendlove, G. A. (1950). *Pub. Health Rep.* 65:466.
9. Gremillion, G. G. (1959). In *Proceedings Second Symposium on Gnotobiotic Technology*. Univ. of Notre Dame Press, Notre Dame, Ind., p. 171.
10. Hanson, R. P., Sulkin, S. E., Buescher, E. L., McD. Hammon, W., McKinney, R. W., and Work, T. H. (1967). *Science* 158:1283.

11. Morris, E. J. (1960). *J. Med. Lab. Techn. 17*:70.
12. Phillips, G. B. (1965). *Causal Factors in Microbiological Laboratory Accidents and Infections* (Misc. Publ. 2). U.S. Army Biological Laboratories, Fort Detrick, Md.
13. Philipps, G. B. (1965).*J. Chem. Educ. 42*:A43, A117.
14. Philipps, G. B. (1967). In *Handbook of Laboratory Safety*, N. V. Steere (Ed.). Chemical Rubber Co., Cleveland, Ohio, p. 384.
15. Philipps, G. B. (1969). *Am. Ind. Hyg. Assn. J. 30*:170.
16. Philipps, G. B., and Jemski, J. V. (1963). *Lab. Animal Care 13*:30.
17. Pike, R. M., and Sulkin, S. E. (1952). *Scien. Monthly 75*:222.
18. Reitmann, M., Frank, M. A., Sr., Alg, R., and Wedum, A. G. (1953). *Appl. Microbiol. 1*:14.
19. Reitmann, M., and Wedum, A. G. (1956). *Pub. Health Repts. 71*:659.
20. Shapton, D. A., and Board, R. G. (Eds.) (1972). *Safety in Microbiology*. Academic Press, New York.
21. Shepard, C. C., May, C. W., and Topping, N. H. (1945). *J. Lab. and Clin. Med. 30*:712.
22. Smadel, J. E. (1951). *Am. J. Pub. Health 41*:788.
23. Sulkin, S. E., and Pike, R. M. (1951). *J. Am. Med. Assn. 147*:1740.
24. Sulkin, S. E., and Pike, R. M. (1969). In *Diagnostic Procedures for Viral and Rickettsial Infections*, E. H. Lennette and N. J. Schmidt (Eds.), 4th ed. Amer. Public Health Assn., New York, p. 66.
25. Wedum, A. G. (1950). *J. Lab. and Clin. Med. 35*:468.
26. Wedum, A. G. (1953). *Am. J. Pub. Health 43*:1428.
27. Wedum, A. G. (1957). *J. Clin. Path. 10*:88.
28. Wedum, A. G. (1949). In *Proceedings Second Symposium on Gnotobiotic Technology*. Univ. of Notre Dame Press, Notre Dame, Ind., p. 105.
29. Wedum, A. G. (1961). *Bact. Rev. 25*:210.
30. Wedum, A. G. (1964). *Pub. Health Repts. 79*:619.
31. Wedum, A. G., and Kruse, R. H. (1969). *Assessment of Risk of Human Infection in the Microbiological Laboratory* (Misc. Publ. 30, 2nd Ed.). Industrial Health and Safety Directorate, Dept. of the Army, Fort Detrick, Md.
32. Wedum, A. G., Barkley, W. E., and Hellman, A. (1972). *J. Am. Vet. Med. Assn. 161*:1557.
33. Whitewell, F., Taylor, P. J., and Oliver, A. J. (1957). *J. Clin. Path. 10*:88.
34. Subcommittee on Arbovirus Laboratory Safety of the American Committee on Arthropod-borne Viruses. (1980). *Am. J. Trop. Med. Hyg. 29*:1359.

35. Richardson, J. H., and Barkley, W. E. (Eds.) (1984). *Biosafety in Microbiological and Biomedical Laboratories*. U.S. Government Printing Office, Washington, D.C. HHS Publ. No. (CDC) 84-8395.
36. Lynch, P. (1982). *Occup. Health and Safety 1*: 30.
37. U.S. Dept. of Health, Education and Welfare, Office of Research Safety, National Cancer Institute and the Special Committee of Safety and Health Experts. (1978). *Laboratory Safety Monograph: A Supplement to the NIH Guidelines for Recombinant DNA Research*, National Institutes of Health, Bethesda, Md.
38. Sansone, E. B., and Twari, Y. B. (1978). *Am. Ind. Hyg. Assn. J. 39*:169.
39. Stanbrook, B. W., and Runkle, R. S. (1986). In *Laboratory Safety: Principles and Practices*, B. M. Miller (Ed.). American Society for Microbiology, Washington, D.C., p. 164.
40. Thomas, S. M. G. (1970). *Occup. Health 22*:281.
41. Occupational Health and Safety Administration. (1981). *Respiratory Protection*. General Industry Standards 20 CFR 1910.134, U.S. Dept. of Labor, Washington, D.C.
42. American National Standards. (1982). *Practices for Respiratory Protection*. Z88.2-1980, American National Standards Institute, Inc., New York.
43. Bureau of Mines. (1972). *Respiratory Protection Devices; Tests for Permissability; Fees*, 30 CFR 11, U.S. Department of Interior, Washington, D.C.
44. Rosenstock, R. (1986). *Profess. Safety 1*:18.
45. Occupational Safety and Health Administration. (1984). *Respiratory Protection* (OSHA 3079). U.S. Government Printing Office, Washington, D.C.
46. Birkner, L. R. (1980). *Respiratory Protection: A Manual and Guideline.* American Industrial Hygiene Association, Akron, Ohio.
47. Krey, H. F., Ludwig, H., and Rott, P. (1979). *Arch. Virol. 61*:283.
48. Kristensson, K., Ghetti, B., and Wisniewski, H. M. (1974). *Brain Res. 69*:189.
49. American Medical Association Council on Occupational Health. (1964). *J. Am. Med. Assn. 188*:397.
50. Kuhn, H. S. (1961). *J. Am. Med. Assn. 178*:1055.
51. Fox, S. L. (1967). *J. Occup. Med. 9*:18.
52. Guthrie, J. W., and Seity, G. (1975). *J. Occup. Med. 17*:163.
53. Rengstorff, R. H., and Black, C. J. (1974). *J. Am. Optomet. Assn. 45*:270.
54. Rose, R. D. (1979). *Chem. Eng. News 57*:84.
55. Ennis, J. L., and Arons, I. (1979). *Chem. Eng. News 57*:4.
56. Steere, A. C., and Mallison, G. F. (1975). *Ann. Inter. Med. 83*:683.

11

Liquid Chemical Germicides

EVERETT HANEL, JR. *NCI-Frederick Cancer Research Facility, Program Resources, Inc., Frederick*

I. INTRODUCTION

A. History

For many centuries, man has used various procedures to control the spread of infectious diseases. There was, of course, no basic understanding of contagion and certainly the role of microorganisms in the infectious disease process was totally unknown. However, the Egyptians and Jews clearly believed that disease can be communicated by touch and fomites. Bulloch, in his book, *The History of Bacteriology*, beautifully covers in detail the developments in knowledge of infectious diseases from the ancient doctrines to the early twentieth century (1). Compounds of mercury were used by Arab physicians in the Middle Ages to prevent sepsis in open wounds (2). In the early 1800s chlorinated lime was used to treat the open sewers in London. This treatment was primarily to reduce odors, although it was felt that harmful miasmas were also reduced. Semmelweis carried out a series of excellent scientific studies starting in 1847 (3) that showed the spectacular reduction in sepsis and death rates that could be achieved by having physicians and medical students wash their hands with chloride of lime on leaving the autopsy room and before examining obstetric patients. Although Lister was not the first scientist to use phenol as a disinfectant, it was hiw work in the 1860s that brought phenol into prominence and made the chemical popular in surgical procedures. He covered open wounds with phenol-soaked dressings and sterilized sutures and instruments. He sprayed a 1:20 aqueous solution of phenol in the operating room during surgery. His results in reducing infections were

so dramatic that these practices were adopted by hospitals throughout
Europe and America (4,5).

B. Sterilization by Heat

Sterilization is an absolute term that means the destruction or elimination of all forms of life. From early recorded history, fire and boiling
water have been used as purifying agents to check the transmission of
diseases. Boiling water is not a sterilizing agent because of its relatively low temperature. However, boiling water is an excellent sanitizer
and disinfectant. In 1810, Nicholas Appert and co-workers in France
(6) developed canning methods for the preservation of foods by heat
and maintaining the foods in a sealed container. In 1884, the Chamber-
land autoclave was developed and a Parisian engineering firm began
manufacturing and marking it (7). Sterilizaiton by steam under pressure is the accepted standard, since all living organisms are destroyed
by heat. Dry heat, as an alternative to steam under pressure, is relatively slow and requires higher temperatures to achieve sterilizaiton.
However, dry heat will penetrate materials not soluble or miscible in
water such as oils and greases and with sufficient exposure time will
sterilize materials in closed containers.

C. Sterilization by Methods Other Than Heat

Ethylene oxide gas has been used for the last 40 years for sterilizing
materials that cannot be subjected to heat or moisture. Ethylene oxide
is toxic and has been shown to cause cancer in laboratory animals.
Some studies indicate that ethylene oxide may be associated with an
increased risk of cancer in man (8,9). OSHA has established controls
on the use of ethylene oxide (10) that cover all aspects of work and
exposure to this material. Formaldehyde gas has been used for the
inactivation of microorganisms since before the turn of the centery.
     Besides heat and gases, other methods of achieving sterility in-
clude: (1) filtration—the development of the membrane filter allows
the absolute removal of all particulates down to the pore size of the
particular membrane. Filtration is in general use to produce sterile
pharmaceutical and biological fluids. The most common pore size used
is 0.45 µm, although some fluids are passed through membrane filters
having a pore size of 0.2 or even 0.1 µm. A depth prefilter consisting
of fiberglass, cotton, or granular on sintered materials are generally
used as prefilters. (2) Ionizing radiation from gamma rays or s-rays
can be used for the destruction of microorganisms. Gamma radiation
has become a useful system for the sterilization of a variety of hospital
and laboratory supplies. (3) Ultraviolet (UV) radiation is germicidal
at a wavelength near 260 nm. Low pressure mercury vapor lamps emit
radiant energy at 254 nm, which is near the maximum wavelength ab-

sorbed by nucleic acids and the maximum wavelength for germicidal action. Although UV is quite effective in reducing the number of microorganisms, especially airborne microorganisms, it is quite difficult to achieve sterility with UV. The UV rays must penetrate the microorganism to achieve inactivation. The short-wave UV has poor penetrating power so that dust, dirt, grease, or clumps of microorganisms may prevent the UV from irradiating the microorganism. Also, if the exposure of the microorganism to UV is not sufficient to cause irreversible damage, then photoreactivation of the cell may occur when the cell is exposed to visible light.

II. LIQUID CHEMICAL GERMICIDES

A. Desirable Properties

The microbiologist is quite dependent upon disinfectants to prevent exposure of himself/herself and co-workers to hazardous microorganisms and to protect experimental materials from contamination. Although the type of disinfectant needed will vary depending upon the microorganism in use and the procedures and equipment being used, there are certain general properties of disinfectants that make them useful. A germicide should have a wide sprectrum of activity and should be rapid in action. A germicide should not be highly toxic or carcinogenic. A germicide should be commercially available, inexpensive, and should not be a strong solvent for plastics, paint, and similar materials. Nonflammability is a desirable feature for a germicide. Noncorrosivity is a good characteristic, but noncorrosive is a relative term as low pH water is somewhat corrosive. In some instances, it may be desirable for the disinfectant to leave a film of disinfectant on the floor or other surface as will occur with phenolic compounds, but in other areas, such as ventilated cabinet where work with tissue cell cultures is being carried out, the investigator will prefer to use a germicide such as alcohol that leaves no residue. Germicides should and most do contain a surface active agent to give good penetrating and cleaning properties.

B. Factors Influencing Germicidal Action

Concentration of Germicide

The concentration of the germicide will vary depending upon exposure time, the organisms to be treated, the characteristics of the germicide and the purity of the media being treated. For example, in a municipal water supply system, chlorine may be added at the plant to 5-10 ppm. The water will be exposed to the chlorine for at least several hours before use, and chlorine concentration will gardually decrease to a range of 0.2 to 1 ppm at the tap of the user. The organisms to be

controlled are primarily enteric bacteria and viruses. Bacterial spores
will probably not be killed by this chlorine concentration, but they
are of minor health concern. The water being treated is relatively
pure so that very little chlorine is adsorbed by extraneous organic
materials.

pH

The pH of the fluid being treated is an important factor for most disinfectants. Chlorine and chlorine dioxide are both excellent germicidal
agents. The germicidal action of chlorine dioxide increases with an
increase in pH, which makes it very suitable for treating high pH water
such as lime-softened water (11,12). On the other hand, chlorine is
known to be more germicidal at lower pH. It has been shown that at
each pH value, the log of the time required for 99% kill plotted against
the log of concentration is a straight line, and the rate of sporicidal
action decreases rapidly as the pH increases (13).

Temperature

The effect of temperature on germicidal activity is dependent upon
the thermal death rate of the microorganism, as well as the temperature
dependency of the germicide. However, for limited ranges of temperature, the mcirobial dealth rate when exposed to a germicide can be
described as a simple chamical reaction representing a first-order (log-
arithmic) reaction, as shown in Figure 11-1. For use at temperatures
considerably below room temperature, one could select a germicide
having a relatively low activation energy, so that the germicide is relatively independent of change in temperature. There have been few
studies relating to germicidal effectiveness at low temperatures. Jones,
Hoffman, and Phillips (15) showed that sodium hypochlorite was an
excellent disinfectant at low temperatures. Using ethylene glycol to
prevent freezing, hypochlorite solutions were effective against *Baccilus
subtilis* var. *niger* from 0 to -40°C. (See Figure 11-2.)

Effect of Organic Material

Most germicides are adversely affected by the presence of organic materials. Organic materials may coagulate, form films, or form clumps
of materials that shield and protect the microorganisms from the disinfectant. Also, the germicide may combine with the organic materials
causing a loss or reduction in germicidal action. Chlorine is usually
cited as a germicide that loses or is reduced in its capacity as a germicide in the presence of organic materials. Chlorine reacts with ammonia
or amino compounds to form chloramines, which reduces the free available chlorine. The chloramines or N-chloro compounds require a higher
concentration and much more time to effect a kill of microorganisms

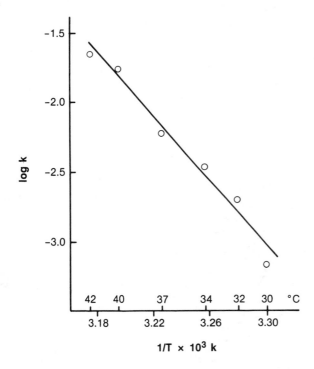

FIGURE 11-2

than does free chlorine. These effects of organic material on chlorine are most noticeable at low levels of chlorine. This effect of organic material on free chlorine concentration may be somewhat deceiving. Chlorine may be used as a germicide at 10 ppm or .001%, and if the break-point is at the 1% level, then chlorine must be added to reach the 10,000 ppm or 1% level at which point further addition of chlorine will result in a linear increase in free chlorine. On the other hand, many germicides are used at the 5% or 8% concentration levels. It is obvious that a 1% reduction in concentration of these higher concentration disinfectants will still leave killing concentrations of the disinfectant.

C. Comparative Resistance of Microorganisms

Microorganisms of the various classes show enormous resistivity differences to heat, gases, and liquid germicides. In general, resistivity increases from vegetative bacteria, to viruses, to fungal spores, to bacterial spores. However, there is a large overlap among these

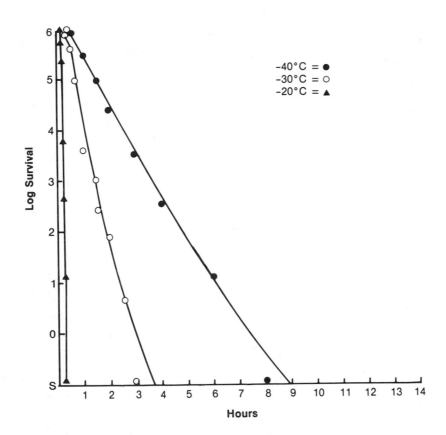

FIGURE 11-1

groups. Viruses, being obligate intracellular parasites, are afforded
a degree of protection within the cell. The rickettsia, also, are primarily intracellular parasites, although a few species are facultatively
extracellular. The rickettsia *Coxiella burnetii* (Q fever) is very resistant to deleterious environmental conditions such as drying and sunlight so that it, to some extent, resembles a bacterial spore. The DNA
viruses are generally more resistant to environmental conditions than
are the RNA viruses, and viruses carrying a lipid outer membrane that
reacts with lipophilic germicides are much more susceptible to germicides
than are the nonlipid (hydrophilic) viruses (16). Fungal spores are
quite resistant to environmental conditions, and fungi, such as *Asper-
gillus niger*, that produce black spores are the most resistant forms
of life. However, as with the other groups of microorganisms, there
is a great variation among the spore formers. *Bacillus stearothermo-
philus* is very resistant to heat, whereas *Bacillus subtilis* species are
more resistant to gaseous and liquid germicides.

The subacute spongiform encephalopathies or slow viruses are transmissible infectious agents best described as highly unconventional viruses. These viruses, especially some unusually resistant strains of scrapie, survive treatment with formaldehyde (17), alcohol (18), chloroform (19), and standard autoclaving time and temperatures (20). Brown, Rohmer, and Gajdusek (20) conclude that autoclaving at 123°C for 60 minutes results in a nondetectable titer of scrapie or CJD virus. Exposure to 1.0 N NaOH for 60 minutes results in a 6.8 log 10 LD50 decrease, and exposure to 2.5% sodium hypochlorite results in a 7.0 log 10 LD50 reduction in titer. Slow viruses are studied in relatively few laboratories, but they do exemplify the great differences in resistivity to adverse environmental conditions that occur among microorganisms.

III. ACTION OF SPECIFIC GERMICIDES

A. Halogens

Chlorine

Chlorine and chlorine compounds are more widely used than any other germicide, especially if one includes the metropolitan water systems that treat drinking water with chlorine. In the laboratory, chlorine is used in the form of sodium or calcium hypochlorite. Hypochlorous acid (HOCl) appears to be the active ingredient, but the hypochlorite ion (OCl) may play a role as well. Chlorine acts rapidly at very low concentrations and is bactericidal at low temperatures. Chlorine produces destructive permeability changes in the cell wall (21), and as time goes on the very reactive, oxidative chlorine forms N-chloro compounds with protoplasm (22) and also acts on the SH groups of one or more enzyme systems (23). Hypochlorite can be obtained as sodium, lithium, or calcium hypochlorite. The advantage of calcium hypochlorite, besides being a dry, stable powder, is the fact that it has an available chlorine equivalent of two chlorine atoms for each hypochlorite ion (OCl), or four chlorine atoms for each molecule of $Ca(OCl)_2$, whereas sodium hypochlorite is equivalent to two chlorine atoms. Available chlorine by weight for calcium hypochlorite is 70-72% or 700,000 ppm. It is obvious that the use dilution is 1 to 100 (7,000 p$\frac{1}{2}$pm), 1 to 1,000 (700 ppm), or even 1 to 10,000 (70 ppm). Calcium hypochlorite has a wide spectrum of germicidal activity so that it is widely used in dairies, food preparation areas, and swimming pools.

There are several commercial germicides, such as zonite, that use sodium hypochlorite as the active ingredient. The hypochlorites have been used for 200 years for bleaching cotton cloth. There are dozens of household bleaches, such as Clorox, that use sodium hypochlorite with available free chlorine in a concentration of 5.25% (52,000 ppm).

These hypochlorite bleaches make excellent wide spectrum germicides. They are generally diluted 1 to 10 (5,250 ppm) or 1 to 100 (525 ppm), and their penetrating power and cleaning action can be enhanced by the addition of a 0.1% nonionic detergent.

Chlorine meets many of the standards for an ideal disinfectant except that it is a strong oxidizing agent and is corrosive to metals, and it readily combines with organic compounds, which makes it more difficult to use effectively in very low concentrations. Some manufacturers of chlorine germicides have made their products less corrosive by stabilizing and buffering the solutions at high pHs, and others have produced chloramines by the reaction of HOCl with an amine, amide, imine, or imide. The organic chloramines are the N-chloro derivatives of sulfonamides, heterocyclic compounds with nitrogen in the ring, condensed amines from guanidine derivatives, and anilides. Chloramine-T (sodium p-toluene sulfonchloramide) is a white crystalline powder containing about 25% available chlorine. Chloramine-T was the active ingredient in Dakin's solution, which was used during World War I for irrigating open wounds to prevent infection and to loosen and remove necrotic tissues. Hypochlorites with free chlorine kill microorganisms much more rapidly than do chloramines. However, the chloramines are less corrosive and since they do not release free chlorine they do not form trihalomethanes with organic chemicals. The trihalomethanes are generally believed to be carcinogens.

Iodine

Iodine is the second halogen that is in widespread use as a germicide. Iodine is a highly reactive element, and because it is so reactive it is an excellent germicide. It is germicidal over a wide pH range and is a broad spectrum germicide including bacterial spores (24). Iodine has been shown to be an excellent tuberculicidal agent (25). Iodine is only slightly soluble in water (1 to 3,000), but it is highly soluble in aqueous solutions containing sodium or potassium iodide, and it is soluble in alcohol and glycerine. Iodine is the only halogen that is a solid at room temperature. Aqueous and tinctures of iodine are the most effective iodine germicides, but they are corrosive, they produce stains, and they are somewhat allergenic and toxic. Consequently, iodine has been mixed, complexed, and combined with a wide variety of compounds to produce scores of commercailly available iodine-based disinfectants. Iodine is complexed with polyvinylpyrrolidine to produce compounds that are less toxic, less corrosive, nonallergenic, and stains are easily removed. Iodine is also complexed with a group of nonionic surface active agents to produce the iodophors, which are very popular disinfectants. Studies have shown that an idophon containing 0.75% available iodine will reduce microflora inoculated onto hands and will reduce the natural microflora of hands during a short exposure (15 seconds) hand wash (26). Although iodophor antiseptic preparations are widely used in hospitals and laboratories as disinfectants, they are FDA-approved only for antisepsis of skin and mucous membranes. There have been several reports of iodophor preparations being intrinsically contaminated with *Pseudomonas cepacia* or *Pseudomonas aeruginosa* (27-29).

B. Phenolic Compounds

Phenol itself is rarely used as a disinfectant, however, there are hundreds of derivatives of phenol, many of which are very useful germicidal agents. Among these derivatives are cresols, resorcinol, hexyl resorcinol, o-phenylphenol, o-benzyl-p-chlorophenol, chlorothymol, 8-hydroxyquinoline (antifungal), and the chlorinated phenols such as pentachlorophenol have been widely used as wood preservatives. Phenolic compounds, such as o-phenylphenol, have low volatility and consequently are especially useful as floor disinfectants since they produce a continuing antibacterial potential to the disinfected surface. Ortho-phenylphenol has been shown to be an effective tuberculocidal germicide (30).

o-Phenylphenol

Ortho-phenylphenol has been used as a preservative in products such as cosmetics since it is relatively nontoxic and nonallergenic to the skin at use concentrations. O-phenylphenol, as with other phenolics, is inactivated by nonionic detergents, and this must be kept in mind when using or formulating these germicides.

The bis-phenols contain two phenolic rings joined together by oxygen, sulfur, or most frequently methylene (CH_2). These compounds are bacteriostatic and fungistatic. Chlorine, sulfur, and hydroxyl groups can be added to those compounds in a large variety of ways. Hexachlorophene was widely used in soap and other products, but the use of hexachlorophene is now severely restricted due to neurological problems associated with its use, especially in burn patients and in hospital nurseries.

The salicylanilides contain two phenolic rings joined by a carbamide NHCO. The carbanilides contain a linkage between the nitrogen atom of the analine moiety and the phenyl carbamide. Both the salicylanilides and carbanilides have been used as skin antiseptics and preservatives in cosmetics, toiletries, and pharmaceuticals.

Phenolic germicides are extremely versatile as they can be formulated in a great variety of ways to meet specific needs. Various phenolic compounds can be combined with alcohols, quaternary ammonium
compounds, anionic detergents, natural soaps, sulfur compounds, and
chlorine and iodine. The phenolic compounds are biodegradable by
certain soil bacteria, or they can be decomposed by adapted strains.
The phenolic germicides are not effective sporicidal agents.

C. Quaternary Ammonium Compounds

The quaternary ammonium compounds are cationic surface active agents
in widespread use in the food industries and in dairy sanitization.
The general formula for the quaternary compounds is as follows:

Quaternary Ammonium Compounds

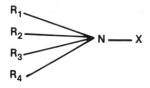

The R1 through R4 are alkyl chains that may be identical or different,
substituted or unsubstituted, saturated or unsaturated, branched
or unbranched, cyclic or acyclic, and they may be aromatic or substituted aromatic groups. The nitrogen atom plus the attached alkyl
groups forms the cation portion, and the anion is usually chloride or
bormide to form the salt. Hugo (31) has made a detailed study of the
mode of action of the quaternary compounds and believes that germicidal action is most likely due to the effect on the cytoplasmic membrane
controlling cell permeability. Due to the high detergency action of
quaternaries, they function as a cleaner as well as a germicide. The
quaternaries are highly bacteriostatic, and this feature led to some
erroneous reports in the 1930s and 1940s. These reports (32) indicated that the quaternaries had a wide spectrum of germicidal activity,
including sporicidal and tuberculocidal activity at extremely low concen-
tractions. Later studies have shown that these first reports were a
result of inadequate neutralization of the quaternary, which was carried over to the subculture medium where it produced stasis rather
than cidal action.

One of the very popular quaternaries is:

Alkyl dimethyl benzyl ammonium chloride

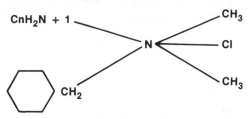

The length of the alkyl chain is important in determining the germicidal potency of the quaternary (33). The C_{14} chain appears to provide the most germicidal action. The quaternary shown above has
the following carbon chains: C_{14} 60%, C_{16} 30%, C_{12} 5%, and C_{18} 5%.
Quaternary compounds can be formulated with phenolic germicides,
with different quaternary compounds or with nonionic detergents for
specific sanitizing needs. Quaternaries are most active in the alkaline
pH range, and they show only a slight loss in germicidal activity in
hard water, even at 800 ppm of $CaCO_3$. Quaternaries are relatively
nonvolatile, have little odor and are stable and active in hot water,
which makes them attractive for use in food processing and dairy sanitization. Also, these compounds are nontoxic when used on intact
skin at 0.1% or less, and they do not act as sensitizers, which enhances
their use as preservatives in cosmetic and pharmaceutical products
(34). The quaternary compounds are nonflammable, noncorrosive to
metals, relatively inexpensive, and possess good detergency action-
properties that make these compounds very useful for surface treatment, especially floor treatment in animal rooms, hospitals, schools,
day care centers, and similar facilities. Quaternary residues are easily
rinsed from surfaces, however, and in some situations, as in animal
rooms, it may be desirable to leave a quaternary film to provide some
continuing bacteriostatic action. Hands may be wetted with a quaternary after scrubbing to provide a bacteriostatic film. Quaternaries
are strongly adsorbed by fabrics, and some hospitals have felt it advantageous to provide a residual bacteriostatic effect to sheets,
blankets, and operating room garments. Diapers have also been treated
with quaternaries to impart a prolonged bacteriostatic action.
    Quaternary compounds are not sporicidal, tuberculocidal, or viru-
cidal against hydrophilia viruses at high concentration levels (35).
At low concentrations, quaternaries are bacteriostatic, tubuculostatic,
sporostatic, fungistatic, and algistatic (36). At standard use concentrations quaternaries are bactericidal, fungicidal, algicidal, and virucidal against lipophilic viruses. These have been reports of quaternary
compounds being associated with disease outbreaks due to gram-nega-

tive bacteria (37) and a report of two *Pseudomonas* species that will grow in benzolkonium chloride solution that contains ammonium acetate (38). The quaternaries are much more active against gram-positive bacteria than against gram-negative.

D. Aldehydes

Formaldehyde

Formalin is a solution of 40 g of formaldehyde gas in 100 ml of water, or 37% by weight. Formalin generally contains 10 to 15% methanol as a preservative. Formalin has long been used to inactivate virus in the preparation of vaccines, since it produces little change in antigenic properties. A 0.2 to 0.4% formaldehyde concentration is usually used with an exposure at this concentration for several days. Twenty percent formalin (8% formaldehyde gas) is sporaicidal, bactericidal, virucidal, tuberculocidal, and fungicidal. A combination of 20% formalin with 70% isopropyl alcohol greatly enhances the germicidal activity of the solution. Formalin is most active in the alkaline range. Formaldehyde is a strong reducing agent and is not corrosive. Therefore, a formaldehyde-isopropyl alcohol mixture is a very useful solution for the disinfection of surgical instruments. Formaldehyde is detectible at about 0.05 ppm, and at 0.5 ppm it produces eye irritation (39). At 0.1-25 ppm, formaldehyde produces upper airway irritation (40), and 5-30 ppm lower airway and pulmonary effects will be noted (41), and at 50-100 ppm pulmonary edema, inflammation, and pneumonia will occur (42). The concentrations causing these effects is quite varied due to great differences in individual susceptibility.

Formalin is a potent germicide in a concentration of 8% in water. Formalin is a high level germicide as its activity is greatly enhanced in a concentration of 8% in 70% isopropanol. It is sporicidal, tuberculocidal, bactericidal, virucidal, and fungicidal. Formalin is noncorrosive to metals, which makes it a good treatment for disinfecting surgical instruments. Formalin is inexpensive, and it is active in the presence of organic materials. Formaldehyde does not penetrate well, and it has a slow germicidal action. Formaldehyde has a pungent, disagreeable odor and at low concentrations is irritating to the eyes. At slightly higher concentrations, formaldehyde produces respiratory irritation and pulmonary distress. There is now sufficient evidence to show that formaldehyde is carcinogenic in experimental animals (43). However, numerous epidemiologic studies do not provide adequate evidence for assessing the carcinogenicity of formaldehyde in man (44). The American Conference of Governmental Industrial Hygenists (ACGIH) 1985-1986, lists a time-weighted average (TWA) for formaldehyde of 1.5 mg/m^3 (1 ppm) and a 15 minute short-term exposure limit (STEL) of 3 mg/m^3 (2 ppm).

Gluteraldehyde

Gluteraldehyde is a saturated dialdehyde related to formaldehyde, and when used in a 22% aqueous, alkaline concentration, it is more active than 8% formaldehyde in 70% isopropanol. Gluteraldehyde is stable when stored in the acid range. The gluteraldehyde is activated by adding sodium bicarbonate to elevate the pH to 7.5 or higher. This alkalinized germicide can be used for about 14 days and then discarded and a fresh solution prepared. Gluteraldehyde is sporicidal, tuberculocidal, and virucidal with contact times of 10 to 30 minutes at room temperature. It is effective in the presence of serum (45) and is noncorrosive to metals. It is the liquid germicide of choice for lensed instruments, inhalation therapy apparatus, surgical instruments, and similar hospital devices. To achieve sterility with these items, a contact time of 6 to 8 hours is recommended. Sensitization to glutaraldehyde occurs much less frequently than sensitization to formaldehyde, and cross-reaction of formaldehyde-sensitive subjects does not seem to occur. Gluteraldehyde is a high-level germicide, but it is quite expensive. It has been approved by the EPA as a sterilant when used as a 2% aqueous alkalinized solution with an exposure time of 6 to 10 hours at room temperature, and it has been recommended by the CDC as an effective cold sterilant for treating respiratory therapy equipment (46). The ACGIH lists a threshold limit value-ceiling (the concentration that should not be exceeded during any part of the working exposure) of 0.7 mg/m^3 (0.2 ppm) for gluteraldehyde.

E. Alcohols

Ethanol

Ethyl alcohol has been used as a disinfectant for almost 100 years. Ethyl alcohol is bactericidal against vegetative bacteria, it is tuberculocidal and virucidal, although it is more virucidal against the viruses with a lipid envelope. It possesses little or no sporicidal action. The antibacterial action of alcohol appears to be due to its denaturing of proteins. Below a concentration of 20% by weight, ethyl alcohol has little germicidal activity. A concentration of 70-80% is the maximum germicidal range (47) for ethyl alcohol. It has long been used as a skin disinfectant before procedures that break the intact skin, such as hypodermic injections, venipunctures, and finger pricks. Ten percent acetone is sometimes added to ethyl alcohol to enhance germicidal action. It is virtually impossible to sterilize the skin, but a significant reduction in skin contamination by transient bacteria, including pathogens, can be accomplished. Alcohols are attractive disinfectants for those working with tissue cell cultures as the alcohol rapidly evaporates and leaves no residue or film that could contaminate their tissue cultures.

Isopropanol

This secondary propyl alcohol is, as expected, the strongest bacteri-
cide of the water-soluble alcohols (48). Although isopropyl alcohol
may be slightly more bactericidal than ethyl alcohol, it has the disadvantage that vapors of isopropyl alcohol are much more toxic than ethyl
alcohol, and these toxic effects are longer lasting (49). Whenever
alcohols are being used as disinfectants, it should be remembered that
alcohols are flammable, and whenever flames, such as a Bunsen burner
or equipment that produce sparks, are present, the alcohol must be
used with great care.

F. Miscellaneous Germicides

Peracetic Acid

Peracetic acid is almost in a class by itself. It is a very powerful,
rapid-acting, universal germicide. Peracetic acid has long been sued
to sterilize gnotobiotic animal-holding chambers and equipment. It
is used in a 2% solution and is received as a 40% concentrate. This
40% solution must be handled with great care as it decomposes with
explosive violence if contaminated with heavy metals or reducing agents
or is heated rapidly. It is flammable and should be stored at refrigeration temperatures, as it slowly decomposes when held at room temperatures. Personnel who handle 40% peracetic acid containers should wear
rubber gloves, a rubber apron, and safety goggles. Once diluted
it degrades rather rapidly, so fresh solutions should be prepared daily.
Peracetic acid is corrosive to metals, and the acetic acid that is released during use is so irritating that respiratory protection will probably be needed.

Hydrogen Peroxide

A commercially available 3% solution of hydrogen peroxide is a stable,
inexpensive, nontoxic, fast-acting bactericidal, virucidal, and fungicidal disinfectant. Concentrations of hydrogen peroxide between 10
to 25% have been shown to be sporicidal (50). The enzyme catalase
degrades hydrogen peroxide to water and oxygen. This effect of catalase is expecially noticed in low concentrations of hydrogen peroxide—
.01 to 1%. Many aerobic and facultative anaerobic bacteria that possess
cytochrome systems produce catalase. *Staphylococcus aureus, Serratia
marceseus*, and *Proteus mirabilis*, which produce catalase, required
exposure times of 30 to 60 minutes to 0.6% hydrogen peroxide to achieve
an eight decimal reduction, whereas *Escherichia coli, Streptococcus
faecalis*, and Pseudomonas species, which do not produce catalase,
experienced an eight decimal reduction in 15 minutes (51). Hydrogen
peroxide should be considered whenever there is a need to disinfect
surfaces such as work benches.

Chlorine Dioxide

Chlorination to produce potable water has been implicated in the production of hazardous trihalomethanes, which are products of halogen reactions with naturally occurring organic substrates. Chlorine dioxide does not produce trihalomethanes. Consequently, the use of chlorine dioxide as a water disinfectant has become widespread. Chlorine dioxide is an excellent germicide. It is equal to and, in some few respects, superior to sodium hypochlorite. However, chlorine dioxide has the considerable disadvantage of having a short half-life in the active state. The disinfectant is activated by adding water, a base, and an activator. Once activated, the solution provides effective sterilizing action for about 24 hours. Chlorine dioxide is considerably more costly than sodium hypochlorite, but chlorine dioxide is an excellent all-purpose germicide.

Mercurial Compounds

Mercurial compounds have been in use longer than any other disinfectant. Mercurial compounds have enormous bacteriostatic action, but in standard use concentrations, these compounds lack adequate cidal action.

IV. SELECTING A DISINFECTANT

In selecting a disinfectant, one must first know what device, surface, or liquid is to be treated. One must decide whether sanitation, disinfection or sterilization is required. Is the treatment directed against a specific bacterial, viral or fungal agent, or is the decontamination to cover all microorganisms that might be present? What concentration of microorganisms might be expected, and what concentrations of proteinaceous material and dirt is likely to be present? Does one wish to have the disinfectant evaporate and disappear like an alcoholic compound, or is it better to have a persistent film remaining as with a phenolic disinfectant or a quaternary ammonium compound? There will be other factors to consider in each specific case in which a disinfecting procedure is to be carried out.

There is often a considerable discrepancy between disinfectant research studies and practical use conditions. Often this difference in results arises because in practical use conditions, the disinfectant does not contact the microorganism. This difference was dramatically brought to our attention when using the disinfectant dunk tank to remove metal containers from a Class III cabinet. Even though a 5,000 ppm hypochlorite solution was present in the tank, containers were exposed for a full five minutes, and the agent in the cabinet was a fragile vegetative microorganism, an occasionally positive sample was recovered from the surface of the container. It was then noted that there were air bubbles on the surface of the container, and it was obvious that the surface under the air bubbles was dry. Specks of grease, rust, or dirt can also protect the microorganism from contact with the disinfectant. It became a standard practice to scrub the surface of the container vigorously with a brush while it was submerged in the disinfectant, and this practice eliminated this problem.

Disinfectants used in the laboratory usually are directed against the microorganism(s) under investigation. Laboratory cultures and preparations usually have titers far in excess of those encountered in nature. Agar, proteinaceous nutrients, and cellular materials can retard or chemically bind the active moieties of chemical disinfectants, which will require longer exposure times and higher concentrations than those shown to be effective in test tube studies. When selecting a disinfectant, one should remember that a disinfectant selected to be used against an organism of a known level of resistance will be effective against organisms lower on the resistance scale. For example, if the disinfectant selected is strongly sporicidal, it will be effective against all other organisms that may be present in the laboratory.

V. MICROORGANISMS OF SPECIAL CONCERN

A. Human Immunodeficiency Virus (HIV)

The first cases of AIDS in the United States were reported in 1981 (52,53). HTLV III/LAV, now known as HIV, is the causative agent of AIDS. Health care workers, clinical laboratory personnel, research personnel, and all workers who handle blood products and/or specimens from AIDS patients are greatly concerned about their risk of exposure and infection. Research workers at the Viral Oncology Unit, Institut Pasteur, carried out some of the first studies on the inactivation of HIV by chemical disinfectants (54). In this study, 25% ethanol or 1% glutaraldehyde was felt to be sufficient to disinfect medical instruments, and 0.2% sodium hypochlorite was recommended for cleaning floors and work benches. A summary of recommendations for preventing infection in the workplace with HIV appeared in the MMWR in November 1985 (55). This CDC document states that several liquid chemical germicides commonly used in laboratories and health care facilities have been shown to kill HIV at concentrations much lower than are normally used. This document also states, "In addition to hospital disinfectants, a freshly prepared solution of sodium hypochlorite (household bleach) is an inexpensive and very effective germicide. Concentrations ranging from 5,000 ppm (a 1:10 dilution of household bleach) to 500 ppm (a 1:100 dilution) sodium hypochlorite are effective depending on the amount of organic material (e.g., blood, mucus, etc.) present on the surface to be cleaned and disinfected."

In November 1986, Tierno published a summary of published studies on disinfectants against HIV (56). These various studies showed that the common disinfectants used in laboratories are effective against HIV when used in the manufacturer's recommended use dilution. Effective disinfectants included sodium hypochlorite, glutaraldehyde, formalin, hydrogen peroxide, ethyl and isopropyl alcohol, Lysol, NP-40 detergent, chlorhexidine gluconate plus 25% ethanol, and a quaternary ammonium chloride. Resnick et al. carried out studies on HIV stability and inactivation by various standard disinfectants. They reported that HIV with an initial infectious titer of approximately 8 $\log_{10}$ tissue culture infectious dose ($TCID_{50}$) per ml can be recovered for more than a week from an aqueous environment at 36-37°C. HIV is reduced at a rate of approximately 1 $\log_{10}$ $TCID_{50}$ per 20 minutes at 54-56°C. Dried HIV at room temperature retains infectivity for more than 3 days. Viral infectivity is undectable and reduced more than 7 $\log_{10}$ $TCID_{50}$ within one minute with 0.5% sodium hypochlorite, 70% alcohol, or 0.5% NP-40, and within 10 minutes with 0.08% quaternary ammonium chloride or with a 1:1 mixture of acetone-alcohol.

The retrovirus HIV is a relatively fragile virus that is readily inactivated by most of the common disinfectants found in the laboratory.

B. *Legionella pneumophilia*

Legionellosis first occurred in epidemic form at the American Legion Convention in Philadelphia in July 1976. *L. pneumophilia* is an ubiquitous microorganism, and in one study (58), *L. pneumophilia* was isolated from 67 different lakes, ponds, rivers, and creeks from five different Southeastern states and one Midwestern state. Although *L. pneumophilia* is widespread in nature, the number of human infections is relatively small. Some 833 legionellosis infections were reported by the CDC to have occurred in 1986 (59). Since legionellosis is frequently a mild disease, it is obvious that many infections are not diagnosed and reported to CDC. Nosocomial legionellosis has been reported from a number of hospitals and is usually associated with the hot water systems. Shower heads, faucet aerators, vacuum breakers, dead-end sections of pipe risers, scale and sediment aggregate areas may harbor *L. pneumophilia*. A study by Veterans Administration Medical Center and University of Pittsburgh scientists (60) showed that clorine, heat, ozone, and UV light are effective in eradicating *L. pneumophilia* from a model plumbing system. The most widely used method for eliminating *L. pneumophilia* from water systems is to raise the water temperature to 60-80°C for several hours and to flush all outlets for at least 30 minutes. This disinfection treatment must be repeated periodically since the system may be recolonized over time. Skaliy et al. (61) showed that chlorine, 2,2-dibromo-3-nitro-lorpopion-amide, and didecyldimethylammonium chloride in 20% isopropanol were very effective in destroying concentrations of 10^5 to 10^6 viable cells per ml when used in standard use concentrations of these disinfectants.

C. Hepatitis B

In a study of laboratory-associated infections, Pike (62) found that brucellosis, with 424 infections and five deaths, was the most frequent laboratory-acquired infection. However, during the past 10 years, numerous studies (63-65) have shown that hepatitis B is a common laboratory-acquired infection and has almost certainly surpassed bru-cellosis as the leading occupational infection of laboratory workers. Hepatitis B tends to occur among those working with blood and its derivatives rather than those working in microbiology. A hepatitis B vaccine has been available since mid-1982, and an improved recombi-nant DNA hepatitis B vaccine will be available for general use by mid-1987. These vaccines should lower the incidence of hepatitis B among laboratory and health care workers.

There are few definitive studies on disinfectants against hepatitis viruses (HBV cannot be cultured in the laboratory). Havens (66) showed that heating at 56°C for 30 minutes did not kill all virus, and Krugman et al. (67) showed that hepatitis B virus in serum (1:10) will be killed in direct transmission studies by exposure to boiling water ($\geqslant 98$°C) for 30 minutes. A Ministry of Health Memorandum published in Lancet (68) stated that a mixture of equal parts of phenol and ether in a concentration of 0.5% is ineffective in killing hepatitis virus. Williams (69) recommended the use of sodium hypochlorite in a concen-tration of 0.5-1.0% for surface disinfection. Neefe et al. (70) carried out a study on the inactivation of infectious hepatitis in drinking water using human volunteers. This study showed that water heavily con-taminated with infectious hepatitis virus and then treated by coagula-tion with aluminum sulfate, treated with activated carbon, and filtered through a diatomaceous silica filter remained infectious for the volun-teers. However, when this treated water was exposed to chlorine for 30 minutes at concentrations of 25 ppm, 7.5 ppm, or 3.25 ppm, none of the volunteers were infected by ingesting this treated water.

A comprehensive review of hepatitis B by Bond et al. from the Hepatitis Laboratories Division of the CDC (71) listed specific decon-tamination strategies and recommendations for sterilization or disinfec-tion of hepatitis B. Since hepatitis B virus cannot be cultured in the laboratory, hepatitis B surface antigen (HBsAg) is used as a model and an indicator of contamination. It is known that the HBsAg is much more resistant to physical and chemical stresses than is the infective particle (HBV). Therefore, a large safety factor is produced when using HBsAg as a model for hepatitis B virus. It is emphasized that it is very important to meticulously clean all surfaces when blood from potential hepatitis B patients may be spilled, such as hemodialysis units

or the clinical laboratory. It was found that Haemosol-moistened cloths
were more effective than water-moistened cloths for cleaning surfaces.
Cleaning before any sterilizing or disinfecting treatment is started is
very important. If possible, materials should be treated by standard
autoclaving, dry heat, or ethylene oxide. In a study on the effect
of sodium hypochlorite solution on HBsAg at concentrations of 5,000,
500, 50, 5, and 0.5 ppm of available chlorine at exposure times of 3,
15, and 30 minutes, at which times sodium thiosulfate was added to
neutralize the chlorine, the CDC investigators obtained the following
results: (1) in this test system 0.5 and 5.0 ppm free chlorine had
no effect on HBsAg in 30 minutes exposure; (2) the 50 ppm concentration gave approximately one-half log reduction in antigen units in 30
min; (3) both 500 and 5,000 ppm available chlorine destroyed HBsAg
reactivity within 3 minutes exposure; (4) undiluted serum plus 5,000
ppm available chlorine inactivated the antigen in slightly more than
20 minutes. Recommended disinfectant and sterilizing procedures
against HBV is summarized in Table 1.

D. Slow Viruses

The virus-induced slow infections are spongiform encephalopathies
that include Creutzfeldt-Jakob disease (CJD), kuru, scrapie, and
transmissible mink encephalopathy. These unconventional viruses
are extremely small and are unusally resistant to heat, germicides,
ethylene oxide, ionizing radiation, and ultraviolet light. All chemical
and physical inactivation studies of the slow viruses have concluded
that extended autoclaving should be used whenever possible (72-74).
A temperature of 121°C with an exposure period of at least 30 minutes
will completely inactivate CJD virus, but some unusually resistant
strains of scrapie (75) may require a 60-90 minute exposure period.
Sodium hypochlorite (household bleach) at a concentration of 25,000
ppm available chlorine in the initial incubation mixture was recommended
by Brown (76) as a potent chemical for inactivation of both CJD virus
and scrapie virus. Also, 1.0 N NaOH with an exposure period of one
hour yielded no detectable virus infectivity in the undiluted inoculum,
dialyzed or not.

VI. SURFACE DECONTAMINATION

A. Work Benches

Laboratories where work with microorganisms or clinical specimens
from animals or human beings is being conducted should decontaminate
work benches at the end of each work day. A quaternary ammonium
compound, Ortho phenylphenol, or 70% ethanol or isopropanol might
be used for disinfecting the work bench if it is known that these

TABLE 1 Recommended Disinfectant and Sterilizing Procedures, Concentrations, and Exposure Times for HBV Contamination

| Items | Disinfection (key no.) | | Sterilization (key no.) |
	Noncritical[a]	Semicritical[b]	Critical[c]
Smooth, hard-surfaced objects	3 (≥30 min) 5-9 (≥10 min)	3 (≥30 min) 5-9 (≥30 min)	1 2 4 5 (12 hr) 6 (10 hr) 7 (10 hr)
Rubber objects	3 (≥30 min)[d] 5-9 (≥30 min)[d]	3 (≥30 min)[d] 5-9 (≥30 min)[d]	1 4
Plastic objects	3 (≥20 min)[d] 5-9 (≥10 min)	3 (≥30 min)[d] 5-9 (≥30 min)	1 4 5 (12 hr) 6 (10 hr) 7 (10 hr)
Lensed instruments		5-9 (≥30 min)	4 5 (12 hr) 7 (10 hr)
Thermometers		5-9 (≥30 min)	4 5 (12 hr) 6 (10 hr) 7 (10 hr)
Hinged instruments	3 (≥30 min) 5-7 (≥10 min) 9 (≥10 min)	3 (≥30 min) 5-7 (≥30 min) 9 (≥30 min)	4 5 (12 hr) 6 (10 hr)
Environmental surfaces* (floors, walls, tables, machine panels, telephones, etc.)	5-9 (≥10 min)		

Key numbers:
Values given are for time at temperature in the most occluded area; see manufacturer's recommendations for compatibility of material with heat.
1. Steam under pressure (autoclaving): 121°C, 15 psig, 10 min; or 134°C, 29 psig, 3 min.

TABLE 1 (continued)

2. Dry air oven: 121°C, 15 hr; 140°C, 3 hr; 160°C, 2 hr; or 170°C, 1 hr.
3. Boiling water (_98°C), 30 min. This temperature for 1 min was shown in a direct transmission study to inactivate HBV in serum diluted 1:10.
   Chemicals may be harmful to some materials, and compatibilities must be established on an individual basis.
4. Ethylene oxide gas, 31/2-16 hr (depending on object). Check manufacturer's recommendations for relative humidity and temperature requirements.
5. Formalin, 20% aqueous (8% formaldehyde, aqueous).
6. Formalin, 20% in 70% alcohol.
7. Alkalinized glutaraldehyde, 2%.
8. Sodium hypochlorite, 500-5,000 ppm (0.05-0.5%)* available chlorine. Commercial household bleach is usually supplied in 5.25% hypochlorite solution.
9. Iodophors, 500-5,000 ppm (0.05%)* available iodine. Commercial products are usually supplied in 1% concentration.

*Concentrations to be used depend upon amount of organic soil present at time of application (71).
[a]Items not usually contacting skin or mucous membranes.
[b]Items frequently contacting skin or mucous membranes.
[c]Items entering skin or mucous membranes.
[d]Depending on thermal stability.
[e]Recognizing the small infective doses and environmental stability of HBV, these surfaces in high-risk environments may easily result in direct or indirect contamination of semicritical or critical items.

disinfectants are effective against the microorganisms being used. Many laboratories have equipment, materials, and supplies far beyond their immediate needs, which makes decontaminating the work benches difficult. Class 3 and 4 agents are worked with only in ventilated enclosures, but the Class 2 agents are intermediate between the 1 and 3 classes. The CDC/NIH Guidelines (77) recommend that work with Class 2 agents be confined to a biological safety cabinet when there is a high potential for creating aerosols or high concentrations or large volumes are used. However, in our laboratories we require that work with moderate risk viruses (78) and Class 2 agents be carried out in closed systems or ventilated cabinets. It is obvious that Class 2 agents such as *Baccillus anthracis* or *Blastomyces dermatitidis* will require the application of high level germicides such as gluteraldehyde 2%, formaldehyde 8%, or sodium hypochlorite 5,000 ppm to obtain effective germicidal action.

B. Floor Treatments

Floors in laboratories, animal rooms, and hospitals are usually heavily
contaminated with the prevailing microflora. Studies over the years
have shown that a disinfectant in a one-bucket cleaning solution is
not important in reducing microbiological contamination. The removal
of soil is the most important aspect of floor contamination (79,80). If
a disinfectant, such as a phenolic disinfectant or a quaternary ammonium
compound, is used for floor treatment, then a two-bucket protocol as
recommended by Vesley and Lauer (81) should be used. The principle
of this protocol is that fresh disinfectant-detergent solution is always
applied to the floor from one bucket, while all spent solution removed
from the floor is wrung from the mop and collected in the second bucket.
When the mop is saturated with the disinfectant-detergent from the
first bucket, it is lightly wrung out into the second bucket and then
applied to the floor. After a 5-minute contact period, the solution
is removed with a wrung-out mop. Other studies (82,83) have shown
that dry treatment with dust suppressant chemicals decreases microbial
colony counts from floors to the same extent as that achieved by chemical germicidal solutions. Dry cleaning of floors is more convenient,
less costly, and is safer as it avoids falls associated with wet surfaces.

C. Water Baths, Sinks, and Floor Drains

A heavy microbial growth will be present whenever water is present,
as in water baths, sinks, and floor drains. Untreated water baths
carry a heavy load of microorganisms that will contaminate the outside
of any vessel placed in the bath. This contamination is of particular
concern if tissue cell culture work is being carried out in the laboratory. It is recommended that a relatively nonvolatile disinfectant such
as a quaternary ammonium compound or a phenolic compound be maintained in the water bath. Sinks, cup drains, and bench drain troughs
are usually moist and therefore will support a luxuriant growth of
microorganisms. Periodically hot water can be run through the sink
or drain trough to achieve a high temperature pasteurization. A more
thorough decontamination can be obtained by filling the sink with water
and adding sodium hypochlorite solution to a level of 1,000 ppm and
allow exposure for 30 minutes before draining. Drains such as cup
sinks and especially floor drains may be seldom used. These drains
must have a liquid added periodically so that the traps do not dry out,
which will allow sewer gases to enter the room. It is recommended
that propylene glycol be poured into seldom used drains to maintain
a trap-seal for a prolonged period of time.

D. Plants and Flowers in Laboratories

Many laboratory workers decorated their window sills and reagent
shelves with a variety of house plants. The damp soil in the flower

pots is teeming with bacteria and fungi. Watering the plants or moving
or working with them will release a significant concentration of microbial aerosol. Another problem with house plants is the frequent presence of mites that feed on vegetation or decaying organic matter in
humus or soil in the flower pot. These anthropods are about .02 mm
in length, which is too small to ordinarily be seen without magnifica-
tion. These arachnids move rapidly and for considerable distances.
They are able to pass through cotton plugs and into Petri dishes carrying bacteria and fungi from culture to culture. Living plants or cut
flowers should not be brought into the laboratory.

E. Biological Safety Cabinets

The laminar-flow biological safety cabinet has about 14 complete air
changes per minute. Since the air flowing downward through the cabinet is virutally sterile, the contamination within the cabinet will be
the result of the worker and materials being handled in the cabinet.
Careful workers will generate little contamination within the cabinet,
whereas careless workers will produce considerable contamination on
the floor of the cabinet and may even splash contaminated liquids on
the walls of the cabinet. At the end of the workday or upon the completion of a project, the interior of the cabinet should be wiped with
a disinfectant. If possible, ethanol should be used as it evaporates
and leaves no residue. However, if more resistant organisms have
been used, then a phenolic compound or an iodophor solution should
be used, followed by ethanol or isopropanol to remove the film. Some
investigators insist on using UV as an aid in disinfecting their laminar-
flow cabinets. Ultraviolet fixtures should not be mounted within the
cabinet in a manner that disrupts the laminar airflow. At the Frederick
Cancer Research Facility, we have designed a counterweight, hinged
fixture that can be swung down to cover the 8" or 10" front opening.
The shiny, stainless steel interior of the cabinet gives good reflectance
of the ultraviolet so that all areas will be bathed with germicidal radiation. However, it should be remembered that 254 nm wavelength emitted
by the germicidal lamp has very limited penetrating power so that dust,
dirt, grease, or clumps of microorganisms may prevent the UV form
irradiating the microorganism. Also, the fact that the germicidal lamp
emits a blue color is no indication as to whether the lamp is emitting
254 nm radiation. A small, hand-held meter should be used to measure
the 254 nm radiation at 6-month intervals. When the radiant energy
drops to 60-70% of rated output for the particular lamp, the lamp should
be replaced. The output of the lamp is greatly affected by dust or
dirt on the surface of the lamp. Therefore, the lamp should be cleaned
with an ethanol-dampened cloth at biweekly intervals. Great care
should be exercised in preventing eye exposure to a germicidal lamp.
The 254 nm radiant energy produces a superficial but extremely pain-

ful conjunctivitis that appears after a several-hour latent period. Symptoms usually disappear after a number of hours, but after a more prolonged exposure, several days may be required for a return to normal.

VII. SKIN ANTISEPSIS

It is not possible to sterilize the skin in either handwashing procedures or in preparing patients for surgical procedures. However, it is important that procedures be used that will remove transient microbes, both pathogenic and saprophytic types, that are not ordinarily part of the resident skin flora. Washing the hands with running water and soap will remove most transient microorganisms, and rinsing with 70% ethanol will further improve the procedure. The use of a soap containing an antiseptic and scrobbing the hands with a surgical brush is a more rigorous method of reducing the transient bacterial load. Tincture of iodine is an excellent disinfectant for treatment of the skin before venipuncture or for skin preparation before surgical procedures. Povidine-iodine and chlorhexidine digluconate (Hibitane[R]) are widely used as antiseptic handwashing agents. Although neither of these antiseptics can be classified as high level germicides, they are very effective as antiseptic handwashing agents. These handwashing agents possess a uniformity of action against different species of vegetative bacteria, they are painless and do not irritate tissues, and they provide a lasting film on the treated skin.

VIII. PRECAUTIONS IN USING MICROBIAL AGENTS

Germicidal agents are used frequently in laboratories to protect personnel and to prevent cross-contamination. The routine use of these microbiocidal materials may produce an attitude of familiarity and inattention that is not commensurate with the true hazards presented by exposure to these materials. Many germicides disrupt the cell wall, and others will interfere with enzymatic systems. Some germicides disrupt nucleic acids and therefore have the potential of being mutagenic, teratogenic, or carcinogenic. Ethylene oxide gas is an alkylating agent that can cuase an increased sister chromatid exchanges or chromosomal aberrations in humans (85). OSHA has established controls on the use of etheylene oxide [IO] that cover all aspects of work and exposure to this germicidal gas. An action level of 0.5 ppm as an 8-hour time-weighted average and a permissible exposure limit over an 8-hour time-weighted average of 1 ppm of ethylene oxide has been established. Some disinfectants, especially the phenolic compounds, can produce skin irritation, depigmentation, and dermatitis.

Formaldehyde gas occasionally produces a hypersensitivity to this gas.
Hexachlorophene is severely limited in its use as it can be neurotoxic,
especially in infants. The fact that microbiocidal agents can produce
health problems such as respiratory irritation, dermititis, and hypersensitivity requires that care be exercised when preparing or using
these microbiocidal agents. Gloves, safety glasses, and a fully fastened
laboratory coat or wrap-around gown should be used when preparing
use dilutions from concentrated materials. When diluting glutaraldehyde, a chemical fume hood or ventilated enclosure should be used
since this compound has a TWA of 0.2 ppm by the ACGIH. Formaldehyde is received in a 37% solution with 15% methanol. Formaldehyde
gas is highly irritating to the eyes, mucous membranes, and respiratory system. It is readily detected when present at 1 ppm. Formaldehyde is carcinogenic in rats after prolonged exposure. Epidemiologic studies have not established formaldehyde as a human carcinogen.
However, it may be reasonably anticipated to be a weak carcinogen in
humans. The TWA for formaldehyde is 1 ppm (ACGIH). A ventilated
hood or enclosure should be used when preparing use dilutions. Gloves,
goggles, and a laboratory coat or surgical gown should be worn when
working with 37% formaldehyde. If working on an open bench, a half-
face respirator with goggles or a full-face respirator should be worn.

Sodium hypochlorite is not toxic at use dilutions, but it is corrosive to metals on prolonged exposure. If pans of items that are covered with hypochlorite solution are to be autoclaved, it is recommended
that the chlorine be neutralized by adding sodium thiosulfate (1 ml
of 5% sodium thiosulfate per ml of 5% hypochlorite ion), which will prevent corrosion within the autoclave.

All workers should remember that alcohols are flammable and when
treating work benches, laminar-flow safety cabinets, or equipment,
there should be no open flames in the area.

IX. SPILL PROTOCOLS

Laboratory personnel should be prepared to deal with the release of
an infectious material, which will occasionally occur when a glass container breaks or some other type of release occurs. When an accident
occurs, the individual involved in the accident should hold his or her
breath and get out of the room. All personnel in the area should be
notified that an accident has occurred. The worker involved in the
spill should remove contaminated clothing and, if available, take a
shower. If the laboratory is under a reduced air pressure, as it should
be, then the laboratory should remain undisturbed for 30 minutes to
allow particles to settle and dilution to take place. If the laboratory
room is not under a reduced air pressure, then clean-up should proceed immediately. If a biological safety officer is available, he or she

should be called and asked to bring a spill kit. The safety officer will don protective clothing including a surgical gown, gloves, shoe covers, head cover, and a full-face respirator. A disinfectant, such as 5% bleach, will be poured around the spill area and gently worked into the center of the spill using a bench brush or paper toweling. About 30 minutes should be allowed for exposure and then materials should be pushed into a dust pan and disposed of into an autoclavable plastic bag. A second bag can be used for the clothing used during the clean-up. The autoclavable bags containing the clean-up materials should be treated in the nearest available autoclave.

X. SUMMARY

The early biological pioneers, such as Paul Koch, Louis Pasteur, Ferdinand Coh, and many others who established the germ theory of disease, also carried out a number of significant studies on germicides. Since the late 1800s, many thousands of studies have been published relating to microbiocidal agents. It is somewhat surprising to note the considerable differences in results obtained by different investigations. However, when one notes the many variables that influence or change final results, the differences in these studies is more understandable. Effective contact between the germicide and the microorganism is essential. Microorganisms may be protected when within an air bubble on the surface of an item submerged in a disinfectant, or a bit of dirt, grease, oil, rust, or a clump of microorganisms may shield the organism. Different strains of a species may exhibit considerable differences in resistivity to a particular germicide, and resistivity may vary with the age of the culture and the components of the medium in which the organism was propagated. Germicidal action only occurs when the germicide interacts with a component of the microorganism that is essential to its structure or metabolism. Germicidal activity is not solely a property of chemical composition but is greatly affected by such factors as the concentration of the microorganisms and of the active ingredient, duration of contact, pH, temperature, the presence of organic matter, and the presence of electrolytes. Some microbiocides are adsorbed to solids. For example, quaternary ammonium compounds will adsorb to talc or kaolin.

There are scores of germicides available under a variety of trade names. Unfortunately, the more active the germicide, the more likely it will possess undesirable characteristics. For instance, peracetic acid is a fast-acting, universal germicide. However, in the concentrated state it is a hazardous compound that can readily decompose with explosive violence. When diluted for use, it has a short half-life, produces strong, pungent, irritating odors, and is extremely corrosive to metals. Nevertheless, it is such an outstanding germi-

cide that it is used in germ-free animals studies despite these unde-
sirable characteristics.

The halogens are high level germicides and are effective over a
wide range of temperatures. The ahlogens have several undesirable
features. They combine readily with protein, so that an excess of
the halogen must be used if proteins are present. Also, the halogens
are somewhat unstable, especially at lower pH levels, so that fresh
solutions must be prepared frequently. Finally, the halogens corrode
metals. A number of manufacturers of disinfectants have treated the
ahlogens to control some of the undesirable features. For example,
sodium hypochlorite reacts with p-toluenesulfonamide to form chloramine
T, and iodine can be complexed with surface-active agents to form
the popular iodophors. These "tamed" halogens are relatively stable,
nontoxic, odorless, and less corrosive to metals. However, these treatments decrease the germicidal effectiveness of the halogens. This
trade-off is acceptable, if the investigator is aware of the difference
in germicidal activity between the treated and untreated halogen and
knows that the treated halogen is adequate for his disinfecting purpose.
At the present time, there are many biochemists working with a variety
of microorganisms in molecular biology, and they are less likely to be
well informed in microbiology and in the use of disinfectants.

If microorganisms are used in a laboratory, then disinfectants will
be needed. If not needed to protect the workers, disinfectants will
still be necessary to protect the research materials from contamination.
Germicides are an integral part of microbiology, and all investigators
must be aware of the effectiveness and the limitations of the germicides
that they plan to use in their work.

REFERENCES

1. Bullock, W. (1938). *The History of Bacteriology*. Oxford University Press, London.
2. Grier, N. (1977). Mercurials—inorganic and inorganic. In *Disinfection, Sterilization and Preservation*, 2nd ed., Seymour Block
   (Ed.). Lea and Febiger, Philadelphia.
3. Semmelweis, I. P. (1847). Höchst wichtige Erfahrungen über
   die Aetiologie der in Gebaranstalten epidemischene Puerperalfieber.
   Ztsch. d.K.-k., gesellsch. d. Aerzte zu Wien, 1847, Jahrg. IV,
   242, 1849 Jahrg. V, Bd.i,64.
4. Lister, J. (1868). Antiseptic system of treatment in surgery.
   Brit. M.J., ii 53, 101, 461, 515.
5. Lister, J. (1869). Observations on ligatures of arteries on the
   antiseptic system. *Lancet i*:51.
6. Appert, N. (1810). L'art de conserver pendant plusieurs annees
   toutes les substances animales et vegetales. Paris.

7. Block, S. S. (1977). Historical Review in *Disinfection, Sterili-
    zation, and Preservation*, 2nd ed. Seymour Block (Ed.). Lez
    and Febiger, Philadelphia.
8. Hogstedt, C., Malmquist, N., and Wadman, B. (1979). Leukemia
    in workers exposed to ethylene oxide. *JAMA 241*:1132-1133.
9. Hogstedt, C., Rohlen, O., Berndstsson, B. S., Axelssen, O.,
    and Ehrenberg, L. (1979). A cohort study of mortality and can-
    cer incidence in ethylene oxide production workers. *Br. J. Ind.
    Med. 36*:276-280.
10. Department of Labor, OSHA, 29 CFR, Part 1910, section 1910.1047.
    Ethylene oxide. *Federal Register 9*(122):25796-25809.
11. Ridenour, G. M., and Armbruster, E. H. (1949). Bactericidal
    effect of chlorine dioxide. *J. Amer. Water Works Assoc. 41*:537-
    550.
12. Noss, C. I., and Oliveiri, V. P. (1985). Disinfecting capabili-
    ties of oxychlorine compounds. *Appl. and Environ. Microbiol.
    50*:1162-1164.
13. Marks, H. C., Wyss, O., and Strandskov, F. B. (1945). Stud-
    ies on the mode of action of compounds containing available chlorine.
    J. Bact. 49:299-305.
14. Kondo, W. (1957). Studies on the mechanism of resistance of
    bacteria. VII. The measure of thermodynamical quantities in the
    disinfection of phenol against *E. coli. Bull. Tokyo Med. Dent.
    Univ. 4*:81-98.
15. Jones, L. A., Hoffman, R. K., and Phillips, C. R. (1968). Spori-
    cidal activity of sodium hypochlorite at subzero temperature. *Appl.
    Microbiol. 16*:787-791.
16. Klein, M., and DeForest, A. (1963). The inactivation of viruses
    by germicides. *Proc. Chem. Spec. Mfg. Assoc.*, 49th Mid Year
    Meeting, pp. 116-118.
17. Pattison, I. H. (1965). Resistance of the scrapie agent to forma-
    lin. *J. Comp. Pathol. 75*:159-164.
18. Hartley, E. G. (1967). Action of disinfectants on experimental
    mouse scrapie. *Nature 213*:1135 (letter).
19. Lavalle, G. C. (1972). Large fraction of scrapie virus resistant
    to lipid solvents. *Proc. Soc. Exp. Biol. Med. 141*:460-462.
20. Brown, P., Rohmer, R. G., and Gajdusek, D. C. (1986). Newer
    data on the inactivation of scrapie virus or Creutzfeldt-Jakob
    disease virus in brain tissue. *J. Infect. Dis. 151*:1145-1148.
21. Baker, R. J. (1959). Types and significance of chlorine resid-
    uals. *J. Am. Water Works Assoc. 51*:1185-1190.
22. Rudolph, A. S., and Levine, M. (1941). Factors affecting the
    germicidal efficiency of hypochlorite solutions. Bull. 150, Eng.
    Exp. Sta., Iowa State College.
23. Knox, W. E., Stumpf, P. K., Green, D. E., and Auerbach, V.
    H. (1948). The inhibition of sulfhydryl enzymes as the basis
    of the bactericidal action of chlorine. *J. Bacteriol. 55*:451-458.

24. Gershenfeld, L., and Witlin, B. (1952). Iodine solution as a sporicidal agent. *J. Am. Pharm. Assoc. 41*:451-452.
25. Gershenfeld, L., Flagg, W., and Witlin, B. (1954). Iodine as a tuberculocidal agent. *Milit. Surg. 114*:172-183.
26. Stiles, M. E., and Sheena, A. Z. (1985). Efficacy of low-concentration iodophors for germicidal hand washing. *J. Hyg. Camb. 94*:269-277.
27. Craven, D. E., Moody, B., Connolly, M. G., Kollisch, N. R., Stottmeier, K. D., McCabe, W. R. (1981). Pseudobacteremia caused by povidone-iodine solution contaminated with *Pseudomonas cepacia. N. Engl. J. Med. 305*:621-623.
28. Berkelman, R. L., Lewin, S., Allen, J. R., Anderson, R. L. et al. (1981). Pseudobacteremia attributed to contamination of povidone-iodine with *Pseudomonas cepacia. Ann. Intern. Med. 95*:32-36.
29. Parrott, P. L., Terry, P. M., Witworth, B., Frawley, L., McGowan, J. E., and Sikes, R. K. (1982). *Pseudomonas aeruginosa* attributed to a contaminated iodophor solution. *Morb. Mort. Weekly Rev. 31*:197-198.
30. Tilley, F. W., McDonald, A. D., and Schaffer, J. M. (1931). Germicidal efficiency of o-phenylphenol against *Mycobacterium tuberculosis. J. Agric. Res. 42*:653-656.
31. Hugo, W. B. (1965). Some aspects of the action of cationic surface-active agents on microbial cells with special reference to their action on enzymes. S.C.I. Monograph No. 19, Surface active agents in microbiology, pp. 67-82. Soc. Chem. Ind. London.
32. Domagk, G. (1935). Eine neue klasse von disinfektionsmitteln. *Dtsch. Med. Wochenschr. 61*:829-832.
33. Cutler, R. A., Cimijotti, E. G., Okolowich, T. J., and Wetterau, W. F. (1966). Alkylbenzyldimethyl-ammonium chlorides–a comparative study of the odd and even chain homologues. C.S.M.A., Proceedings of the 53rd Annual Meeting, pp. 102-113.
34. Finnegan, J. K., and Dienna, J. B. (1954). Toxicity of quaternaries. *Soap Sanit. Chem. 30*:147-153.
35. Smith, C. R., Nishehara, H., Golden, F., Hoyt, A., Guss, C. O., and Kloetzel, M. C. (1950). The bactericidal effect of surface-active agents on tubercle bacilli. U.S. Public Health Rept. No. 48, pp. 1588-1600.
36. Hueck, H. J., Adema, D. M. M., and Wiegmann, J. R. (1966). Bacteriostatic, fungistatic, and algistatic activity of fatty nitrogen compounds. *Appl. Microbiol. 14*(3):308-319.
37. Dixon, R. E., Kaslow, R. A., Mackel, D. C., Fulkerson, C. C., and Mallison, G. F. (1976). Aqueous quaternary ammonium antiseptics and disinfectants. *JAMA 236*:2415-1417.
38. Adair, F. W., Geftic, S. G., and Gelzer, J. (1969). Resistance of *Pseudomonas* to quaternary ammonium compounds. 1. Growth in benzolkonium chloride solution. *Appl. Micro. 18*:299-302.

39. Fairhall, L. T. (1957). *Industrial Toxicology*, 2nd ed. Williams and Wilkins, Baltimore.

40. U.S. Dept. HEW, PHS, CDC, NIOSH. (1976). Criteria for a recommended standard. Occupational exposure to formaldehyde. NIOSH publication No. 77-126. Washington, D.C., U.S. Gov't Printing Office.

41. Hendrick, D. J., and Lane, D. J. (1977). Occupational formalin asthma in hospital staff. *Brit. Med. J. 1*:607-608.

42. Porter, J. A. H. (1975). Acute respiratory distress following formalin inhalation. *Lancet 2*:603-604.

43. International Agency for Research on Cancer (IARC). (1982). Monograph on the evaluation of the carcinogenic risk of chemicals to humans. Supplement 4, Lyon, France.

44. International Agency for Research on Cancer (IARC). (1982). Monograph on the evaluation of the carcinogenic risk of chemicals to humans. Vo. 29. Lyon, France, pp. 345-389.

45. Borick, P. M., Dondershire, R. H., and Chandler, V. L. (1964). Alkalinized glutaraldehyde, a new antimicrobial agent. *J. Pharm. Sci. 53*:1273.

46. Centers for Disease Control. (1973). Methods of prevention and control of nosocomial infections: Recommendations for the decontamination and maintenance of inhalation therapy equipment. National Nosocomial Infections Study, Quarterly Report, Third quarter, 1972. Issued Aug. 1973, pp. 12-17.

47. Price, P. B. (1950). Re-evaluation of ethyl alcohol as a germicide. *Arch. Surg. 60*:492.

48. Tanner, F. W., and Wilson, F. L. (1943). Germicidal action of aliphatic alcohols. *Proc. Soc. Exp. Biol. Med. 52*:138-140.

49. Wise, J. R. (1969). Alcohol sponge baths. *N. Eng. J. Med. 280*:840.

50. Swartling, P., and Lindgreen, B. (1968). The sterilizing effect against *Bacillus sublitis* spores of hydrogen peroxide at different temperatures and concentrations. *J. Dairy Res. 35*:423-428.

51. Schaeffer, A. J., Jones, J. M., and Amundsen, S. K. (1980). Bactericidal effect of hydrogen peroxide on urinary tract pathogens. *Appl. and Envir. Microbiol. 40*:337-340.

52. Friedman-Kien, A., Laubenstein, L., Marmor, M., et al. (1981). Kaposi's sarcoma and *Pneumocystis pneumonia* among homosexual men—New York City and California. *MMWR 30*:305-308.

53. Gottlieb, M. S., Schranker, H. M., Fan, P. T., et al. (1981). *Pneumocystis pneumonia*—Los Angeles. *MMWR 30*:250-253.

54. Spire, B., Montagnier, L., Barre-Sinoussi, F., and Chermann, J. C. (1984). Infactivation of lymphadenopathy associated virus by chemical disinfectants. *Lancet 2*:899-901.

55. Anonymous. (1985). Summary: Recommendations for preventing transmission of infection with human T-lymphotropic virus Type

III/lymphadenopathy-associated virus in the workplace. *MMWR* 34:681-695.

56. Tierno, P. M., Jr. (1986). Preventing acquisition of human immunodeficiency virus in the laboratory: safe handling of AIDS specimens. *Lab. Med.* 17:696-698.

57. Resnick, L., Veren, K., Salahuddin, Z. S., Tondreau, S., and Markham, P. D. (1986). Stability and inactivation of HTLV-III/LAV under clinical and laboratory environments. *JAMA 255:* 1887-1891.

58. Fliermans, C. B., Cherry, W. B., Orrison, L. H., Smith, S. J., Tison, D. L., and Hope, D. H. (1981). Ecological Distribution of *Legionella pneumophilia. Appl. Environ. Microbiol. 41:* 9-16.

59. MMWR. (1987). Summary–cases specified notifiable diseases, United States. *MMWR 35(53):810.*

60. Muraca, P., Stout, J. E., and Yu, V. L. (1987). Comparative assessment of chlorine, heat, ozone, and UV light for killing *Legionella pneumophilia* within a model plumbing system. *Appl. Environ. Microbiol.* 53:447-453.

61. Skaliy, P., Thompson, T. A., Gorman, G. W., Morris, G. K., McEachern, H. V., and Mackel, D. C. (1980). Laboratory studies of disinfectants against *Legionella pneumophilia. Appl. Environ. Microbiol. 40:*697-700.

62. Pike, R. M. (1976). Laboratory-associated infections: Summary and analysis of 3,921 cases. *Hlth. Lab. Sci. 13:*105-114.

63. Lauer, J. L., Van Drunen, N. A., Washburn, J. W., and Balfour, H. H. (1979). Transmission of hepatitis B virus in clinical laboratory areas. *J. Infect. Dis. 140:*513-516.

64. Grist, N. R. (1976). Hepatitis in clinical laboratories. *J. Clin. Path. 29:*480-483.

65. Polakoff, S. (1986). Acute viral hepatitis B: Laboratory reports 1980-1984. *Br. Med. J. 293:*37-38.

66. Havens, W. P., Jr. (1945). Properties of the etiologic agent of infectious hepatitis. *Proc. Soc. Exper. Biol. Med. 58:*203.

67. Krugman, S., Giles, J. P., and Hammond, J. (1971). Viral hepatitis virus: Effect of heat on the infectivity and antigenicity of the MS-1 and MS-2 strains. *J. Infect. Dis. 122:*432-436.

68. Ministry of Health Memorandum prepared by Medical Officers. (1943). Homologous serum jaundice. *Lancet 1:*83.

69. Williams, S. V., Huff, J. C., Feinglass, E. J., Gregg, M. G., Hatch, M. H., and Matson, J. M. (1974). Type B viral hepatitis in hospital personnel. *Am. J. Med.* 57:904-911.

70. Neefe, J. R., Baty, J. B., Reinhold, J. G., and Stokes, J. (1947). Inactivation of the virus of infectious hepatitis in drinking water. *Am. J. Pub. Health* 37:365-372.

71. Bond, W. W., Petersen, N. J., and Favero, M. S. (1977). Viral

hepatitis B : Aspects of environmental control. *Health Lab. Sci.*
14:235-252.

72. Brown, P., Rohwer, R. G., Green, E. M., and Gajdusek, D.
    C. (1982). *J. Infect. Dis.* 145:683-687.

73. Gajdusek, D. C., Gibbs, C. J., Asher, F. M., Brown, P., Diwan,
    A., Hoffman, P., Nemo, G., Rohwer, R., and White, L. (1977).
    N. Eng. J. Med. 297:1253-1258.

74. Jarvis, W. R. (1982). Precautions for Creutzfeldt-Jacob disease.
    Infect. Control. 3:238-239.

75. Walker, A. S., Inderlied, C. G., and Kingsbury, D. T. (1983).
    Conditions for the chemical and physical inactivation of the K.
    Fu. strain of the agent of Creutzfeldt-Jakob disease. *AJPH 73*:
    661-665.

76. Brown, P. (1986). Newer data on the inactivation of scrapie
    virus or Creutzfeldt-Jakob disease virus in brain tissue. *J.
    Infect. Dis. in Concise Communications Section 153*:1145-1148.

77. CDC/NIH Publication No. (CDC)84-8395, 1st Edition (1984).
    Biosafety in microbiological and biomedical laboratories. U.S.
    Gov't Printing Office 0-487-328.

78. National Cancer Institute Safety Standards for Research Involving
    Oncogenic Viruses. 1974. HHS publication No. (NIH) 75-790,
    Office of Research Safety, NCI.

79. Ayliffe, G. A. J., Collings, B. J., and Lowbury, E. J. L. (1966).
    Cleaning and disinfection of hospital floors. *Br. Med. J.* 2:442.

80. Finegold, S. M., Sweeney, E. E., Gaylor, D. W., et al. (1962).
    Hospital floor decontamination: Controlled blind studies in evalu-
    ation of germicides. In *Proceedings of the Second Interscience
    Conference on Antimicrobial Agents and Chemotherapy*, Sylvester,
    J. C. (Ed.). ASM, pp. 250-258.

81. Vesley, D., and Lauer, J. (1986). In *Laboratory Safety*: *Prin-
    ciples and Practices*. Miller, B. (Ed.). ASM, pp. 182-198.

82. Schmidt, E. A., Coleman, D. L., and Mallison, G. F. (1984).
    Improved system for floor cleaning in health care facilities. *Appl.
    Environ. Microbiol.* 47:942-946.

83. Vesley, D., Klapes, N. A., Benzow, K., and Le, C. T. (1987).
    Microbiological evaluation of wet and dry floor sanitization systems
    in hospital patient rooms. *Appl. Environ. Microbiol.* 53:1042-1045.

84. Jong, S. C. (1987). Prevention and control of mite infestations
    in fungus cultures. *ATCC Newsletter* 7:1, 7.

85. Landrigan, P. J., Meinhardt, T. J., Gordon, J., Lipscomb, J.
    A., Burg, J. R., Mazzuckelli, L. F., Lewis, T. R., and Lemen,
    R. A. (1984). Ethylene oxide: an overview of toxicologic and
    epidemiologic research. *A.J. Ind. Med.* 6:103-115.

12

Infectious Waste Management

JAMES L. BOYLAND   *Biotech Services, Inc., Clayton, Missouri*

JUDITH GORDON   *Gordon Resources Consultants, Inc., Reston, Virginia*

ROBERT A. SPURGIN   *BFI Medical Waste Systems, Santa Ana, California*

DANIEL F. LIBERMAN   *Massachusetts Institute of Technology, Cambridge, Massachusetts*

I. INTRODUCTION

The purpose of this chapter is to provide information for institutions to use in developing logical and environmentally sound infectious waste management programs. The Resource Conservation and Recovery Act of 1976 (RCRA) granted the United States Environmental Protection Agency (EPA) the authority to define and regulate hazardous waste (1). In RCRA, five characteristics of hazardous waste (toxicity, corrosivity, reactivity, ignitability, and infectiousness) were described.

   The first EPA proposal for hazardous waste regulations was published December 18, 1978 in the *Federal Register* (2). This proposal included definitions and treatment/disposal requirements for infectious waste (2). After a series of public hearings and review of the public hearings and review of the public comments, the EPA published Phase I of the hazardous waste regulations in the *Federal Register* on May 19, 1980 (3). Phase I included certain final hazardous waste regulations as well as new proposals or modifications of other regulations. Infectious waste regulations were deferred to Phase II of the hazardous waste regulatory development program (3). Subsequently, the EPA decided not to continue with the development of infectious waste regulations.

   In response to requests for guidance, the EPA issued the *Draft Manual for Infectious Waste Management* in September of 1982 (4). The reasons for issuing this draft were to share the accumulated information, to provide comprehensive guidance for infectious waste management, to provide a source of technical information to state or municipal agencies for regulatory purposes, and to provide a potential focal point for exchange of information on infectious waste management and treatment technology. Continued requests for updated information prompted the EPA to modify the Draft Manual; it was reissued in 1986 as the *EPA Guide for Infectious Waste Management (5)*.

II. DEFINITION OF INFECTIOUS WASTE

There is no universally accepted definition for infectious waste. Regulatory agenices, hospitals, and academic and industrial laboratories have different perspectives, which reflect their specific objectives. This has led to an inconsistency in the terminology used to define these wastes. For example, the terms pathological, biomedical, biohazardous, medical, toxic, viable, and biocontaminated have all been used to describe this class of waste.

For the purposes of this chapter, infectious waste will be defined as waste that is capable of causing an infectious disease. This definition is based on the factors that are necessary for induction of disease. These factors include:

Presence of a pathogen of sufficient virulence
Dose
Portal of entry
Resistance of host

In order for a waste to be infectious and a hazard, it must have pathogens or biologically active material present in sufficient concentration or quantity so that exposure of a susceptible host to the waste could result in disease. In addition, a portal of entry must be available for exposure to occur.

A. Designation of Infectious Wastes

On the basis of this definition of infectious waste, it is advisable to designate as infectious those wastes that probably contain pathogenic agents that--because of their virulence, concentration, and quantity-- may cause disease in those exposed to the wastes.

Wastes should certainly not be tested for the presence of pathogens as a means of identifying waste that is infectious. Because many microorganisms grow only under very specific conditions and some pathogens (e.g., those causing hepatitis) cannot be cultured, negative cultures do not constitute evidence that no pathogens are present. There is no justification for the risk of handling infectious waste in order to select a representative sample that can be cultured, nor for the expense of providing all possible culture conditions for the waste sample.

The following types of waste may be considered infectious waste and managed accordingly:

Isolation wastes*
Cultures and stocks of infectious agents*
Human blood and blood products*
Pathological wastes*
Contaminated wastes from surgery and autopsy
Contaminated laboratory wastes
Contaminated sharps*
Dialysis unit wastes
Contaminated animal carcasses, body parts, and bedding*
Discarded biologicals*
Contaminated food products and other products
Contaminated equipment

The asterisks denote those categories of waste that, according to the most recent EPA recommendations, should be designated as infectious waste (5). Because of unknown risks in wastes, differences in the definition of infectious waste in the various state laws, and considerations of prudent practice, best judgment should be exercised in classifying each type of waste as infectious or noninfectious.

Clearly these types of waste are not necessarily always infectious. While some pathological wastes, sharps, discarded biologicals, surgical wastes, etc., may not always be contaminated with infectious agents, they should always be handled in accordance with management practices that minimize any hazard. Any special or unique problems associated with the particular waste (e.g., limbs, organs, vaccines, sharps, allergy diagnostic material) should also be addressed.

B. Types of Infectious Waste

Isolation Wastes

Isolation wastes are all those generated in the care of and by hospitalized patients who are isolated to protect others from their communicable diseases. Wastes from hospital patients in protective isolation (i.e., isolation that protects patients from the diseases of others) are not infectious unless such a patient is also isolated because of illness with a communicable disease. EPA has recommended that isolation wastes be managed in accordance with the guidelines of the Centers for Disease Control (CDC) (6).

Cultures and Stocks of Infectious Agents

All discarded cultures and stocks of infectious agents should be handled as infectious waste because this material usually contains large quantities or concentrated growths of microorganisms. This type of infectious waste is often generated in medical, pathological, microbiological, and research laboratories. It includes cultured specimens from patients, stocks maintained for research, and wastes from the production of certain pharmaceuticals.

Human Blood and Blood Products

All human blood and blood products are potentially infectious because of the possible presence of pathogens of bloodborne disease, disease which may or may not have been diagnosed. The principal diseases of concern today are hepatitis B and acquired immunity deficiency syndrome (AIDS), although other diseases are also transmissible through bloodborne pathogens (e.g., malaria, congenital rubella, and dengue). Because all human blood and blood products are not tested routinely for every possible bloodborne pathogen, all of these materials--when discarded--should be managed as infectious waste. Generators of this type of infectious waste include medical laboratories, blood banks, dialysis centers, and pharmaceutical companies.

Pathological Wastes

Pathological wastes are the body tissues, organs, and body parts that are removed during biopsy, surgery, and autopsy. All such waste has the potential hazard of causing disease because of the possible presence of pathogens for diagnosed or undiagnosed disease. Therefore, it is prudent to manage all pathological wastes uniformly as infectious waste.

Contaminated Wastes from Surgery and Autopsy

Waste generated during the surgery and autopsy of septic ("dirty") cases or of patients with infectious diseases may be contaminated with pathogens. Therefore, these wastes should be managed as infectious waste.

Because of possible difficulties in identification of septic cases and the potential for unidentified or unrecognized infections, some hospitals manage as infectious all contaminated wastes from surgery and autopsy. It is not essential to adopt such an approach, but this approach should also be considered during evaluation of the risks and benefits of the various alternatives for management of contaminated wastes from surgery and autopsy.

Wastes in this category include suction canisters, tubing, sponges, lavage tubes, drainage sets, underpads, surgical gloves, and soiled dressings and drapes.

Contaminated Laboratory Wastes

The category of contaminated laboratory waste includes all wastes that were in contact with pathogens during laboratory work and that are not included in another infectious waste category such as cultures or sharps. These wastes are generated in all types of laboratories including medical, pathological, microbiological, pharmaceutical, and other commercial, research, and industrial laboratories. Wastes that are generated in medical and pathological laboratories during the culturing of patient specimens are hazardous because antibiotic-resistant organisms are often present in these cultures. Contaminated wastes generated during the culturing of pathogens in research, commercial, and industrial laboratories are usually infectious because they were in contact with pure cultures containing high concentrations of microorganisms.

Research and industrial applications of various biotechnologies, especially recombinant DNA, generate classes of waste which must be managed properly. At this time there is divergence of opinion among experts in the field about the extent and degree of the potential hazard posed by wastes associated with biotechnology. In the interests of safety, such wastes from research as well as from commercial/production areas should be managed as infectious waste.

Wastes in this category of infectious waste include culture dishes; devices used to transfer, inoculate, and mix cultures; and paper and cloth items that were in contact with specimens or cultures.

Contaminated Sharps

All discarded sharps that were used in patient care should be classified and managed as infectious waste because of the possibility of undiagnosed bloodborne disease (e.g., hepatitis B and AIDS). In addition, all sharps that were contaminated by exposure to pathogens (e.g., in the laboratory) should be managed as infectious waste.

The dual hazard of contaminated sharps must be recognized. This category of waste has the potential for infliction of injury and induction of disease. The hazards of disease and injury from contaminated sharps can be minimized by the establishment of good management practices.

Waste sharps that are not contaminated still pose the hazard of injury. From considerations of safety, simplicity, and efficiency, it is best to manage uniformly all waste sharps--contaminated as well as uncontaminated--in accordance with the practises established for contaminated sharps.

The category of waste sharps includes discarded hypodermic needles, syringes, Pasteur pipettes, broken glass, and scalpel blades.

Dialysis Unit Wastes

Dialysis unit wastes are those wastes that were in contact with the blood of patients undergoing hemodialysis. Because patients undergoing hemodialysis receive frequent blood transfusions and have a high incidence of hepatitis, each such patient should be regarded as a potential hepatitis case even though previous tests for the disease might have been negative. Therefore, in the interests of prudent management, all such wastes should be classified and managed as infectious. Wastes in this category include disposable dialysis tubing and filters, sheets, towels, gloves, aprons, and lab coats. Sharps from dialysis units should be managed in accordance with the practices established for the handling of all sharps.

Contaminated Animal Carcasses, Body Parts, and Bedding

Contaminated carcasses, body parts, and the bedding of animals that were intentionally exposed to pathogens should be classified and managed as infectious waste. Wastes of this type are generated--for example--during research, the production of biologicals, and the in vivo testing of pharmaceuticals.

Discarded Biologicals

This category consists of waste biologicals (e.g., vaccines) that were produced by pharmaceutical companies for human or veterinary use. These products are being discarded because a manufacturing lot failed quality control, a lot was recalled, the material is outdated, or the product is being removed from the market. These discarded materials should be managed as infectious waste because pathogens may be present.

Manufacturing wastes that are generated during the production of biologicals are classified as other types of infectious waste. These categories include stocks and cultures of infectious agents, sharps, contaminated laboratory wastes, and animal carcasses.

Contaminated Food and Other Products

Food and other products that are being discarded because of contamination with infectious agents should be classified and managed as infectious waste. Wastes in this category include foods, food additives, cosmetics, and drugs that are not safe for consumption or use because they are contaminated. Canned food that is recalled because of the known or suspected presence of *Clostridium botulinum* toxin should be managed as infectious waste; management procedures should be designed to prevent exposure to the toxin, dispersal of the toxin in the environment, and access to the contaminated food.

Contaminated Equipment

Contaminated equipment and equipment parts that are to be discarded should be classified as infectious waste. This category of infectious waste includes equipment that became contaminated through use in patient care, in medical and microbiological laboratories, in research with infectious agents, and in the production and testing of various pharmaceuticals.

III. GENERATORS OF INFECTIOUS WASTE

While various industrial, institutional, and research facilities are sources of infectious waste, the largest generators of infectious waste are:

Health care facilities
Academic and industrial research laboratories
The pharmaceutical industry
Veterinary facilities
The food, drug, and cosmetic industries

These industries are subject to various regulations, guidelines, and standards that pertain to waste management. *Health care* facilities generate infectious waste in caring for individuals who are ill from or carriers of pathogens. Another factor is that hospitals are reservoirs of antibiotic-resistant strains of bacteria. The Department of Health and Human Services has promulgated regulations that apply only to hospitals and medical facilities undergoing construction or renovation if the facilities receive federal funds. These regulatory requirements specify that federal hospitals, long-term care facilities, and outpatient surgical facilities are to have an incinerator or access to one for the destruction of pathological and infectious waste.

Although there are no other federal regulations for the disposal of infectious wastes from medical care facilities, it is important to note that some states do have regulatory requirements for infectious waste disposal. Generators of infectious waste in these states must comply with these regulations.

Three guidelines have been published by federal agencies that contain recommendations on the disposal of infectious waste from medical care facilities. These are the *EPA Guide for Infectious Waste Management* (5), the *CDC Guideline for Handwashing and Hospital Environmental Control, 1985* (7), and the *CDC Guideline for Isolation Precautions in Hospitals* (6).

Hospitals that seek accreditation must meet the standards of the

Joint Commission on Accreditation of Hospitals (JCAH)--the organization that sets standards for, inspects, and accredits hospitals. The
JCAH standards are published annually in the *Accreditation Manual
for Hospitals* (8). Among these standards is Standard #PL.6: Hazardous Materials; it requires "a system to safely manage hazardous materials and wastes"--with infectious wastes specified as one type of hazardous waste to which the standard applies.

   Academic and industrial research laboratories that work with pathogens or that use specific biotechnologies are generators of infectious
wastes. Various guidelines published by federal agencies include
recommendations for the management of infectious waste from research
laboratories. Biosafety guidelines for microbiological and biomedical
laboratories were prepared by the Centers for Disease Control and
the National Institutes of Health (9). The *Guidelines for Research
Involving Recombinant DNA Technology* were developed by the National
Institutes of Health (10); they are mandatory only for institutions that
receive NIH funding for research projects that involve recombinant
DNA molecules. Some state and local jurisdictions have converted these
guidelines into mandatory requirements that apply to all recombinant
DNA research/production activities, academic as well as industrial.
EPA has recommended that in all work with recombinant DNA molecules,
i.e., research as well as commercial activities, the NIH guidelines for
waste treatment included in the *Laboratory Safety Monograph* (11)
should be followed.

   The pharmaceutical industry generates infectious wastes during
the production of pharmaceuticals such as biologicals and antibiotics.
The testing of these and other products such as antitumor drugs for
efficacy and safety by means of in vitro and in vivo studies results
in infectious waste generation. Regulations of the United States Food
and Drug Administration that pertain to the manufacture of pharmaceuticals and of blood and blood components specify only that manufacturing wastes should be disposed of in a safe and sanitary manner.

   Veterinary facilities also constitute a source of infectious waste.
Regulations of the United States Department of Agriculture pertain
to the treatment and disposal of diseased agricultural animals and of
contaminated animals in the food supply. However, there are no regulatory requirements for the disposal of other diseased animals such
as those from research laboratories.

   The food, drug, and cosmetic industries occasionally generate
infectious waste. Examples of such infectious waste are contaminated
lots of food additives, food products, drugs, and cosmetics that must
be discarded because they cannot be used. These products are under
the jurisdiction of the Food and Drug Administration; this agency has
no regulatory requirements for the disposal of contaminated products.

IV. INFECTIOUS WASTE MANAGEMENT

A. Introduction

Each institution should have a comprehensive waste management plan that is developed under the sponsorship of the administration. This plan should address the infectious, radioactive, chemical, and multiply hazardous (e.g., infectious and radioactive, infectious and toxic, infectious and radioactive, and carcinogenic) wastes as well as the general solid waste. As part of the institutional plan, each laboratory or department within the institution should have detailed, written instructions for the management of the kinds of waste that it generates. The EPA has recommended that each facility establish an infectious waste management plan that will ensure proper handling of infectious waste from the point of discard, through treatment of the waste to render it noninfectious, to final disposal of the treated waste (4,5).

From the management aspect, the infectious waste management plan should designate one person who is to have overall responsibility for infectious waste management at the facility. The plan should specify those persons (laboratory supervisors, operating room supervisors, principal investigators, etc.) who are to be responsible for implementation of the plan. Each of these individuals should have the responsibility as well as the authority to obtain compliance with the provisions of the management plan. The plan should also delegate to specific persons the responsibility for the management of wastes with mixed hazards through evaluation of the different hazards, determination of the relative degrees of risk, and establishment of prioities based on the greatest hazard.

Every facility that generates infectious waste must train its employees in the proper procedures for infectious waste management. Such education is essential for *all* employees who generate or handle infectious wastes regardless of the employee's role--whether technical or custodial, whether as supervisor or supervised. Training is essential for new employees, but it is also necessary for all infectious waste handlers whenever management practices are changed. Refresher courses are useful and important for all personnel; they serve as reminders of established procedures, increase the level of awareness, and help in promoting the infectious waste management plan.

The infectious waste management plan must be clear and unambiguous. It should specify the selected management options and procedures for the facility as a whole and for the individual units that generate infectious waste. The plan should detail the practices and procedures that are to be used with the different types of infectious waste. Protection from the hazards of infectious waste is possible only through careful development of a management plan followed by strict adherence to all the elements of the plan.

B. Selection of Management Options

Management options should be evaluated on the basis of a number of factors including the types and quantities of infectious waste generated, the types and applicability of equipment available for treatment on-site, the availability of off-site treatment alternatives, physical constraints, regulatory and institutional constraints, and cost considerations.

During the evaluation of the various treatment methods, a facility should take into consideration the appropriateness of each method for treating the types of infectious waste that it generates. The selection of management options should be based on the major component(s) of the infectious waste stream, not on the minor constituents. The degree of hazard of each type of infectious waste should also be a factor in the evaluation process. If one treatment option is not suitable for all the types of infectious wastes that are generated, then one principal treatment technique should be selected with other methods specified as necessary for particular types of waste. At many facilities, a combination of treatment techniques or management options is best. For example, laboratory cultures are often steam sterilized while pathological and other infectious wastes are incinerated, either at the facility or at an off-site location.

The next step in selection of management options should be consideration of the types of treatment equipment available to meet particular needs. A variety of techniques and devices can be used for treatment of infectious wastes. Selection of treatment equipment should be based on consideration of various factors, including technical capability to treat the waste effectively, applicability to the types of infectious waste being generated, versatility (i.e., suitability for treating more than one type of waste and for uses other than waste treatment), capacity relative to the quantities of infectious waste generated, reliability, and the ease of operation.

Another important factor that affects the selection of waste management options is the availability of treatment equipment (both on-site and off-site). All the treatment equipment does not have to be situated on-site at the generating facility. All or some of the infectious wastes could be transported to an off-site treatment location in accordance with the regulatory requirements of the particular state.

There are various forms of the off-site treatment option. The treatment equipment could be situated at another institution or at a special treatment facility. Some treatment facilities are owned and operated by the infectious waste generators whom they serve, whereas others are independent, commercial operations. The off-site treatment option can be especially advantageous to generators of small volumes of infectious waste because it provides an alternative to capital expenditures for treatment equipment that would probably never be fully utilized.

Various regulations at the federal, state, and local levels are relevant to infectious waste management, and regulatory constraints must be assessed when management options are being evaluated. Some states and localities have infectious waste regulations. Other regulations do not pertain specifically to infectious waste, but they are relevant to the selection of infectious waste treatment methods and equipment. Among these are the regulations governing air quality and water quality.

There is no federal regulation of infectious waste incinerators, but some state regulations do govern the use and operation of these incinerators. There are also other air quality regulations that could affect the use of incineration as a method of treating infectious waste. In the last decade, many incinerators of various types have been closed down. This action was due primarily to problems in meeting air quality regulations, especially the federal standard for particulate emissions. Although there are no regulations pertaining to emissions from infectious or pathological waste incinerators, the operation of these incinerators has been restricted in some localities because of air quality considerations. Energy conservation policy has also reduced the use of some infectious waste incinerators because of the amount of fuel required to maintain proper incineration temperatures. It should be noted that compliance with the general air quality regulations does not necessarily indicate proper incineration procedures. The criterion for infectious waste incineration should be that no viable microorganisms are recoverable from the stack gas or the ash. This is best ascertained through the use of bacterial spores during incinerator testing.

Water quality regulations apply to other techniques for treating infectious waste. For example, heat sterilization systems must meet the regulatory requirement governing thermal discharges. Chemical treatment and wet grinding systems must be compatible with the requirements of the local wastewater treatment plant regarding wastewater constituents.

Hazardous waste regulations may also apply to residues from infectious waste treatment. For example, incinerator ash may be regulated as a hazardous waste. Chemical treatment could produce a hazardous waste because of altered pH or the presence of residual chemical in the waste.

A final factor in the selection of management options is prevailing community attitudes. These are expressed officially as local laws, ordinances, and zoning restrictions and unofficially as public opinion that can affect the official legal positions. Community attitudes are especially relevant in siting issues such as the siting of incinerators (on-site as well as off-site) and of off-site treatment facilities. It is important to evaluate the different site alternatives and to consider the relevant factors such as the nature of the site itself, the type of hauling and treatment equipment that will be used, and the potential

impacts on the community from odors and emissions, dust, noise, and
the increased volume of traffic.

V. ELEMENTS OF INFECTIOUS WASTE MANAGEMENT

A. Designation of Infectious Wastes

Infectious waste management begins with the designation of the specific
types of infectious waste that are generated at the facility. Once this
designation has been made, those persons responsible for infectious
waste management can know which types of infectious waste are being
generated and where within the facility. They can then use this information to formulate effective and efficient waste management policy
and procedures.

B. Segregation of the Infectious Waste Stream

In order to ensure proper treatment of the infectious waste and to
prevent expenditures for unnecessary special handling of the general
solid waste, infectious waste should not be combined with solid waste
or other hazardous waste. Infectious wastes should be segregated
from other waste streams at the point of waste generation, that is the
point of discard. In addition, types of infectious waste will have to
be segregated into two or more groups if more than one treatment technology is to be used at the facility to ensure that each type of infectious waste will receive the proper treatment. Similarly, waste that
is sent off-site for treatment should be segregated and kept separate
from waste that will be treated within the facility.

C. Packaging and Labeling

Containers

The principal criterion for selection of infectious waste containers
should be ensuring containment of the waste so as to protect the waste
handler, other personnel, patients, and the public from injury and
disease that could be caused by contact with the waste. Infectious
waste should be discarded directly into appropriate containers. The
factors to be considered in selection of containers are the type of waste;
the methods of waste collection, handling, and movement; storage conditions; and the treatment that will be used.

Types of Infectious Waste Containers

    1. Plastic bags. Plastic bags should be used as the principal
container only for nonsharp, notwet infectious wastes. Plastic bags
can also serve as the liners for secondary containers. All plastic bags

should be tear-resistant and leakproof. They should be able to contain the waste during its handling, movement, and treatment, and be strong enough to contain the waste without breaking and spilling.

When the waste is to be treated by steam sterilization, the steam must be able to penetrate the bags and the waste. The plastic bags should be autoclavable (i.e., penetrable by steam and remaining intact throughout the treatment process), or they should disintegrate upon exposure to steam (an outer container therefore being necessary to contain the waste).

Plastic bags made from polyvinyl chloride (PVC) plastics should be used only when this material is required to contain, or protect from, a particular hazard. PVC plastic bags should not be used when the waste will be incinerated because chlorine gas will be released during incineration. Chlorine is harmful in various ways: it corrodes the incinerator, it can react with water vapor to form hydrochloric acid, and it has been implicated in free radical formation.

Plastic bags used for infectious waste should be easily identifiable. Color coding is an acceptable practice, and red or orange are universally recognized as the colors for containers of biohazardous material.

2. Other materials. Metal, rigid plastic, corrugated carton, fiberboard, and glass are generally suitable materials for infectious waste containers. Although they offer the advantages of greater rigidity and strength than plastic bags, they have disadvantages as well. For example, glass is breakable, and, therefore, it should be used only when it is protected from breakage and when any spillage can be contained within a leakproof outer container. Corrugated carton and fiberboard containers are not leakproof, and they must be used only with plastic bag liners.

During selection of containers made of these materials, consideration must also be given to the type of treatment that will be used. If the waste is to be steam sterilized, steam must be able to penetrate the container and the waste, but this may not be possible with some of these materials. If the waste is to be incinerated, excessive use of glass and metal containers should be avoided because these materials can interfere with incinerator operation.

3. Containers for liquids. Liquids should be placed within leakproof containers for movement to the site of treatment. In order to minimize or control spills, it is advisable to limit the size of the containers and to provide a leakproof outer container if there is any possibility of spills. The use of an absorbent packing material around the primary container can be considered.

4. Sharps containers. Special containers should be used for all discarded razor blades, scalpels, syringe needles, broken glass, etc. Handling of untreated sharps should be minimized, and sharps should be placed directly into the special containers without needles being clipped or recapped.

Sharps containers should be rigid, puncture-resistant, and leak-proof. They should be designed to prevent needle penetration through the walls of the container when the container is dropped or handled roughly by collection and treatment personnel.

Plastic bags are not appropriate for sharps, and sharps should never be placed into plastic bags. Heavy-wall corrugated cardboard is generally an unsuitable material for sharps containers because it provides only marginal protection from penetration by needles. In addition, liquids can wet and weaken the cardboard, thereby reducing its puncture resistance.

Sharps containers that are to be steam sterilized must have vent holes that allow steam to enter the container. These holes should not be taped closed; therefore, sharps containers with open vent holes must be maintained in an upright position prior to treatment to prevent the leakage of fluid.

Considerations for on-site waste treatment. The following factors should be considered in selecting containers for on-site treatment:

The containers must protect employees from contact with the waste. Of special importance is protection from contact with sharps.
The containers must remain intact throughout collection, handling, movement, storage, and placement within the treatment equipment.

Special considerations for off-site waste treatment. The principal considerations in the selection of containers for waste that will be treated off-site are similar to those listed above for wastes that are treated on-site. In addition, when waste is moved out of the facility, it is subject to additional handling, and, therefore, additional (or different) packaging is usually required to ensure that the waste containers remain intact.

In general, there are two types of outer containers that can be used for off-site transport of infectious wastes: single-use containers and reusable containers.

1. Single-use containers. Single-use containers are usually corrugated fiberboard cartons or fiber drums. To ensure that these containers have sufficient strength to remain intact during handling and transportation, they should meet or exceed the specifications of the National Freight Classification (12) for the particular item. Especially important are the specifications for minimum bursting test—275 pounds per square inch (psi) for fiber boxes and 700 psi for fiber drums. These containers are not leakproof and, therefore, they should always be used with plastic-bag liners. Each plastic bag should be individually tied or sealed, and the container itself should securely taped closed or sealed.

2. *Reusable containers*. Reusable containers for infectious waste
are usually made of either heavy plastic or noncorroding metal (e.g.,
stainless steel, chrome-plated steel, plastic-coated metal). Such containers are leakproof, easily cleaned and disinfected, and durable.
Furthermore, because these materials are noncorroding, the strength
and leakproof qualities of the containers will not be compromised by
rusting.

Reusable containers should either be lined with plastic bags before loose waste is placed in them, or the containers should be filled
with bagged waste that is tied or sealed in plastic bags. These practices will ensure that the infectious waste remains contained and that
workers will not be exposed to loose infectious waste at the treatment
facility and while handling the emptied containers.

Reusable containers must be cleaned and decontaminated before
they are used again. This should be done at the treatment facility.

Labels and Markings

It is important that containers of infectious waste be easily identifiable.
This can be accomplished through the use of specific colors, symbols,
or wording, or by any combination of these. Some state regulations
have specific labeling requirements, and some also specify the terminology that must be used.

Certain colors, markings, and terms are generally recognized as
denoting infectious or biohazardous waste. The colors red, orange,
and red-orange specifically identify biohazards and are generally used
as the colors for plastic bags and other containers for infectious waste.
The international biohazard symbol is often used to identify infectious
waste containers. Terms that are in common use include infectious
waste, biohazard, biohazardous waste, contaminated waste, and hazardous medical waste.

The labels and markings can either be incorporated into the container itself (e.g., color coding or printed symbols or wording), or
they can be on signs or labels that are affixed to the containers.
(When reusable containers are used, the labels must be removed before
the container is returned for reuse.)

Labeling also serves other functions in addition to identifying a
waste as infectious. It should be used to identify and segregate the
different waste streams so that the components of each stream will be
handled properly and will receive the appropriate treatment. This
is important when different treatment methods are used for the different types of infectious waste.

When infectious waste is transported away from the facility for
off-site treatment, information on the waste containers should include
the name and address of the generator. The names, addresses, and
telephone numbers of the hauler and the treatment facility are also

needed; this information could be included on the container label or
in the shipping documents (manifest). In addition, the name and telephone number of an emergency contact should be provided for use
in the event of a spill or accident.

D. Handling and Movement of the Infectious Waste

The purpose of developing handling procedures is to minimize exposure of waste handlers and others and to ensure that the waste will
ultimately receive the proper treatment. These goals can be accomplished by:

Maintaining the integrity of the waste containers during their handling
        and movement
Disinfecting collection carts and reusable waste containers
Minimizing the potential number of persons who might be exposed to
        the waste
Implementing a system for delivery of the different infectious waste
        containers to the proper site for appropriate treatment

Waste Collection

The principal consideration in the selection of waste collection vehicles
and procedures is to ensure that the packages and containers of infectious waste remain intact so that exposure to the waste is minimum.
Therefore, it is important that each container of infectious waste be
closed, tied, or sealed by an appropriate means before it is moved
from the point of waste generation. The waste containers may then
be collected in any convenient way that does not compromise the integrity of the packaging.
     Plastic bags require special handling because they are not as
strong as other containers and are more readily torn and broken by
rough handling.
     Infectious waste containers should be collected in leakproof carts
that have sides or barriers to minimize spillage. These carts should
be clearly labeled and easily identified. They should be used only
for infectious waste collection and movement. Cart design should allow
easy cleaning. The carts should be cleaned, disinfected, and maintained on a regular schedule.

Transport of Waste Within the Facility

The transport of infectious waste within the facility should be carefully planned. The objectives are to provide timely collection of the
waste, to minimize possible contact with the waste, and to ensure delivery to the appropriate treatment/disposal site. In planning collection schedules and designing routes for waste movement within the

facility to the site of treatment, consideration should be given to the activity levels at the facility during the course of the day and to the population density (personnel and visitors) at different hours and in different sections of the facility.

Compaction of Waste

Infectious waste should not be compacted before treatment. Compaction destroys containment, creates aerosols and dust, and may result in the release of liquids, thereby increasing the potential for worker exposure. In addition, compaction increases the density of the waste which may reduce the effectiveness of subsequent treatment, be it steam sterilization of incineration.

E. Storage

Infectious waste is best treated as soon as possible after it is generated. However, in the event that storage is necessary, storage times should be as short as possible, and storage conditions should minimize possible exposure to the waste and prevent any increase in the numbers of potentially harmful organisms associated with the waste.

Storage areas should be secure, with entry limited to those persons who have been trained to handle infectious waste. The infectious waste storage area should be clearly identified with signs (e.g., the biohazard sign). Refrigeration may be necessary to prevent rapid multiplication of infectious organisms. The storage area should be kept vermin-free to prevent the spread of vectorborne disease, and it should be cleaned and disinfected on a regular schedule.

The waste that is placed in storage should be properly packaged and contained so that the waste will not spill. Containers should be clearly labeled with information about type of waste, generation date, and designated treatment. A log of the movement of waste containers in and out of storage is essential in tracking the contents of the storage area and in establishing priorities for transport of waste for treatment as soon as treatment capacity becomes available.

F. Treatment Techniques

Infectious waste is treated so as to render it noninfectious. Two treatment technologies are generally used for the treatment of infectious waste--incineration and steam sterilization. Other types of treatment (e.g., use of ethylene oxide, gamma radiation, microwaves, and dry heat) can be effective. However, they have various problems and constraints, and these technologies are rarely used for infectious waste treatment in the United States.

After treatment, most infectious waste may be combined with the general waste stream from the facility for final disposal. For some wastes, however, additional treatment may be necessary before disposal, depending on the type of treatment that was used for infectiousness. For example, after steam sterilization, needles and syringes would still be usable and body parts would be recognizable. Another example is wastes with multiple hazards; for these wastes, additional treatment may be required because treatment processes that eliminate one hazard may be ineffective for the other hazard(s) present (e.g., steam sterilization does not destroy or change antineoplastic drugs present in the waste).

For proper management, treated infectious waste must be distinguishable from the untreated waste to ensure that all the waste is treated before disposal. Some treatment methods produce obvious changes in the appearance of the waste--e.g., incineration burns the waste while steam sterilization produces color changes on indicator strips and crumples many plastic bags. If the treatment process has no effect on waste appearance, some method of distinguishing treated from untreated waste must be used. Options include removing or obliterating biohazard symbols and labels, and compacting or grinding the treated waste.

Treated infectious waste is usually disposed of by the burial in a sanitary landfill of the treated waste or the residue from incineration. Other options for the disposal of specific types of infectious waste after treatment include the pouring of liquid wastes down the drain to the sanitary sewer system and the grinding and flushing to the sewer system of certain solid wastes (e.g., pathological). Such use of the sanitary sewer system must be in accordance with local sewage treatment regulations and the agreement between the facility and the local sewer authority.

Incineration

Incineration destroys infectious waste through the process of high temperature combustion, leaving a residue ash that is nonhazardous. The effectiveness of incineration is a function of incinerator design, operating temperatures, and dwell times. It is also affected by the type of waste. For example, effective incineration of pathological waste requires lower temperatures and long combustion times, whereas high-temperature incineration and long dwell (gas retention) times are needed for effective destruction of chemotherapy wastes.

Types of incinerators. Three different incinerator designs are commonly used for the incineration of infectious wastes. These are:

The single-chamber incinerator with an afterburner in the stack
The two-chambered controlled-air incinerator

   1. The single-chamber incinerator. The single chamber incinerator is usually referred to as a "pathological" or "Type 4" incinerator.
It normally operates at a relatively low temperature in the 1200 to
1400°F range. Because of the low operating temperatures and short
dwell time (gas retention time), this type of incinerator has marginal
destruction capability. It can be suitable for incineration of pathological wastes, but it is not suitable for the incineration of wastes that
contain plastics or chemotherapy wastes.

   2. The single-chamber incinerator with an afterburner. An afterburner located in the incinerator stack normally operates at temperatures of 1600 to 1800°F, thereby providing additional incineration of
the combustion gases. This enhanced combustion efficiency is limited,
however, by the short retention time of the gases in the afterburner,
which is usually less than 0.1 second.

   3. The two-chambered controlled-air incinerator. In this type
of incinerator, waste is burned in the primary chamber with less-than-
sufficient air for complete combustion (starved air incineration). The
incinerator then completes the combustion process using excess air
in the secondary chamber. Operating temperatures are usually 1600
to 1800°F in the primary chamber and 2000 to 2200°F in the secondary
chamber. Gas retention time in the secondary chamber is a minimum
of 0.5 second, but it can be as long as 2.0 seconds. This type of incinerator is best for infectious waste incineration because the high
combustion temperatures and longer retention times are the conditions
needed for effective combustion of most kinds of infectious waste.
(Temperatures and dwell time should be at the upper ends of the
ranges when antineoplastic wastes are being incinerated.)

   Incinerator operation and control. Various automatic controls on
the incinerator enhance incinerator operation and help to ensure that
the incinerator is operated in accordance with the specified procedures
and under the designated conditions. These controls include the following:

Automatic waste feed systems that control the rate at which waste is
     fed into the primary combustion chamber
Lockout controls that prevent charging of waste into the incinerator
     when the unit is not operating within its specified operating param-
     eters
Burndown controls that fire the burners in both chambers for a pre-
     set time and then fire the burners in the secondary chamber for
     a specified period of time after those in the primary chamber have
     shut down

Various monitoring devices to monitor combustion temperatures, air
    flow, fuel, stack gas flow and content, opacity of stack gases, etc.
Linkages between the monitors and the control panel to sound an alarm
    and to stop incinerator operation in the event of a malfunction

    Regardless of automation, monitors, and controls, incineration
is a complex and complicated process. In order to achieve safe and
effective incineration, incinerators must be operated only by well-
trained operators.
    There are no federal regulations that apply specifically to infec-
tious waste incinerators. Some states and localities have regulated
infectious waste incineration, and the requirements may specify testing, operating conditions, etc. In the absence of regulatory requirements, certain characteristics should always be regarded as signs of
a poorly operating incinerator; these include the emission of smoke
and the presence of unburned material in the ash.

Steam Sterilization

Steam sterilization utilizes pressurized steam at 250 to 270°F (121 to
132°C) to kill pathogenic organisms that are present in the infectious
waste.
    The steam sterilization process does not destroy the waste itself
(as does incineration), and the entire load of treated waste has to be
disposed of after treatment. There is usually some decrease in waste
volume (the result of air removal in the sterilizer) and some increase
in its weight (the result of steam wetting the waste).

    Types of steam sterilizers. Pressurized vessels are used for steam
sterilization, and the equipment is variously known as steam sterilizers,
autoclaves, and retorts. Pressurized vessels are needed to maintain
the steam pressure so that the waste is exposed to sufficiently high
temperatures for a long enough time to sterilize the waste.
    Steam sterilizers differ in type of construction material (stainless
or other steel), the presence or absence of a steam jacket, and degree
of automation. The main functional difference is the mechanism for
removal of air from the sterilization chamber; this step in the sterilization process can be accomplished by gravity displacement (of air by
steam) or by vacuum.

    Operation and control of steam sterilizers. The effectiveness of
the steam sterilization process is determined by two variables--tempera-
ture and time, i.e., the temperature of the steam and the duration
of exposure to the treatment temperature. The single most important
factor in steam sterilizer operation is removal of the air from the sterilization chamber and from the waste. Complete removal of air is essen-

tial for the steam to penetrate throughout the waste so that the waste
will be at treatment temperature for sufficient time to achieve kill.
Factors that affect air removal include waste type and quantity, waste
packaging, load density and configuration, and container type, size,
and shape.

Because there are so many variables, an arbitrary set of operating
conditions that specifies chamber temperature and cycle length will not
ensure that removal of air will be complete and that all the waste will
be at treatment temperature for the required minimum amount of time.
Standard operating procedures (SOPs) for the steam sterilization process must be developed in order to ensure the effectiveness of treatment. These SOPs could be developed for a single worst-case load
and be used for each run of the sterilizer. Alternatively, a separate
set of SOPs could be developed for each type of infectious waste that
will be steam sterilized; each type of load and the operating conditions
for each load would be standardized. The SOPs for steam sterilization
should be developed using biological indicators placed throughout each
load to monitor the effectiveness of the treatment process. At present,
biological indicators that consist of spores of *Bacillus stearothermo-
philus* are the most effective monitors of the steam sterilization process.
Temperature indicator strips that change color indicate only that the
outside of the container was exposed to a certain temperature; they
do not indicate internal temperatures or exposure times and, therefore, do not indicate that the waste has been sterilized. Although
such indicator strips should not be used for monitoring treatment effectiveness, they are useful for showing the waste has been processed.

Periodic monitoring of routine subsequent operations provides a
means of checking on treatment effectiveness. If monitoring were to
show that sterilization is not being achieved, this would be indication
of a problem--either the equipment is malfunctioning or the SOPs are
not being followed.

State infectious waste regulations generally specify that steam
sterilization is an acceptable treatment technology for infectious waste.
Some of these regulations also include specific requirements for the
steam sterilization process. Regardless of the presence or absence
of regulatory requirements, operations for the steam sterilization of
infectious waste should be monitored at regular intervals using biological indicators to ensure that the treatment is effective.

G. The Off-Site Treatment Option

As an alternative to on-site treatment, off-site treatment options are
available in many areas of the country. Off-site treatment may be
provided at and by another facility, at a jointly owned treatment facility located at another site, or by a commercial treatment facility. Regardless of the nature of the off-site treatment operation, this alterna-

tive offers both advantages and disadvantages for the generator of
infectious waste.

Advantages and Disadvantages of Off-Site Treatment

The principal advantages of off-site treatment are the delegation of
waste treatment to a specialty operation and potential cost savings.
Waste treatment could be best performed by an operation specializing
in provision of that service. Savings could be realized in total expenditures for waste treatment, i.e., in the costs of equipment (capital,
operation, and maintenance costs) and of labor to handle the waste
and operate the equipment.

The principal disadvantage of off-site treatment is that the generator loses control over the waste while still retaining liability for the
safe and proper management, treatment, and disposal of the waste.
For some generators, this disadvantage is slight when reliable off-
site treatment is available. For others, this disadvantage may be
severe enough to outweigh the benefits and to justify continued on-
site treatment.

Evaluation of Off-Site Treatment

A decision on adopting the off-site treatment option should be made
only after careful evaluation of all the relevant factors. This evaluation should include consideration of the following factors:

*State and local regulatory requirements for infectious waste manage-
    ment* (Is off-site treatment allowed? What are the requirements?)
*Availability and capacity of storage space and treatment equipment at
    the generating facility* (Is it adequate? Is it in compliance with
    regulatory requirements?)
The reputation and references of the broker/contractor (Does the
    operation have a record of providing reliable service that is in
    compliance with all regulatory requirements? Is the management
    committed to providing quality service? Is the management ex-
    perienced in hauling and treating infectious waste?)
The type of treatment that will be provided (Is it appropriate for the
    types of infectious waste generated? Will treatment be provided
    for all types of infectious waste generated by the facility?)
The service that will be provided (Will it be a full-service program?
    Will it meet all the generator's needs?)
*The contractor's specifications for containers and packaging and the
    protocols for waste pick-up and transportation* (Are they in com-
    pliance with regulatory requirements? Are they suitable for use
    at the generating facility? Are the drivers qualified and properly
    licensed?)

The contractor's use of storage sites and transfer stations (Are these sites permitted in accordance with all relevant regulations? Are they secure and properly managed?)

The treatment facility and treatment operations (What kind of treatment is provided? Is the facility properly licensed and permitted in accordance with all relevant regulatory requirements? What treatment equipment is used? Is there sufficient equipment to provide reliable service? Is the equipment properly operated and maintained? Are operations monitored to ensure treatment effectiveness?)

Disposal of the by-products of waste treatment (Where and how is the treated waste disposed of? Are disposal operations in compliance with all regulatory requirements? What are the contractor's contractual arrangements with the disposal facility, that is, are they sufficiently long term?)

Chain of custody (What type of control document is used to track and account for the waste? Does the system provide "cradle-to-grave" custody?)

The contractor's emergency response procedures, contingency plans, and treatment back-up arrangements (Do they address all likely and reasonably foreseeable contingencies? Are there back-up agreements with other facilities? Do the back-up plans provide for prompt treatment of the waste at other reliable treatment facilities?)

The contractor's insurance (Is it comprehensive? Is it sufficient?)

The contractor's service fee structure (Are the charges reasonable for the services provided? Does the fee schedule specify charges for regular and special services? How do the charges compare with the total costs for on-site treatment?)

Regulatory Requirements

Regulatory requirements pertaining to infectious waste vary among the states. Whatever they may be, they must be complied with by the generating facility and also by the treatment facility. If the off-site treatment facility is located in a different state, compliance with two different sets of regulations may be necessary, and any contradictions in requirements would have to be resolved.

Generators should know the specific regulatory requirements that pertain to generators and to off-site treatment facilities--they may be the same or different. There may be regulations that pertain to such aspects of the operation as:

Transportation (e.g., requirements for truck size and type, permitting and licensing, signs and placards, disinfection of reusable containers and of vehicles, driver qualifications)

Storage sites and transfer stations (e.g., requirements for permitting,
    storage conditions, security)
Treatment facilities (e.g., requirements for permitting, security, equip-
    ment size and type, operations, monitoring, contingency plans,
    personnel training)
Disposal of treated waste (e.g., type of landfill)
Records (e.g., requirements for manifests, records of operations and
    monitoring, duration of record keeping)

The generator must ascertain that the off-site operations are in full
compliance with all relevant regulations.

H. Contingency Plans and Emergency Procedures

There can be difficulties in implementing any established plan for infectious waste management. It is, therefore, important to prepare
contingency plans and emergency response procedures so that they
will be in place and be ready for implementation whenever needed.

Contingency Plans

Contingency plans are needed to provide alternatives for every phase
of the facility's waste management plan. They would be used whenever
the designated management policy and procedures cannot be implemented. With contingency plans already in place, it will not be necessary to scramble for alternative arrangements when an emergency
arises. The need for contingency plans can occur at any time and
for any of a variety of reasons. It can result from any factor ranging
from delayed receipt of a shipment of plastic bags, to labor difficulties, to the malfunctioning of treatment equipment.
    The most critical need is for alternative treatment arrangements
that can be activated whenever the facility's on-site treatment equipment is inoperable. The plans for such a contingency should include
use of other on-site treatment equipment (if available and appropriate
for the types of waste) and provisions for longer-term storage until
the equipment is back in operation (if such storage is permitted under
the regulations and if space is available). There should also be back-
up arrangements for off-site treatment of the facility's infectious waste;
these back-up agreements should include arrangements with other facilities or with commercial contractors for as-needed service providing
transportation and treatment of the waste.

Emergency Response Procedures

As with contingency plans, it is essential that emergency response
procedures for managing accidents with and spills of infectious waste
be developed and in place before they are needed. The elements of
an emergency response plan are as follows:

Development of spill containment and cleanup procedures for each type
of infectious waste and for wastes with multiple hazards

Placement of clean-up equipment and protective gear at easily accessible locations

Designation of an emergency response coordinator

Assignment of persons responsible for the cleanup, with assignments
based on type of waste or area of the facility

Training of responsible personnel in emergency response procedures

Specification of procedures for accident reports, accident assessments,
corrective action, and follow-up medical surveillance (if deemed
necessary)

I. Disposal of Treated Waste

The purpose of treating of infectious waste is to render the waste noninfectious. Therefore, infectious waste that has been properly treated
should be considered "ordinary" solid waste and should be managed
accordingly. Like solid waste, treated infectious waste or the treatment residue (e.g., incinerator ash) could be disposed of in a licensed
landfill. The liquid portion of treated infectious waste could be poured
down the drain into the sanitary sewer.

Regulatory Factors

It is important to note that landfills and sewer systems are under local
jurisdiction, and, therefore, the disposal of treated waste is subject
to the specific provisions of local ordinances. For example, agreements
with the local sewer authority govern the nature of the wastewater
that may be discharged into the sanitary sewer system by the facility.

There may also be some relevant state regulatory requirements
that pertain to infectious waste. These regulations may stipulate disposal requirements for treated as well as untreated infectious wastes.

Special Considerations

Another critical factor that affects the disposal of treated infectious
waste is the total effect of the treatment process on the waste. That
is, treatment may make the waste noninfectious, but it still may not
be suitable for disposal as solid waste. This may happen when the
waste has certain physical characteristics (e.g., pathological wastes
and sharps) and when the waste has additional hazards that are not
affected by the treatment for infectiousness. (Obviously, these considerations should be a factor in the selection of treatment technology
for these types of waste.)

Pathological wastes. For aesthetic reasons, recognizable body
parts should not be placed into a landfill. Therefore, if a treatment

process renders pathological waste noninfectious but does not alter its physical appearance, it would not be suitable for landfill disposal and additional treatment would be necessary before disposal. Incineration of pathological waste provides complete treatment, whereas steam sterilization would have to be followed by additional handling and additional treatment.

Sharps. With sharps, also, treatment for infectiousness may or may not be sufficient, depending on the treatment technology. Additional treatment for sharps is necessary if they remain intact after treatment. It is essential that, before disposal, the treated sharps first be rendered nonreusable (e.g., by compaction, incineration, or grinding). Because of public health factors, risk, and liability considerations, intact sharps--whether infectious or sterilized-- should never be shipped from a facility for disposal.

When sharps are sent off-site for treatment, however, different considerations prevail. In this circumstance, it is important to consider the risk and liability factors. Risk and liability can be minimized through the policy and procedures that are established for sharps containers, packaging, handling, and shipment off-site to the treatment facility.

Wastes with multiple hazards. The management of wastes with multiple hazards is especially complex because treatment for infectiousness, if it is compatible with the other hazards, may not reduce the other hazards at all. Therefore, if these wastes are treated for infectiousness, disposal in a sanitary landfill may still be prohibited because of the hazardous nature of the waste. Such waste must be disposed of in accordance with all relevant regulatory requirements.

J. Training

In order to attain full implementation of the infectious waste management plan, responsibility for the various aspects of waste handling and treatment must be assigned to specific personnel, and they have to be trained for these tasks. The training program should include the following elements:

The rationale for an infectious waste management program
The overall infectious waste policy of the institution
Organization of the management plan
Delegation of responsibilities and establishment of a chain of command
Assignment of personnel to specific tasks
Detailed training for the assigned tasks including specified procedures
    and use of equipment
Training in emergency response procedures and responsibilities

Training classes should be conducted by qualified instructors. The training program should include initial training, followed by periodic refresher courses. Training should be mandatory for all personnel who generate, manage, or handle infectious waste.

K. Records

Record keeping is an essential part of a waste management program. Records serve a number of useful purposes. They:

Quantify amounts of infectious waste being generated (data that are needed in reviews of waste management policy)
Document waste treatment operations
Provide a log of monitoring data
Document equipment inspection and maintenance activities
Provide a log of employee participation in training sessions
Provide data on accidents, spills, and employee injury
Document follow-up of employee injury and exposure including medical treatment and surveillance
Document shipment of waste to off-site treatment facilities

These types of records can be used to create the data base that is needed for risk management. Furthermore, the documentation provided by carefully kept records can be a key element in limiting liability that might otherwise be attributed to mismanagement of the facility's infectious wastes.

The different types of records need not all be maintained for the same length of time. In some states, a minimum period of time is dictated by regulatory requirement; in other states, it is at the discretion of the facility. Obviously, records that might be useful in defense against litigation should be kept for longer periods.

VI. RELEVANT REGULATIONS

Regulations at the federal, state, and local levels impose requirements on the management of infectious waste. Some of these regulations pertain to infectious waste directly, while others affect indirectly the treatment and the disposal of infectious waste.

A. Federal Regulations

Two acts passed by Congress--the Resource Conservation and Recovery Act (RCRA) (1) and the Comprehensive Environmental Response, Compensation and Liability Act (CERCLA) (13)--control hazardous waste disposal in the United States. Both of these acts granted to the EPA the authority to promulgate regulations deemed necessary to ensure the safe management and disposal of wastes and to protect human health and the environment from the ill effects of improper waste disposal.

RCRA Regulations

The regulation of infectious wastes under RCRA was discussed above in some detail. In summary, this class of waste remains unregulated as a hazardous waste on the federal level. The EPA has issued two guidance manuals on infectious waste management (4,5). No decision has yet been reached by EPA as to when, or if, final hazardous waste regulations will be promulgated for infectious waste.

CERCLA (Superfund) Regulations

CERCLA, commonly referred to as the Superfund legislation, provides for the cleanup of inactive or abandoned waste disposal sites that present substantial danger to the public health or welfare or the environment because of the presence of hazardous substances at the site. In CERCLA, the term "hazardous substance" is defined broadly and includes more substances than the RCRA-listed hazardous wastes.

The EPA has compiled the National Priority List (NPL), a list of sites that are candidates for Superfund cleanup actions. It should be noted that several municipal landfills are already on the Superfund cleanup list, and the number will probably increase.

Under CERCLA, the EPA has the authority, by means of court order, to assume control of an NPL site. The EPA can then institute immediate and long-term remedial actions for the containment or removal of the hazardous substances that are posing the threat.

There is no statute of limitations CERCLA, and some of the remedial actions now underway are at sites where hazardous wastes were deposited several decades ago. If a company or facility is found to be responsible for any of the pollution at a site, it can be held financially liable for the entire cost of the site cleanup under the joint and several liability provision of CERCLA. If infectious wastes were to be regulated under RCRA, they would become subject to CERCLA. An infectious waste generator could then be required to pay for the cleanup of a site (e.g., a landfill) if the generator had ever placed untreated infectious waste into that landfill.

B. State Regulations

Under RCRA, each state may assume the authority to regulate hazardous waste. This authority is granted to the state by the EPA when EPA determines that the state's regulations are at least as stringent as the federal regulations. Most states have now been granted the authority to run their own hazardous waste programs.

A number of states have issued infectious waste regulations under the hazardous waste program. There is much variation in these regulations and in their requirements. For example, state regulations differ in their definition of infectious wastes. Also, infectious waste is classified as hazardous waste in some states and as special waste in others. When infectious waste is classified as hazardous, it is subject to the full gamut of the hazardous waste regulations in some states, but only to certain of the hazardous waste requirements in other states.

C. Status of Infectious Wastes

At present, infectious waste is not regulated by the EPA as a hazardous waste under RCRA. Because infectious waste does not meet any of the other Superfund definitions of "hazardous substance," it is not covered by the CERCLA regulations. If the EPA were to regulate infectious waste as a hazardous waste, infectious waste would then be regulated under Superfund also. Infectious waste could be considered a source of pollution some day, and landfills containing infectious waste might be subject to Superfund cleanup action. Infectious waste generators could then be held liable for the costs of cleanup.

There is more immediate concern in states that have already regulated infectious waste as hazardous waste. In these states, any state Superfund regulations that are promulgated would also apply to infectious waste. Generators of infectious waste would then be potentially responsible for pollution arising from a disposal site in which their infectious waste was deposited.

RCRA, CERCLA, the regulations promulgated under these two acts, and subsequent litigation have established that the liability for hazardous waste remains always with the generator of the waste. Disposal of hazardous waste in a landfill does not relieve the generator of responsibility for the waste. The implications are serious for generators of infectious waste, because waste that is managed improperly today could result in future liabilities.

REFERENCES

1. PL 94-580. The Resource Conservation and Recovery Act. 94th Congress, October 21, 1976.
2. U.S. Environmental Protection Agency. (1978). Hazardous Waste Guidelines and Regulations: Proposed Rules. *Federal Register* 43:58946-59028.
3. U.S. Environmental Protection Agency. (1980). Hazardous Waste and Consolidated Permit Regulations. *Federal Register* 45:33066-33588.
4. U.S. Environmental Protection Agency. (1982). *Draft Manual*

for Infectious Waste Management. EPA document # SW-957. U.S. Government Printing Office, Washington, D.C.

5. U.S. Environmental Protection Agency. (1986). *EPA Guide for Infectious Waste Management.* EPA document #530-SW-86-014. National Technical Information Service, Springfield, Virginia.

6. Garner, J. S., and Simmons, B. P. (1983). *CDC Guideline for Isolation Precautions in Hospitals.*

7. Garner, J. S., and Favero, M. S. (1985). *CDC Guideline for Handwashing and Hospital Environmental Control.*

8. Joint Commission on Accreditation of Hospitals. (1986). *Accreditation Manual for Hospitals*, 1987 Edition. Chicago.

9. Centers for Disease Control and National Institutes of Health. (1984). *Biosafety in Microbiological and Biomedical Laboratories.* U.S. Government Printing Office, Washington, D.C.

10. National Institutes of Health. (1978). *Guidelines for Research Involving Recombinant DNA Molecules.*

11. National Institutes of Health. (1978). *Laboratory Safety Monograph.*

12. National Freight Classification #100-L. Item 222 (Single Wall Corrugated Fiberboard Boxes) and Item 291 (Semiliquid Fiber Drums).

13. PL 96-510. Comprehensive Environmental Response, Compensation, and Liability Act of 1980. 96th Congress, December 11, 1980.

13

Safety Considerations in the Use and Disposal of Chemotherapy Agents

FRANKLIN PEARCE   *Memorial Sloan-Kettering Cancer Center, New York, New York*

I. INTRODUCTION

Treatment of cancer with cytotoxic agents (chemotherapy) has developed to the point where long-term remissions can be induced in patients with specific types of cancer, such as Hodgkin's disease, testicular cancer, and myelogenous leukemia. Chemotherapy in combination with surgery and/or radiation has increased survival times in other types of cancer as well.

Chemotherapeutic drugs are designed to cause cellular death. Although they do not distinguish between normal and cancerous cells, the cellular death rate is proportionally greater in the cancer cell than in the normal cell because of the more rapid proliferation of the latter. As these drugs interact with nucleic acids and protein synthesis, they may be mutagenic, carcinogenic, or teratogenic (1,2).

Many of these agents are acutely toxic. Some may permanently damage the nervous system, reproductive system, liver, or other organs. Many cause reversible effects such as nausea and vomiting (3,4), immunosupression, and bacteriological or viral infections (5-7).

Despite these hazards, the therapeutic benefits of chemotherapy far outweigh the potential risk to the patient. Perhaps because of the benefit or the frequency of use, hospital personnel have become complacent and have overlooked the adverse effects that many of these drugs may have on the persons who prepare, administer, or dispose of them.

There are no epidemiologic studies that demonstrate conclusively that small amounts of cytotoxic agents will have delayed carcinogenic effects. Among health care workers, the studies that have been performed have design deficiencies, such as lack of exposure information regarding specific drugs, differences in preparation protocols, or the failure to include statistically significant numbers of subjects in studies (8). What is clear is that secondary tumors may be induced in patients treated with chemotherapy agents (9).

A number of in vitro, animal, and even human studies have implicated cytotoxic drugs in chromosome damage, teratogenesis, and cancer induction. Testicular and ovarian dysfunction including permanent sterility have been demonstrated in patients (respectively) who have received single or combination therapy. Congenital malfunctions have been attributed to 5-fluoro-uracil (10).

A study of fetal loss associated with cytotoxic agents among nurses in a Finnish hospital is consistent with this potential (11). Clearly, then, chemotherapeutics are chemicals that must be managed with some care. They are toxic, and, therefore, the drugs administered to the patients must be carefully evaluated. These agents are also potentially hazardous to those who handle them, therefore care must be used in the preparation, administration, and disposal of excess and waste material.

II. MANAGEMENT OF CHEMOTHERAPY AGENTS

As parenteral injection is the only practical method of administering chemotherapy agents, this chapter will deal exclusively with preparation, administration, and disposal as it pertains to this method of treatment.

A. Policies

Written policies concerning transport, storage, preparation, administration, and disposal of chemotherapy agents should be written and followed by all departments concerned. These must be consistent with governmental regulations and professional standards and should take into account the "right to know" laws currently being promulgated. Material Safety Data Sheets on each drug used should be on file. These may be obtained from the manufacturers, who are required by law to make them available to the user.

B. Receiving and Storage

It stands to reason that workers must be protected from accidental exposure to cytotoxic drugs from the time the agents are received until their ultimate disposal.

1. Shipping containers should be clearly labeled as to their contents. Receiving departments must be trained to inspect all such packages for damage and to notify the appropriate persons if damage is noted.
2. Storage in pharmacy areas should be in clearly labeled bins to provide secondary containment in the event of spillage or breakage. The storage bins should be no higher than eye level to reduce the chance of breakage. Access to the area should be restricted to authorized personnel only.

C. Containment Cabinetry

In large oncology centers and hospitals, chemotherapy agents are usually prepared in the pharmacy by trained personnel, while in many smaller hospitals and centers these agents may be prepared by physicians or nurses in patient care or staff areas, which may be inadequately ventilated.

Many cytotoxic agents will undergo a number of materials transfer steps prior to patient administration. Even when care is taken, the opportunity for personnel exposure through inhalation or direct skin contact exists.

Aerosols

The withdrawal of needles from drug vials, drug transfers that involve the use of syringes/needles the breaking opening of ampules, and the expulsion of air from a drug-filled syringe all can produce aerosols that may place the worker at risk.

Aerosols generated by these activities expose not only the worker immediately involved, but also staff and patients in the surrounding area. To avoid this type of exposure it is necessary to work in an enclosed and properly designed ventilated work area. Due care must be exercised to ensure that no smoking, drinking, applying cosmetics, or eating where these drugs are prepared, stored, or used take place, as these practices greatly increase the chance of exposure.

Surface contamination control. Respiratory protection, skin protection, and training in the proper handling of these agents are essential to minimize worker exposure. This is especially true in the pharmacy, where the opportunity for exposure still exists even when protective clothing and gloves are worn and careful aseptic technique is used.

Containment devices. Horizontal and vertical clean air benches have been used to provide a clean environment for the preparation of injectable drugs. These units were designed to direct the airflow

over the work surface toward the user, thus protecting the product
being manipulated from external contamination. The design resulted
in an increase in the likelihood of exposure to the preparer of the
drugs and other persons in the immediate area to aerosols generated
during preparation procedures. The use of a Class II laminar flow
biological safety cabinet reduces the potential exposure of workers
(see below).

Biological safety cabinets: These units provide protection for
both the operator and the product. The units are designed so that
room air enters the front intake grill, and travels under the work surface through an exhaust HEPA filter, and 70% flows through a supply
HEPA filter, then vertically toward the work surface, where it splits
toward the front intake grill and a rear slot, and is recirculated. HEPA
filters trap particles as small as 0.3 μ in size with a 99.97% efficiency.
Thus, the air that passes over the work surface as well as the air that
is exhausted to the outside is essentially particle-free.

In contrast to other types of equipment, a biological safety cabinet can malfunction without the operator's knowledge. The airflow
may be unbalanced, or there may be holes or breaks in the filter.
Therefore, these units must be certified at least annually or whenever
the cabinet is moved. The HEPA filters are extremely brittle and may
crack or tear away from the frame if stress is put upon them, such
as that produced by moving the cabinet.

Ideally, the safety cabinet should be placed in a separate room
used only for chemotherapy preparation. If this is not possible, then
the cabinet should be placed in a rear corner of the pharmacy, away
from doors, windows, and high traffic areas. Persons walking behind
the cabinet operator can cause air currents that may compromise the
front air barrier. If the cabinet is left running at all times, dust particles will not accumulate within the cabinet (11).

III.   PREPARATION OF CHEMOTHERAPY AGENTS

A.   Dose Preparation

The following sequence of steps is suggested to minimizing worker
exposure in the preparation and administration of chemotherapy agents:

1.   The interior walls, view screen, and work surface of the safety
     cabinet must be washed with 70% alcohol before each production
     run.
2.   Place a sterile plastic-backed absorbent pad on the work surface.
     Plan your work so that clean materials are on one side, and mate-
     rials to be discarded are placed on the other side. Be sure that
     all the necessary materials are placed within the cabinet, as your

hands should not be removed from the cabinet while the work is in progress.

3. Let the cabinet air wash over your supplies for about 5 minutes before beginning your work. During this time, wash your hands and don your closed-front sterile gown and latex gloves. These gloves should be changed every half hour.

4. All work should be done at least 6 inches in from the front edge of the work surface. The tops and necks of vials should be swabbed with 70% alcohol prior to puncture. When working with a glass ampule, tap the ampule so that all material drains from the top to the bottom. Wrap the ampule in an alcohol-dampened gauze prior to opening it to reduce aerosol generation. It is important to avoid touch contamination of the needles and vials.

5. Prevent positive pressure build-up in the vial by withdrawing an equal volume of air from the vial before adding the diluent. Use only syringes and ivsets with Leur Lock® fittings. Syringes should be of sufficient capacity so that they will not be filled to maximum volume when the agent is measured into them.

6. Final drug measurement should be done prior to removal of the syringe from the vial. If further adjustment is necessary or air bubbles must be removed from the syringe, wrap a sterile gauze over the needle to prevent aerosolization.

7. Prime all iv sets within the safety cabinet before the drug is added to the iv set. All sets and syringes should be wiped clean with 70% alcohol before they are removed from the safety cabinet.

8. All iv sets or syringes should be labeled with the following information:

   Patient name and location
   Name and amount of medication
   Name and volume of diluent
   Date and time the solution was prepared
   Initials of the preparer
   Rate of administration
   Expiration date of the compounded solution
   "Chemotherapy, Dispose of Properly"

9. Dispose of all materials properly. See "Disposal" in this chapter.

10. Wash interior walls and work surface with 70% alcohol at the end of each production run and at the end of each shift.

11. The cabinet work surface should be removed at least monthly and the cabinet area below the work surface cleaned. The cabinet should be turned off and barrier garments worn when the latter operation is being done.

As indicated previously, many chemotherapy agents are acutely cytotoxic. Skin contamination should be treated in the same manner as exposure to other caustic substances. Preparation areas should be equipped with a sink, a liquid soap dispenser, and an eyewash station. A written protocol concerning exposure should be available, and medical treatment should be obtained as soon as possible after exposure.

B. Administration

Basic Considerations

Chemotherapy should be administered only by those who have received appropriate training. In fact, a number of medical centers now have chemotherapy nurses, who are the only persons allowed to administer these agents. These nurses have been trained in proper administration technique, significant adverse effects of the drugs, their mechanisms of action, and appropriate instructions to be given to patients. They are given specific instruction as to how the drug is to be introduced into an iv infusion line and procedures to follow in the event of malfunctions.

Procedure

1. Assemble all the materials and equipment needed, such as alcohol wipes, plastic-backed absorbent pad, empty bottles for excess drug, a plastic container for used needles and syringes, and a sealable plastic bag for disposal of supplies.
2. Wash hands and don a closed-front cuffed gown and latex gloves.
3. Examine infusion sets for signs of leakage.
4. Place the plastic-backed absorbent pad under the injection site. If the iv line must be bled, bleed it into a gauze pad inside a sealable plastic bag.
5. If the chemotherapy agent is an iv piggyback, allow the primary iv solution to infuse at the prescribed rate after the piggyback is empty.
6. Discard all chemotherapy agents, supplies, and equipment according to protocol. Remove your gloves and wash your hands.

C. Patient Care

Patient Wastes

A number of chemotherapy agents are excreted in the urine, feces, or vomitus of patients receiving these drugs within the first 48 hours after administration (see Table 1). Persons caring for these patients should be extremely careful to avoid contact with excretions during

TABLE 1 Safe Handling of Cytotoxic Agents: Routes of Drug
Elimination for Commonly Administered Cytotoxic Agents*

Drug	Route of excretion
Bleomycin	50% of dose in urine by 24 hours. 20-40% of this amount is active drug. Tissue inactivation is rapid and occurs in liver and kidney.
Carmustine (BCNU)	30% in urine after 24 hours. Some metabolized by the liver. 60-70% of dose recovered in urine as metabolites by 96 hours.
Cisplatin	Urine; 27-45% excreted in first 5 days, initially as unchanged drug.
Cyclophosphamide	Exclusively excreted in urine; 15% unchanged, the rest as metabolites.
Cytarabine (Cytosar)	90% of a given dose eliminated in urine by 24 hours, some in the bile; 4-10% unchanged and 86-96% inactive deaminated metabolites.
Dacarbazine (DTIC)	Urine excretion accounts for 30-45% of any dose administered (Maximized after 6 hours).
Daunomycin	Primarily biliary; some in urine, 14% recovery after 7 days.
Doxorubicin (Adriamycin)	Primarily biliary with 10-20% of a dose in feces after 24 hours and 50% by 7 days; 5-6% intact in urine after 5 days.
VP-16 (Etoposide)	44-60% in urine with 67% of that excreted as unchanged drug, the rest as metabolites; recovery in feces ranges from 2-16% over 3 days.
5-Fluorouracil	Given rapid IV injection: 60-80% excreted via lungs as carbon dioxide. 15% of a dose found in urine by 6 hours with 90% excreted in the first hour. Given by continuous IV infusion over 24 hours: 90% excreted as carbon dioxide and 4% in urine.

TABLE 1 (continued)

Drug	Route of excretion
Methotrexate	Excreted unchanged in the urine. 40-50% of a small dose and 90% of a large dose within 48 hours. Most within 8 hours.
Mitomycin C	Primary elimination is by liver metabolism. 10-30% of a dose excreted in the urine within a few hours of administration.
Procarbazine	Liver biotransformation. Renal excretion, after 24 hours, up to 70% of a dose is recovered in the urine with less than 5% of the urine fraction as unchanged drug.
Vinblastine	Excreted intact in both bile and urine.
Vincristine	Primarily by liver into bile and feces with 1/3 of a dose found in feces within 24 hours, and 2/3 in 72 hours. 12% of a dose excreted in urine in 72 hours.

*Compiled by Lyn Cain Zehner and Joann Neuffer, Alexandria Hospital, Alexandria, VA. By permission.

this time period. Gloves and disposable gowns should be worn and discarded as chemotherapy waste after each procedure. Human wastes should be handled in such a manner as to minimize aerosol production. Urinals and bedpans should be emptied slowly and from a low height to avoid splashing. Rinsing of these articles should be done using a slow stream of water. Hands must be thoroughly washed after glove removal.

Clothing

Bed clothing should be protected from contamination as much as possible by placing an absorbent plastic-backed pad under the patient. Contaminated linen should be placed in labeled laundry bags, and laundry personnel should wear gloves and disposable gowns when handling these articles. Nursing and housekeeping personnel may be exposed

to chemotherapy agents if they are not made aware of the potential
hazard and not trained to take precautions.

Plumbers and other maintenance personnel who must repair bathroom fixtures should be advised that the patient is on chemotherapy.
Institutions must provide sufficient training and proper equipment
for tradespeople.

D. Disposal

Waste and Waste Containers

Waste chemotherapy agents and materials that come into contact with
them are easier to discard if they are managed as a completely separate
waste stream. Dedicated containers can be placed at the points of
origin within the facility. Two types of containers at each site should
be considered.

Solid Waste. Two 5-gallon containers should be used, one labeled
"dry chemo-waste" and the other labeled "wet chemo-waste." Wet
chemo-waste is defined as parenteral sets or vials still containing
liquid, whereas dry chemo-waste is defined as empty parenteral
bottles or bags, gauze squares, gloves, gowns, etc. Both containers
are lined with a 10-mil plastic bag.

Liquid Waste. Two 30-gallon fiber drums, also lined with a 10-
mil plastic liner, should be labeled wet or dry chemo-waste. "Wet
chemo" drums should have a 6-inch layer of some combustible liquid
absorbent in the bottom, such as sawdust.

Waste Liners

Contaminated materials should be placed in plastic bags and then placed
into the appropriate 5-gallon container. When these containers are
filled, the liner is tied off, removed from the can, and placed into a
30-gallon fiber drum. When this is filled, the liner is again tied off,
so, in effect, each object in the 30-gallon drum has been triple-bagged.
The drum is covered, sealed, and labeled for collection by a licensed
hazardous waste contractor. The drum should not contain metal parts
so that the entire drum can be incinerated without ever having to be
opened.

In those areas where chemotherapy is given mainly by syringe
rather than with an iv set, it is advantageous to install a large (8-
gallon) plastic sharps container. Syringes, needles, and small drug
bottles can be discarded directly into this container. Once full, these
containers can then be placed in a 30-gallon fiber drum for disposal.

IV. EMERGENCY RESPONSE

A. Spill Control

General Comments

In the real world, accidents and spills will occur. There are two types
of spills/accidents to consider. The first would be a small-scale spill,
e.g., if a small vial is broken in shipment or in the pharmacy, while
the second is a larger-scale spill, such as parenteral bottles falling
from an iv rack. Spills of this latter kind frequently occur in the patient's room.

If your hospital uses plastic rather than glass parenteral sets,
the larger spills will be minimal. Filled plastic bags can frequently
be dropped from a height of 6 feet without breaking.

Spill Kits

There are a number of commercially available spill kits, which contain
disposable gloves, a disposal bag, and absorbent wipes, spill control
pillows, etc. In those instances where the kits contain face masks
and safety glasses, care should be exercised to ensure that the masks
provide appropriate protection. It is important to make sure that the
kit you wish to use is adequate with respect to the amount of liquids
to be picked up. You can test the adequacy of a specific kit by pouring a liter of water on the floor and determining whether or not it ful-
fills your particular requirements. In general, kits containing spill
control pillows will clean up larger volumes, but they also take up more
storage space to meet your needs. It may be necessary to add additional supplies to a commercial kit.

Training. Once a kit is selected or developed, it should be distributed to areas where they are needed, such as pharmacies and nursing stations. As with any other piece of equipment or new procedures,
in-service training should be given to potential users, pharmacists,
nurses, and/or key housekeeping personnel.

Accidents and Spills

Small scale. Problems arising from vials broken in shipment can
be minimized if the shipping container is examined carefully before
it is opened. If the container shows signs of damage, place the container in a biological safety cabinet. The worker should put on a gown
or lab coat and gloves and then carefully open the outer container.
Each wrapping layer should be checked for signs of damage. If a broken
vial is found, place the entire package in a sealable plastic bag for
proper disposal. If undamaged vials are present, they can be washed
with 70% alcohol while still within the cabinet. All waste paper and
any liquid waste is disposed of properly.

Spills of small vials are easily handled and may be handled without resorting to a commercial spill kit. Clear other persons from the area and don gloves, gown, safety glasses, and a disposable dust and mist respirator. If the drug is in powder form, place a very wet plastic-backed wipe over the spill material, then carefully wipe the area. If pieces of glass shards must be removed, place your gloved hand into a small plastic bag, carefully pick up the pieces, then invert the bag to contain the pieces. The spill area must be wiped at least three times with clean wet wipes. The floor or surface should then be washed.

Larger scale. Broken parenteral bottles pose a different problem, which is best addressed through preplanning.

The chemotherapy spill kit should be used on spills of larger volumes. Nonessential persons should be cleared from the area. The person who is to perform the cleanup dons gloves, dust and mist respirator, safety glasses, and disposable gown. As breakage of parenteral bottles containing chemotherapy agents compounds the spill by producing large, very sharp glass shards, the first step is to remove these, using the scoops. Using the absorbent wipes or pillows, begin the cleanup at the edge of the spill, and work towards the center. Discard the wipes or pillows as they become saturated. Work carefully, as there will be many small glass shards adhering to the wipes or pillows.

Once all excess liquid has been absorbed, pour a minimal amount of water on the spill area from a low height to minimize aerosolization. Using a clean wipe, absorb this water, and repeat this step two more times. Discard all materials used as wet chemo-waste, and have the floor washed by housekeeping personnel.

B. General Spill Procedures

Spills should be cleaned up immediately by a properly protected person trained in the appropriate procedures. Broken glass should be carefully removed. A spill should be identified with a warning sign so that other persons in the area will not become contaminated.

Personnel Contamination

Overt contamination of gloves or gowns, or direct skin or eye contact should be treated as follows:

1. Immediate removal of the gloves or gown.
2. Wash the affected skin area immediately with soap (not germicidal cleaner) and water. For eye exposure, immediately flood the affected eye with water or isotonic eyewash designated for that purpose for at least 5 minutes.
3. Obtain medical attention immediately.

Cleanup of Small Spills

Spills of less than 5 ml or outside a safety cabinet should be cleaned immediately by personnel wearing labcoats, gowns, and double surgical latex gloves and eye proteciton (14).

1. Liquids should be wiped with absorbent gauze pads, while solids should be picked up with wet absorbent gauze. The contaminated surfaces should be cleaned three times using a detergent solution followed by clean water.
2. Any broken glass fragments should be placed in a small cardboard or plastic container and then into an appropriate chemo-waste disposal bag, along with used absorbent pads and any noncleanable contaminated items.
3. Glassware or other contaminated items that are reusable should be placed in a plastic bag and washed in a sink with detergent by a trained employee wearing labcoats or gowns and double surgical latex gloves.

Cleanup of Large Spills

The spread of fluid or powder associated with spills that exceed 5 ml should be limited by gently covering the contaminated area with absorbent sheets or spill-control pads or pillows (14). If a powder is involved (greater than 5 grams) the material should be covered with a damp cloth or towel (5). One must be very cautious so that secondary aerosols will not be created.

Personal protection. Protective apparel should include a respiratory protection respirator when there is a danger of airborne powder or an aerosol being generated. The dispersal of aerosols into the immediate vicinity and the possibility of inhalation are serious matters and must be treated as such.

Chemical inactivators. Chemical inactivators should not be applied to the absorbed drug, because they may produce hazardous by-products.

Criteria for cleaning. All contaminated surfaces should be thoroughly cleaned three times with a detergent solution and then washed with clean water. All contaminated absorbents and other materials should be disposed of in the chemotherapy waste bag.

Spills in Safety Cabinets

Decontamination of all interior hood surfaces may be required after the *above procedures have been followed.* If the HEPA filter of a safety cabinet is clogged, the unit must be labeled "Do not use–contaminated," and the filter must be changed and disposed of properly as soon as possible by trained personnel wearing protective equipment. Protective goggles should be cleaned with an alcohol wipe after the cleanup.

Spill Kits

Spill kits, clearly labeled, should be kept in or near preparation and administration areas. It is suggested that kits include a respirator, chemical splash goggles, two paris of gloves, two sheets (12 × 12) of of absorbent material, 250-ml and 1-liter spill control pillows, a small scoop to collect glass fragments and two large waste-disposal bags.

Storage Areas

Access to areas where chemotherapy agents are stored should be limited to authorized personnel. Such areas should be posted with a large warning sign, a list of all drugs covered by policies, and a sign detailing spill procedures. Facilities used for storing agents should not be used for other drugs and should be designed to prevent containers from falling to the floor. Warning labels should be applied to all containers, as well as the shelves and bins where these containers are permanently stored.

Receiving Damaged Packages

Damaged cartons should be opened in an isolated area by an employee wearing the same protective equipment as is used in the preparation of chemotherapy agents.

1. Broken containers and contaminated packaging mats should be placed in a puncture-resistant receptacle and then in chemotherapy waste disposal bags, which should be closed and placed into appropriate containers, both of which are described under "Waste Disposal."
2. The appropriate protective equipment and waste disposal material should be kept in the area where shipments are received, and employees should be trained in their use and the risks of exposure to chemotherapy agents (15).

Transport

Within the medical facility, drugs should be securely capped or sealed and packaged with impervious packing material for transport.

1. Personnel involved in transporting chemotherapeutics should be cautioned and trained in the necessary procedures should a spill occur, including sealing off the contaminated area and calling for appropriate assistance.

2. All drugs should be labeled with a warning label and clearly identified as cytotoxics. Transport methods that produce stress on
contents should not be used to transport these materials.

V. TRAINING AND INFORMATION DISSEMINATION

A. Personnel Training

All personnel involved in any aspect of the handling shipment and receiving personnel, physicians, nurses, pharmacists, housekeepers,
employees, etc. transporting or storing chemotherapy agents or drugs
must receive information on these agents. The information should include the known risks, relevant techniques and procedures for proper
handling, the proper use of protective equipment and materials, spill
procedures and applicable medical policies (including those dealing
with pregnancy and with staff actively trying to conceive children).

B. Evaluation of Staff Performance

Knowledge and competence of personnel must be evaluated periodically.
It would seem appropriate to perform the initial evaluation after the
first orientation or training session, and then yearly, or as often as
a need is perceived. Evaluation may involve direct observation of an
individual's performance on the job. We have found that "mock" solutions can be used to evaluate preparation techniques; flourescein,
which will fluoresce under ultraviolet light, provides an easy mechanism for detection of poor technique.

C. Information

Each area where these materials are used or prepared should maintain
a looseleaf, index card, or computerized file containing information
about treatment, chemical inactivators, solubility, and stability of the
agents used in the area. The file should be available to employees,
and any special instructions for the handling of specific drugs should
be included. If drugs are administered in a centralized area, such
as an oncology floor, a copy of this file should be available there.
Summaries of relevant procedures should be posted in the appropriate
work areas. A complete policy and procedures manual should be
developed and made available to all employees who are re-quired to
come in contact with these agents.

REFERENCES

1. Harris, C. C. (1976). *Cancer* 37:1014-1023.
2. Sieber, S. M., and Anderson, R. H. (1975). *Adv. Cancer Res.* 22:57-69.
3. Crudi, C. B. (1980). *Nat. Intra. Therapy J.* 3:77-80.
4. Reynolds, R. D., Ignoffo, R., Lawrence, J., et al. (1982). *Cancer Treat. Rep. 66*:1885.
5. Rudolph, R., Suzuki, M., and Luce, J. K. (1979). *Cancer Treat. Rep. 63*:592-537.
6. Slickson, A. A., Mottaz, J., and Weiss, L. W. (1975). *Arch. Dermatol. 111*:1301-1306.
7. Levantine, A., and Almeyda, J. (1974). *Br. J. Dermatol. 90*: 239-242.
8. Council on Scientific Affairs. (1985). *JAMA 253*:1590-1592.
9. Harris, C. C. (1979). *J. Natl. Cancer Inst. 68*:275-277.
10. Crudi, C. B. (1980). *Nat. Intra. Therapy J.* 3:77-80.
11. Neal, A. D., Wadden, R. A., and Chiou, W. L. (1983). *Am J. Hosp. Pharm. 40*:597-601.
12. National Sanitation Foundation Standard 49 for Class II (Larimar Flow) Biohazard Cabinetry. (1983). *National Sanitation Foundation*, Ann Arbor, Mich.
13. U.S. Dept. of Health and Human Services. (1978). Title #003087/RA. National Audiovisual Center, Washington, D.C.
14. Office of Occupational Medicine. (1986). *OSHA Instruction Pub. 8-1.1.*
15. Zimmerman, P. F., Larsen, R. K., Barkley, E. W., and Gallelli, J. F. (1981). *Am. J. Hosp. Pharm. 38*:1693-1695.

14

Medical Surveillance Program

ROSE H. GOLDMAN *Cambridge Hospital, Cambridge, Massachusetts*

I. GENERAL PRINCIPLES

A medical program appropriate for employees in biomedical laboratories
should be based upon an understanding of the work processes and
potential exposures and upon carefully thought out program goals.
Such a program might include a variety of evaluations, such as preplacement examinations (PPE); periodic monitoring evaluations (PME);
prolonged or unusual illness tracking; specific illness or exposure-
related evaluations, surveys or epidemiologic studies; immunization
programs; and exit evaluations. In addition, the program might include an assessment of workers for conditions that place them at ex-
traordinary risk to themselves or co-workers during routine job conditions or exposures. It is critical to remember that, except for acute
exposure-related examinations, most of the medical program is directed
toward preventing disease. This is accomplished through carefully
targeted medical surveillance.
   Medical surveillance of workers involves the collection, assimilation,
and use of biological monitoring, medical screening, or other health
data for developing strategies for the prevention of disease (1). Biological monitoring and medical screening are terms frequently used
in relationship to medical surveillance. Biological monitoring generally
refers to tests of body fluids or skin to indicate an exposure, whereas medical screening is usually concerned with assessing health effects
(preferably at a preclinical stage). The purpose of medical screening
is to identify the early signs or symptoms of disease resultant from

workplace exposure in order to detect a disease process early in its course, when it is reversible or more easily treatable (1).

Example of biological monitoring in the setting of a biomedical laboratory include serum testing for evidence of infectious exposure (e.g., hepatitis B surface antigen) or urine testing for a metabolite of a cytotoxic agent. Medical screening would look for early health effects, such as abnormalities of liver function in suspected HBV infection, or signs and symptoms related to the biological activity of an end product.

For the most part, medical screening is a form of secondary prevention. More primary means of preventing occupational illness preclude the initiation of the disease process. In the biomedical laboratory this could be accomplished either by eliminating or limiting exposures, as with the use of hazard control ventilation and good work practices, or by protecting individuals through immunization programs. Environmental sampling and biolgoical monitoring assess exposures and the adequacy of primary control measures. Medical screening, too, can serve as a check of the efficacy of procedures for biological and physical containment in protecting worker's health.

The effectiveness of the medical surveillance program may be limited by a number of factors, including the small size of the groups under study; variability in individual exposures; a potentially long induction-latency period from the beginning of exposure to the appearance of adverse health effects (e.g., cancer); uncertainty as to the adverse effects that may occur; uncertainty as to the connection between an exposure and an effect; and the lack of sensitivity or specificity of many screening tests or outcome measures to detect early changes of disease. In designing an appropriate medical program, one must first clearly define the goals to be achieved and an appropriate protocol to reach those goals (2). The primary goal of a pre-placement examination (PPE), for example, is to identify any medical condition that might put the worker at an increased risk to himself or others as a result of certain job exposures or activities. When such a condition is identified, then an appropriate placement or accommodation can be made. Another important goal is to provide a baseline assessment of health status that can be used for comparison to future testing.

Goals for periodic monitoring evaluations (PME) for biomedical laboratory workers might include:

1. To detect evidence of exposure to biohazards or presence of subclinical infections (e.g., with a PPD test for tuberculosis or serum titres for viral infections); to detect evidence of exposure to chemical toxins.
2. To detect early clinical signs or symptoms of disease.
3. To provide data indicating the need for evaluation of control measures.

4. To detect any changes in the health of the employee since the time of the PPE or the last PME that might indicate the need for a change in job placement or in the work process.
5. To detect any patterns of disease in the workforce that might indicate any underlying work-related problems.

Another important goal of the medical surveillance program is to fulfill any regulatory requirements for protecting workers' health. Optional goals for either the PPE, or the PME in some settings, are the provision of recommended primary care periodic health tests, which are aimed at early recognition of treatable nonoccupational diseases such as hypertension or breast cancer.

For medical evaluations to be useful, they must be targeted to the potential risks and hazards of the laboratory. For the program to be effective, there must be regular analysis of the collected data. Thus the design of an appropriate program involves not only close attention to the general goals but also careful scrutiny of job descriptions and potential exposures.

II. POTENTIAL HAZARDS IN THE BIOMEDICAL LABORATORY

The potential occupational hazards associated with biomedical laboratories will vary according to the microbial species, genetically active materials, products, reagents, processes used, and type of laboratory setting. Potential exposures in biomedical or recombinant DNA (biotechnology) laboratories include biological agents, fermentation (or final) products, chemicals or solvents, carcinogens, radioactive materials, nonionizing radiation (e.g., laser), animal handling, and musculoskeletal stresses (3). Despite these many variations, one can group potential hazards into several categories.

A. Biological Agents

Historically there have been many published reports of laboratory-acquired infections (4-9). In a cumulative series that reported a total of 3,921 cases of laboratory-associated infections (4-6), fewer than 20% of the cases could be attributed to documented accidents, such as mouth pipetting and the use of needles and syringes. Exposure to infectious aerosols was considered to be a plausible but unconfirmed source of infection for the nonaccident-related reported cases. Studies of Harrington and Shannon (7) showed laboratory personnel to have higher rates of tuberculosis, shigellosis, and hepatitis than the general population. Recent reviews have analyzed infectious agents according to infection potential and proposed microbiological practices,

laboratory facilities, and safety equipment necessary to control the
hazard (9-11).

More recently, there have been reports of HIV seroconversion
following needlestick injuries and mucosal contact in health care workers (12-14). Two laboratory workers at the National Institutes of Health
working with a highly concentrated strain of HIV have become infected
with the virus, apparently as a result of workplace exposure (15,16).

There are also incidents and concerns about noninfectious diseases
(e.g., cancer) that could be transmitted by biological agents. A very
unusual case of needlestick transmission of human colonic adenocarcinoma was reported in an otherwise healthy laboratory worker (17).

Concerns have been raised at the Institute Pasteur in Paris following the discovery of three cases of cancer among staff of two adjoining
laboratories studying oncogenes and other mutagenic substances. Subsequently two additional cases were found in the same building (18).
A scientific commission has been appointed to investigate this cluster
of cases, but in the interim, France's social security agency has classified one of the cases as an "occupational disease." The International
Agency for Research on Cancer (IARC) responded to these incidents
by proposing a worldwide retrospective epidemiologic study of the
molecular biology research population.

Although these various reports, retrospective studies, and anecdotal information suggest that laboratory personnel are at some increased risk of being infected by the agents they handle, actual incidence rates of infection are typically not available (9). To reduce
the risks of infection in laboratory personnel, it is necessary to utilize
proper work practices and methods of containment as described in a
previous chapter of this book and in other resources (9,10). It is
important to note that the risk assessment and biosafety levels recommended (9) presuppose a population of immunocompetent individuals
(see section on individual risk factors).

B. Chemicals

Solvents and chemical reagents are used extensively in the laboratory
and are an integral part of clinical as well as research and development
efforts, such as fixation of tissues; isolation or modification of nucleic
acids and proteins; and extraction, separation, and purification of
biotechnology products. The medical surveillance program must be
structured so that it can identify symptom complaints or other findings
suggestive of toxicity or excessive exposure. It should also recognize
employees who may have medical conditions that place them at an increased risk for developing toxic effects from specific chemicals.

Solvents commonly used in the laboratory, such as phenol, methanol, chloroform, xylene, and ethers, can all produce an acute syndrome
characterized by decreased alertness, lightheadedness, headache,

nausea, and drying and irritation of the skin and mucous membranes. The problem of xylene poisoning in laboratory workers, for example, has been well documented in a survey by the California Association of Cytotechnologists (CAC) (19).

Formalin (formaldehyde) is another fixative common in the pathology laboratory and associated with irritation of the mucous membranes (eyes, nose, and throat irritation). Formaldehyde has also been found to be a carcinogen in laboratory animals (20).

In addition to the risk posed by exposure to single toxic agents in the laboratories, workers are often exposed to a mixture of chemicals, which may produce reactions that are then additive or synergistic.

Certain chemical reagents used in the laboratory can cause sensitization by skin contact or inhalation or can precipitate allergic reactions in persons already sensitized. Allergic responses may include asthma, cough, rhinitis (nasal congestion), contact dermatitis, or hives. Some examples of known sensitizers used in the biomedical laboratory include hydrazine, piperazine, formaldehyde, ethylenediamine, and antibiotics (21). In addition, sensitization by small amounts of allergen in the laboratory might be followed by an acute immediate type of hypersensitivity reaction upon exposure to the same or a related material in a nonoccupational situation, as when a taking orally, or applying to the skin a medication containing the agent (22). For example, sensitization to neomycin, which is sometimes used in tissue culture media, might lead to contact dermatitis if the worker applied a topical antibiotic containing neomycin. Another type of reaction could occur if a worker sensitized to an agent in the laboratory then ingested an immuno-chemically related medication and then developed a systemic (generalized) contact-type dermatitis (22).

For these reasons, known sensiting agents should be handled with special care to avoid inhalation or skin contact. Although rubber gloves can be worn for protection against reagents or detergents, the glove can also be the source of allergic reactions due to exposure to rubber additives, which are allergens. In such instances this problem may be avoided by wearing cotton-lined rubber gloves.

It is important for the medical surveillance program to be familiar with the properties of the chemicals used and to set up appropriate medical monitoring and/or training and emergency response programs where indicated. As an example, if a laboratory were using an agent that can react to release hydrogen cyanide gas (e.g., cyanogen-bromide), the medical program should then provide information on the first aid response to cyanide poisoning (i.e., use of amyl nitrate and oxygen).

Carcinogens and mutagens find their way into the laboratory in a variety of ways, as in the use of certain solvents (such as benzene) or agents used during DNA sequencing (ethidium bromide). Substances can be classed as to their carcinogenic potential, e.g., as known carcinogen in humans, suspect carcinogen, or animal carcinogen as

described by IARC (23). Since known thresholds may not exist, carcinogens and mutagens should be handled under strictly controlled
conditions in order to minimize exposure (19).

Biolgoical monitoring of workers handling carcinogenic agents is
still a controversial area. Tests have been employed that assess employees' urine for mutagenicity or metabolites of the carcinogenic agents
or that examine the blood for sister chromatid exchange. Although
implying exposure, these tests may not be very specific and are still
difficult to interpret when applied on an individual basis (24). In the
future more specific monitoring of agents might be available through
the use of urinary monitoring of DNA adducts of specific agents. Accidental exposure to a carcinogen or mutagen may have even greater
implications for a pregnant employee or those planning families (see
section on risk factors).

C. Radioisotopes

Radioisotopes are important reagents used in biomedical research. The
major isotopes used include I^{125}, P^{32}, S^{35}, C^{14}, and H^3. The Nuclear
Regulatory Commission (NRC) has specific requirements related to
the use of radioisotopes and also has regulations related to the monitoring of occupationally exposed personnel for external or internal
radiation dose, or both (25). Three devices used for monitoring external radiation dose are film badge dosimeters, thermoluminescent
dosimeters (TLD), and pocket ionization dosimeters. Methods of internal dose monitoring will depend on the distribution and elimination
of the radioisotope from the body. Some (such as P^{32}) can be monitored through urine analysis, while others need to be followed through
direct counting of a particular organ or the whole body. Since iodine
is taken up by the thyroid, the thyroid gland can be scanned with an
NaI scintillation detector and analyzer to measure the amount of radio-
iodine present. A permanent record of radiation dose received by
personnel is required by NRC regulations. Since the embryo and fetus
are more susceptible to the effects of radiation than the adult, a pregnant employee working with radioactivity needs to consider the potential for exposure and possible dose, and whether or not to continue
to work with radioactivity during the gestation period. For protection
of the unborn child, NRC recommends that an occupationally exposed
woman who is pregnant not receive more then 0.5 rem during the 9-
month gestation period (26).

D. Product Hazards

Biotechnology employees working in the fermentation and purification
processes encounter risks similar to those of workers in pharmaceutical
and chemical industries, that is, exposure to the final products that

are physiologically active. Because of the physiologically reactive
nature of many of these biological and pharmaceutical products, exposures to even minute quantities of some may cause medically significant
side effects. In fact, various illnesses associated with drug manufacture have been reported, such as adrenocortical suppression in workers employed in manufacturing synthetic glucocorticosteriods (27),
and gynecomastia in males and intermenstrual bleeding in female workers
employed at a plant formulating estrogen containing oral contraceptives
(28). Biotechnology companies are producing a wide variety of products, including interferon, growth hormone, insulin, various cell line
stimulating factors, enzyme inhibitors, etc., which could potentially
exert effects in exposed employees. The medical surveillance program
should examine the feasibility of assessing workers for any specific
effects related to the products.

In addition to the possibility of experiencing the effects of the
biological activity of an agent, biotechnology workers also face the
risk of allergic reactions due to sensitization to proteins and peptides
generated during the course of fermentation and extraction.

E. Allergic Risks of Animal Handling

The development of allergies to laboratory animals, termed "laboratory
animal allergy" syndrome (or LAA), has been well described and recently reviewed (29). The manifestations of LAA include cough, wheezing (asthma), eye watering, itchy eyes and skin, sneezing, or skin
rash and can develop in less than one to several years after initial
contact. The prevalence of LAA has been reported to range between
11-30% (30), with one prospective study reporting a first-year incidence of 15% (31). Workers can develop allergies to the dander of
animals or to (aerosolized) urine proteins. Another source of sensitizing agents in the animal facility can be found in the animal food or
bedding, which might contain supplements such as antibiotics, arsanilic
acid, ethylenediamine, and piperazine. The role of atopy (i.e., presence of other IgE-mediated allergies, such as hayfever and asthma)
as a predisposing factor to the development of LAA is controversial
(29,32,33). However, it is important to note that a substantial number of nonatopic individuals can also be sensitized. Thus, it is important to reduce exposures in the animal care facility through such measures as filter-top cages, well-designed ventilation systems, and employee use of protective clothing and gloves.

F. Physical (or Musculoskeletal) Stresses

Principles of ergonomics, such as proper positioning of employees at
work stations (well-fitting chairs and tables) apply in the laboratory
setting. Glass washers can develop strained shoulders or hand ten-

donitis from awkward postures and repetitive motions. For example, we saw an employee with shoulder strain caused from cleaning many large Erlenmeyer flasks with a hand brush. She also had to lift and carry trays of glassware. We have also seen a tissue culture technician who developed tendonitis fo the wrist from repeated twisting of media bottle caps. Materials handlers who load and unload boxes are at increased risk for back strain. These problems can be eliminated by analyzing jobs and adjusting processes in order to avoid stressful positions. In the case of the dishwasher mentioned above, the problems were eliminated with the addition of a mechanical dishwasher and a cart on wheels for transporting the glassware.

III. CONDITIONS INCREASING THE RISK OF ADVERSE HEALTH OUTCOME

Conditions which might place some individuals at an increased risk of adverse health outcomes as a result of performing certain laboratory procedures include diseases or drugs that alter host defenses, allergies, inability to receive a specific vaccination, and reproductive system concerns (see below).

It is important to recognize these risk factors and then to evaluate the nature of an employee's work. In this way one can determine whether or not some change or accommodation is necessary to eliminate the risk for that employee. It may be found that the employee will have to be removed from that particular job or task. These decisions need to be made on a case-by-case basis with input from the employee, management, the occupational medical service, and, in some instances, the employee's physician.

A. Deficiencies of Host Defenses

In the light of the possibility of laboratory-acquired infections, it is desirable that, in addition to environmental measures to maximize safety in dealing with dangerous agents, the individuals involved have unimpaired defenses against infection, or be provided additional protective measures, if indicated. In fact, the risk assessments and Biosafety Levels recommended in the agent summary statements of the CDC-NIH Biosafety in Microbiological and Biomedical Laboratories (9) presuppose a population of immunocompetent individuals.

Some workers may face increased risks for certain infections because they have conditions that alter host defenses at body surfaces or impair the functioning of the immune system. Host defenses at the skin are disrupted by diseases such as chronic dermatitis, eczema, and psoriasis. Skin disease not only leads to an increased risk of skin infection but also to an increased risk of systemic infection transmitted

through the broken skin surface. Such workers may have to wear
gloves at all times or avoid handling certain organisms.

Proteciton against infection by pathogenic organisms at the level
of the surface (or mucosa) of the gastrointestinal tract is partially
afforded by the presence of normal bacterial flora. Antibiotic treatment
that suppresses normal flora or inflammatory bowel conditions (such
as colitis or Crohn's disease) that disrupt the mucosal surface may
interfere with the protective capabilities of the gastrointestinal tract.

The immune system, which consists of antibody-mediated (B cell),
cell-mediated (T cell), phagocytosis, and complement defends against
a constant assault by viral, bacterial, fungal, protozoal, and nonrepli-
cating agents (34). Each system may act independently or in concert
with one or more of the others. The ineffective operation of any one
aspect of the immune system either due to a congenital or acquired
problem, particularly predisposes individuals to certain types of infections, as listed below:

1. Antibody (B cell) immunodeficiency disorders consist of diseases
    with decreased immunoglobulines, which leads to recurrent bac-
    terial infections particularly with streptococcus pneumoniae and
    haemophilus influenzae.
2. Defective cell-mediated immunity (T cell) leads to an increased
    susceptibility to infection with fungal, protozoal, and viral orga-
    nisms that may present as pneumonia, or chronic infection of the
    skin, mucous membranes, or other organs.
3. Phagocytic disorders predominantly predispose individuals to in-
    creased susceptibility to bacterial infections.
4. Complement abnormalities and deficiencies can be associated with
    increased susceptibility to bacterial infections.

Most primary immunodeficiencies usually manifest in childhood with
devastating effects on health, and thus would not be a consideration
in adults at the workplace. Persons with selective IgA deficiency,
the most common immunodeficiency disorder, can live to adulthood in
which the most frequent presenting symptoms are recurrent sinopul-
monary viral or bacterial infections. Allergies, autoimmune disease,
and gastrointestinal disease can also be associated with IgA deficiency,
although many persons remain relatively asymptomatic.

Secondary or acquired immunodeficiencies occur more commonly
(see 34). Some of the more common temporary causes of immunodeficiency that one might see in the working population include steroid
treatment for medical conditions such as asthma or inflammatory bowel
disease, acute viral infection, and pregnancy. More longstanding problems with immunodeficiency might arise to a mild degree in workers
with diabetes mellitus or certain connective tissue diseases, or to a
more severe degree in those receiving cancer chrmotherapy or those

with HIV infection. In certain cases measures can be taken to additionally protect immunocompromised workers from infection. For example, a worker without a spleen who thus has an increased risk of infection with encapsulated organisms could receive the Pneumovax immunization against pneumococcus. In more serious cases of immune deficiency, employees may need to be removed from working with any potentially infective organisms.

B. Reproductive Hazards

Concerns about occupational risks to the reproductive system may involve questions about infertility in either sex; exposures during pregnancy that might result in adverse outcomes such as spontaneous abortion or birth defects; male exposures that could result in adverse pregnancy outcomes through mechanisms such as damage to the sperm, transmission of toxic agents in seminal fluid, or contact of the pregnant woman with her partner's contaminated clothing; or breastfeeding as a source of toxic exposure to the child after birth (35).

The laboratory may contain certain biological, chemical, and/or physical hazards that could be toxic to the reproductive system. Anesthetic gases, radiation, ethylene oxide, cytotoxic drugs, and certain solvents are some agents established or suspected as causing adverse pregnancy outcome from human studies (35). Chemical agents suspected of causing adverse birth outcomes identified in animal experiments include chloroform, glycol ethers, and xylene (35).

Concerning biological agents, although some infections can result in infertility in either sex (e.g., mumps, chlamydia, and gonorrhea), it would be unlikely to acquire these infections in the laboratory setting. More commonly, concerns are directed to the potential congenital infection of a child as a result of a pregnant employee acquiring a work-related infection. Infectious agents can be described according to the level of risk of congenital transmission (see 36). Clinical laboratory workers with no direct patient contact probably would be more likely to confront CMV, rubella, HBV, and toxoplasmosis. Since HIV has been associated with congenital infections in the community, concerns would be raised about a potential occupational exposure in a pregnant employee working with HIV. Concerns may also be raised when a pregnant employee is handling a pathogenic virus in which the reproductive toxicity is not specifically known.

In addressing the job placement or work content of a pregnant laboratory employee or employees (male or female) attempting to conceive, one needs an approach that considers several factors: the agent and what is known about its level of risk, the means available to prevent exposure, and availability of other jobs (35). The employee should be completely informed of the potential risks. Decisions need to be made that consider the needs of the pregnant employee and the health of the fetus, and the decision-making process should involve the employer, employee, and the employee's physician.

C. Allergies

In the biomedical laboratory setting, workers can develop allergies to proteins (biological products derived form raw materials, fermentation products, or enzymes), chemicals, as well as to the dander or (aerosolized) urine proteins of animals. Also of concern are allergies to substances contained within a vaccine that might interfere with the ability to be immunized, which may be considered an important protective measure for certain laboratory work. As an example, allergy to baker's yeast (*Saccharomyces cervisiae*) might prevent receipt of recombinant DNA HBV vaccine.

If the medical surveillance program identifies an individual with any condition that increases the risk of an adverse health outcome. the response usually has to be individualized and depends on the level of increased risk, the nature of the hazards, and opportunities for limiting exposure. Responses to such an individual might include a change in the work process, reassignment of specific tasks, job transfer, or additional protective measures.

IV. PROGRAM DESIGN

Having defined the program's goals and identified potential hazards and protective measures, one can then determine the content and participants of the medical surveillance program. A preplacement evaluation might be more extensive, including history, physical examination, and laboratory testing for individuals handling biohazards agents (particularly those requiring Biosafety 3 or above) or particularly toxic chemicals. A more limited evaluation might be indicated for those working with less hazardous agents (9). The program should also be in compliance with federal, state, and local legislation for the protection of the welfare of the employee.

At the time of hiring it is reasonable to give all individuals at a laboratory or company a questionnaire that includes an occupational and medical history. Administrative staff, or those working with low level hazardous infectious agents, who have no other specific risk factors may need no further evaluation, except possibly a serum sample for storage. Others who might require a more complete type of evaluation include:

Scientists working with recDNA requiring Biosafety level 3 or 4 containment

Individuals working with or potentially exposed to Biosafety level 3 or 4 infectious agents or infected animals

Individuals handling teratogenic, carcinogenic, and mutagenic chemi-
    cals
Individuals working with oncogenic and teratogenic microbial agents
Animal caretakers, maintenance workers, custodial and housekeeping
    staffs, or others who work in areas where potential exposure to
    hazardous materials can be expected
Personnel working in areas shared by laboratory and clerical personnel,
    where overcrowding exists, or where containment systems are
    lacking
Individual identified by questionnaire who have some condition that
    places them at an increased risk of handling certain agents

A. Contents of the Medical Surveillance Program

Preplacement Evaluation

The preplacement examination serves as a baseline exam and should
include a medical and occupational history as a minimum and physical
examination and other testing as indicated.

Medical history. A complete medical history should be taken that
particularly emphasizes certain areas, such as host defenses, allergic
conditions, immunization history, and past significant infectious diseases. It may also be useful to inquire about the presence of selected
symptoms (e.g., persistent nasal congestion, headache, cough, diarrhea,
other gastrointestinal complaints) that might be used in the future
for tracking early manifestations of illness.

Occupational history. The occupational health history is the corner-
stone of the occupational health examination (37,38). The initial occupational history includes a job profile (past jobs), a listing of symptoms, illnesses, or injuries related to past jobs or exposures, a description of the new job along with specific questions relevant to the requirements or potential hazards involved in that work, and a listing of significant community and home exposures.

Physical examination. When a general physical examination is performed, particular attention should be paid to evidence of altered host
defenses or allergy (e.g., lymphadenopathy, hepatosplenomegaly,
skin rash, wheezing).

Laboratory testing. Ideally, laboratory testing should be targeted
to exposures. In the case of biohazards there are few specific tests.
As a result, it may be a prudent practice to store a baseline serum
sample that could be used as a reference in evaluating an unforeseen
problem in the future. If some employees are handling viruses, in

some circumstances it might be advisable to obtain a serum sample for baseline titre levels and then recheck periodically for conversion to a positive test. Such a conversion or marked titre rise after starting work suggests that an exposure may have occurred at work, and thus an evaluation of control measures is needed.

Biological monitoring for solvent exposure (e.g., urinary metabolites) is rarely indicated in this setting. Such biological monitoring is less useful, in general, because individual results cannot be specifically linked to health effects, although means of group results might be more useful and interpretable (39). Air sampling for solvents (in which air concentrations are linked to permissable levels and health effects) would generally be preferable. As mentioned earlier, efficacy of monitoring for carcinogen exposure is not yet established in most cases, but new techniques are being developed that may be more specific and useful.

A complete blood count with differential is a general test which may indicate compromised host defenses or suggest presence of infections. Additional testing (such as liver function tests) might be indicated, depending upon what other biological or chemical agents will be handled.

Urinalysis or urine dipstick can demonstrate proteinuria, reflecting kidney disease, or glucose, reflecting diabetes mellitus. These conditions could have implications concerning host defense capabilities.

It is rarely necessary to do a chest radiograph in an otherwise healthy person without pulmonary symptoms, who has a normal physical exam, and negative PPD. If the PPD has recently converted to a positive, then a chest radiograph is ordered to detect the presence of active tuberculosis. Periodic chest radiographs in asymptomatic employees with longstanding positive PPDs have not been found to be useful (36,40).

Pulmonary function testing is not done routinely unless there are specific indications, such as the assessment of fittness to wear a respirator.

Preplacement patch testing generally has not been proven to be an effective means of preventing allergic reactions among animal handlers.

It is also important not to confuse tests for general health maintenance with those that are targeted to the occupational setting. For example, an electrocardiogram would not be an indicated occupational medical surveillance test given the usual risks encountered in biomedical laboratories.

Preventive measures. After assessing an individual and potential exposures at the preplacement evaluation, recommendations can be made for indicated preventive measures such as immunizations, further medical screening tests, or job accommodations or restrictions.

Immunizations: If there is a risk of exposure to biologic agents for which immunizations are available, employees who have no contraindications should be immunized (see summary list of agent-specific vaccines). The preplacement examination is a good time to review the immunization history of every laboratory worker to be sure he or she is up to date or has had basic immunizations such as tetanus, diphtheria, and rubella.

Tuberculosis: If working in a clinical medical laboratory or other setting where there is risk of tuberculosis exposure, employees can be screened for tuberculosis by yearly skin test (purified protein derivative (PPD) if not previously found to be positive.

Accommodations/restrictions: Recommendations for job accommodations and/or restrictions should be appropriate and avoid exclusions or restrictions based on inaccurate presumptions of risks or future disabilities.

Periodic Monitoring Evaluations

A Periodic Monitoring Examination (PME) consists of an interval occupational and medical history, biological monitoring or medical screening tests (if indicated), targeted physical examination (if indicated), and review of interval monitoring results. When possible, the interval evaluation should be structured to look for effects of potential exposures, for example allergic symptoms, complaints or laboratory tests related to the pharmacologic effects of end-products, by-products, media components, or inactivated biological agents, and signs or symptoms of infectious exposure. Serum can be drawn and stored at appropriate intervals (e.g., yearly or every second year). In order to detect a significant change in viral titres, one would need to compare results from serum drawn at the time of the periodic exam to that from the baseline exam, with both samples being tested at the same time. Monitoring for radiation exposure or dose is generally checked more frequently, such as monthly.

Exit Evaluations

It is prudent to perform an "exit" evaluation at the time an employee leaves the company and laboratory. This evaluation, in general, would resemble the PPE and contain an interval history. It could be modified if the person had undergone either the PPE or PME within 6 months. In some cases, it would be appropriate to draw a serum sample for storage.

Agent-Specific Surveillance

A reporting system may prove useful when tracking symptoms related to a specific infectious etiologic agent or to effects of final products. For example, reliable surveillance of personnel for laboratory-associated rickettsial infections could lead to the early treatment of rickettsial disease and prevent the occurrence of severe acute disease (9).

Accidental Exposure/Incident Reporting

Incidents such as needlesticks or other accidental exposures should
be reported to the medical program to determine if additional tests,
treatment, or follow-up is needed. The incident should also be reported to the biosafety staff in order to determine what measures are
needed to avoid a future similar accident. Protocols for handling anticipated accidents, such as needle sticks and spills, should be developed and employees trained accordingly. Laboratory personnel can
be trained as "first responders" for various medical emergencies (10).

Medical incidents should be reported and recorded in compliance
with legal requirements, e.g., OSHA.

B. Recordkeeping and Result Notifications

Workers should be completely informed of the results of their occupational medicine evaluations. The personnel office or the biosafety officer
need only receive information about those medical conditions or re-
strictions/accommodations that relate specifically to individuals' fitness
for work as recommended by the AOMA Code of the Ethical Conduct for
Physicians Providing Occupational Medical Services.

It is advisable to periodically review the data collected in the medical surveillance program. For larger laboratories one could, for example, examine prevalence rates of symptoms or abnormalities in different job or exposure groupings. The results of the medical monitoring
data could then be compared to environmental monitoring data.

REFERENCES

1. Halperin, W. E., and Frazier, T. M. (1985). Survillance for
   the effects of workplace exposure. *Ann. Rev. Public Health 6*:
   419-432.
2. Goldman, R. H. (1986). General occupational health history
   and examination. *J. Occup. Med. 28*:967-974.
3. Landrigan, P. J., Harrington, J. M. and Elliot, L. J. (1984).
   The biotechnology industry. In *Recent Advances in Occupational
   Health*, J. M. Harrington (Ed.). New York, Churchill Living-
   stone, pp. 3-13.
4. Sulkin, S. E., and Pike, R. M. (1951). Survey of laboratory-
   acquired infections. *Am. J. Public Health 41*:769-781.
5. Pike, R. M., Sulkin, S. E., and Schulze, M. L. (1965). Continuing importance of laboratory-acquired infections. *Am. J.
   Public Health 55*:190-199.
6. Pike, R. M. (1976). Laboratory associated infections: Summary
   and analysis of 3921 cases. *Health Lab. Sci. 13*:105-114.

7. Harrington, J. M., and Shannon, H. S. (1976). Incidence of tuberculosis, hepatitis, brucellosis, and shigellosis in British medical laboratory workers. *Br. Med. J. 1*:759-762.
8. Wedum, A. G. (1964). Laboratory safety in research with infections aerosols. *Pub. Health Rep. US 79*:619-633.
9. U.S. Dept. of Health and Human Services, Centers for Disease Control and National Institutes of Health. (1984). Biosafety in Microbiological and Biomedical Laboratories. Publication no (CDC) 84-8395. Washington, D.C.: U.S. Govt. Printing Office.
10. Miller, B. M. (Ed.). (1986). *Laboratory Safety*: *Principles and Practices*. Washington, D.C.: American Society for Microbiology.
11. Liberman, D. F. (1984). Biosafety in biotechnology: A risk assessment overview. *Developments in Industrial Microbiology 25*:69-75.
12. Stricof, R. L., and Morse, D. L. (1985). HTLV-III/LAV seroconversion following a deep intromuscular needlestick injury. *N. Eng. J. Med. 314*:115 (letter).
13. Oksenhendler, E., et al. (1986). HIV infection with seroconversion after a superficial needlestick injury to the finger. *N. Eng. J. Med. 315*:582 (letter).
14. CDC. (1987). Update: Human immunodeficiency visur infections in health-care workers exposed to blood of infected patients. *MMWR 36*:285-289.
15. VNA Occupational Safety and Hazards Reporter. 9/9/87:565.
16. VNA Occupational Safety and Health Reproter. 10/14/87:824.
17. Gugel, E. A., and Sanders, M. E. (1986). Needle-stick transmission of human colonic adenocarcinoma. *N. Eng. J. Med. 315*: 1487 (letter).
18. McGraw-Hill's Biotechnology Newswatch. (1986). *6*:1-2.
19. 29 CFR Part 1910: Health and Safety Standards. (1986). Occupational exposures to toxic substances in laboratories; Proposal Rule. *Federal Register 51*:26660-26684.
20. IARC. (1982). *Monographs on the Evaluation of the Carcinogenic Risk of Chemicals to Humans*. *Some Industrial Chemical and Dyestuffs*, Vol. 29, IARC. LYON, pp. 345-389.
21. AIHA Biohazards Committee. (1985). *Biohazards Reference Manual*: *Allergies*. Akron: American Industrial Hygiene Association, pp. 13-19.
22. Fisher, A. A. (1976). Allergic dermatitis medicamentosa; The "systemic contact-type variety." *Cutis. 18*:637-639,642.
23. IARC. (1982). *Monographs in the Evaluation of the Carcinogenic Risk of Chemicals to Humans*. *Chemicals, Industrial Processes and Industries Associated with Cancer in Humans*, Supplement 4. IARC, LYON.
24. McDiarmid, M. A., and Emmet, E. A. (1987). Biologic monitoring and medical surveillance of workers exposed to anti reoplastic agents. *Sem. Occup. Med. 2*:109-117.

25. NRC regulations. 10 CRF 20. 202, personal monitoring require-
    ments.
26. U.S. Nuclear Regulatory Commission. (1975). Regulatory guide
    8.13. Instructions concerning prenatal radiation exposure. Pen-
    sion 1, November 1975. Office of Standards Development, U.S.
    Nuclear Regulatory Commission.
27. Newton, R. W., Browning, M. C. K., Igbal, J., Piercy, N.,
    and Adamson, D. G. (1978). Adrenal cortical suppression in
    workers manufacturing synthetic glucocorticords. *Br. Med. J.*
    1:73-75.
28. Harrington, J. M., Stein, G. F., Rivera, R. V., deMorales, A.
    V. (1978). The occupational hazards of formulating oral contra-
    ceptives–a survey of plant employees. *Arch. Environ. Health*
    33:12-14.
29. Bland, S. M., Evans, R., and Rivera, J. C. (1987). Allergy
    to laboratory animals in health care personnel. *Occup. Med.*:
    State of the Art Reviews 2:525-546.
30. Agrup, G., Belin, L., Sjostedt, L., and Skerfving, S. (1986).
    Allergy to laboratory animals in laboratory technicians and ani-
    mal keepers. *Brit. J. Ind. Med. 43*:192-198.
31. Davies, G. E., Thompson, A. V., Niewda, Z., et al. (1983).
    Allergy to laboratory animals: a retrospective and prospective
    study. *Brit. J. Ind. Med. 40*:442-449.
32. Cockcroft, A., McCarthy, P., Edwards, J., and Anderson, N.
    (1981). Allergy in laboratory animal workers. *Lancet 1*:827-
    830.
33. Slovak, A. J. M., and HILL, R. N. (1987). Does atopy have
    any predictive value for laboratory animal allergy? A comparison
    of different concepts of atopy. *Brit. J. Ind. Med. 44*:129-132.
34. Ammann, A. J. (1987). *Immunodeficiency Diseases in Basic and*
    Clinical Immunology, Stites, D. P., Stubo, J. D., and Wells,
    J. V. (Eds.). Norwalk, Conn. and Los Altos, Calif.: Appleton
    and Lange, pp. 317-355.
35. Welch, L. S. (1986). Decision making about reproductive haz-
    ards. *Seminars in Occu. Med. 1*:97-106.
36. Sheretz, R. J., and Hampton, A. L. (1986). *Infection Control*
    Aspects of Hospital Employee Health in Preventional Control of
    Nosocomial Infections. Werzel, R. P. (Ed.). Baltimore: William
    and Wilkins, pp. 175-204.
37. Goldman, R. H., and Peters, J. M. (1981). The occupational
    and environmental health history. *JAMA 246*:2831-2836.
38. Occupational Environmental Health Committee of the American
    Lung Association of San Diego and Imperial Countries. (1983).
    Taking the occupational history. *Ann. Inter. Med. 99*:641-651.

39. Bureau of National Affairs. (1984). Report of ACGIH Biological
     Exposure Indies Committee. Occupational Safety & Health Re-
     porter, 5/31/84, 1376-1378.
40. Reeves, S. A., and Noble, R. C. (1983). Ineffectiveness of
     annual chest roentgenograms in tuberculin skin test-positive
     hospital employees. *Am. J. Infect. Control* 11:212-216.

15
Chemical Health Risks
in Biohazards Management

LAWRENCE M. GIBBS*   *Yale University, New Haven, Connecticut*

I. INTRODUCTION

The only way to eliminate health and safety risks associated with bio-medical research and development (R&D) activity is to eliminate R&D activity itself. While this statement may sound ludicrous, there is no way to completely eliminate risk from our lives. In reality, a risk-free world would most likely be quite dull and boring. However, this does not infer that we should make no attempt to minimize the risks associated with biomedical research to an acceptable level.

We are constantly exposed to risks throughout our lives, from the moment of birth until death. Whether walking across the street, riding in an automobile, flying in an airplane, participating in sports activities, eating, drinking, smoking, etc., we are constantly exposed to risks which can produce adverse health effects.

The same is true of risks due to chemicals. There is no way to eliminate completely the risks due to the use and handling of chemicals. Nor should this necessarily be the goal we strive to achieve. We can, and should, strive to reduce/minimize the identified chemical health risks to an acceptable level. There are few, if any, who would disagree with this concept. The difficulty lies in determining not only what levels of chemical exposure constitute an acceptable risk, but also includes who determines the acceptibility of the risk. The acceptable level determination is often debated and argued about within the professional health and safety community and often challenged by the employees and the public.

Current affiliation: Department of Environmental Health and Safety, The University of Connecticut, Storrs, Connecticut

The ultimate determination of acceptable risk associated with an event is the responsibility of each institution, facility, or organization. Each organization is accountable and eventually must assume the liability and responsibility for protection of their most valuable asset, their employees, from the workplace environmental risks. The purpose of this chapter is to identify some factors to be considered in evaluating the risks associated with the use of chemicals in facilities where biohazard management is also of concern.

II. CHEMICAL RISK FACTORS

Chemical risk factors can be classified as chemical safety risks and chemical health risks, depending on the type of potential hazard posed to personnel and facilities. Chemical safety risks are associated primarily with those properties of chemicals that can produce adverse outcomes in short periods of time and are external to the worker's body. Fires, burns, explosions, runaway reactions, etc., are examples of chemical safety risks. Assessment of chemical safety risks includes evaluation of such properties as the flammability, corrosivity, reactivity, and explosion hazard. Evaluation of these types of properties are necessary for an assessment of the total risk due to chemicals. In research laboratories, chemical safety risk factors are often underemphasized by the laboratory staff. They are usually more concerned about the chemical health risks. This is a problem because the chemical safety risks may present greater risk to personnel and the institution than the chemical health risks.

A. Chemical Health Risk

Many of the chemicals used in the laboratory are known to be toxic (1). New and untested substances that present unknown hazards are also frequently encountered. Proper evaluation of the health hazard potential requires an understanding of the types of toxicity, routes of possible exposure, and a recognition of the major classes of toxic chemicals (Tables 1 and 2).

Toxicity

The toxicity of a chemical refers to its ability to damage or interfere with the metabolism of living tissue. This may be damage to an organ system, such as the liver, caused by excessive ethyl alcohol consumption or overexposure to many of the halogenated solvents, kidney disfunction caused by overexposure to chloroform, disruption of a biochemical process, such as the blood-forming mechanism as caused by benzene, or disruption of an enzyme system by cyanide gas at some

TABLE 1 Major Classes of Organic Solvents

Chemical class	Homolog	Examples
Saturated aliphatic solvents (alkanes)	$C_n H_{2n+2}$	Hexane
Unsaturated aliphatic solvents	$C_n H_{2n}$	Hexane
Aeromatic hydrocarbon solvents		Benzene Benzo(a) pyrene Benzo(a) anthracene
Alcohols	R-OH	Methanol

Glycols	R—C—C—R
	   | |
	  OH OH
		Ethylene glycol

Aldehydes	H
	   \\
	     C = D
	   /
	R
		Formaldehyde
		Glyceraldehyde
		Acrolein
		Acetaldehyde

Ketones	R
	   \\
	     C = D
	   /
	R
		Acetone
		Methyl-n-butyl ketone

Carboxylic acids	O
	       //
	R—C
	      \\
	       OH
		Acetic acid
		Trichloro acetic acid

Esters	O
	       //
	R—C
	      \\
	       OR
		Ethyl acetate

Ethers	R—O—R	Ethyl ether Isopropyl ether
Phenols (aromatic alcohols)		Phenols
Halogenated alkanes		Methyl chloride Methylene chloride Vinyl chloride

TABLE 1 (continued)

Chemical class	Homolog	Examples
Amines		
Primary	R–N⟨ $\begin{smallmatrix}H\\H\end{smallmatrix}$	Methyl amine
Secondary	R–N⟨ $\begin{smallmatrix}R\\H\end{smallmatrix}$	Dimethyl amine
Tertiary	R–N⟨ $\begin{smallmatrix}R\\R\end{smallmatrix}$	Trimethyl amine
Amides	R–C⟨ $\begin{smallmatrix}O\\NH\end{smallmatrix}$	Formamide
Nitriles (alkyl cyanides)	RC ≡ N	Acetonitrile Acrylonitrile

site in the body removed from the site of contact. The systemic damage
caused by chemicals is not capricious or random. A chemical does not
affect one set of functions in one person and a different set in another.
They can have an effect on several different functions within an individual, and individual variability may affect the sensitivity of threshold for dosage. The toxicity of a chemical is a property of that particular chemical. Just as the flammability, boiling point, and density
are physical properties, and solubility is a chemical property, toxicity
may be regarded as a biochemical property of the chemical. However,
until they reach and produce toxic effects inside the body, toxic substances are not harmful.

Toxicological effects of chemicals are dependent upon a number
of factors. One of the most important is exposure. Exposure is determined by the dose-time relationship, that is, how much chemical is
absorbed into the body and over what period of time or how often.
Two distinct types of toxicity may result from exposure: acute toxicity
and chronic toxicity.

Acute toxicity. The acute toxicity of a chemical refers to its ability
to do systemic damage as a result of a one-time exposure to relatively

TABLE 2 Pesticides

Chemical class	Examples
Organic phosphates	TEPP (tetraethyl pyrophosphate), Parathion, Dichlorous, Diazinon, Dimethoate, Malathion
Carbamates	Propoxur (Baygon), Carbaryl (Seven), Aldicarb (Temik)
Organochlorines	DDT (dichlorodiphenyltrichloro ethane), Chlordane, Heptachlor, Aldrin, dieldrin, Endrin, Endosulfan, Kepone (chlordecone)
Herbicides	2,4-D (2,4-dichlorophenoxy acetic acid, 2,4,5-T (2,4,5-trichlorophenoxy acetic acid), Dinitrophenols, Paraquat

large amounts of the chemical (1). The effects associated with acute toxicity exposures are usually prompt or only slightly delayed. These effects can include serious chemical burns, inflammation, allergic responses, or damage to the eyes, lungs, or nervous system. Some chemicals, such as cyanide or sulfide gases, are extremely dangerous in this respect—small amounts can cause death or serious injury very quickly. Acute toxicity is characterized by rapid absorption of the substance, and the exposure tends to be sudden and severe. Some substances, such as chlorine or ammonia, give considerable warning, while others, such as carbon monoxide, are not readily detected and therefore are more insidious. In laboratories, acute exposures most often result from the accidental spillage or release of large quantities of the material. Typical incidents illustrating an acute exposure might include the breakage of a one-liter container of concentrated hydrochloric acid, the inadvertant mixing of an acid with a cyanide or sulfide salt on the lab bench, the failure of a compressed gas regulator, the heating of volatile, toxic chemicals in the laboratory room, etc. All of these incidents would release significant amounts of toxic material into the room environment in a short period of time. Other actions, such as the mouth pipetting of a cyanide solution or skin contact with aniline, can cause a large exposure with acute toxic response. Acute toxicity usually refers to the administration of all of the material within a short time period, usually the working day. The symptoms of an acute exposure are also quickly manifested. The time may range from immediate reaction, as with a cyanide exposure, to reactions such as dermatitis, which may not show up for some time after the exposure. Normally, symptoms will appear within 24 to 48 hours after the expo-

sure. Such symptoms range from headache, nausea, and reddening
of the skin to unconsciousness, shock, or even sudden death.

Chronic toxicity. Chronic toxicity refers to the ability of a
chemical to do systemic damage to the body as a result of repeated
exposures, during a prolonged period of time, to relatively low levels
of the chemical (1). These toxicological effects may involve cumulative
damage to many different organs or parts of the body. Some are re-
versed on elimination of exposure to the chemical, but some are nearly
irreversible, especially after much damage has occurred. The symptoms produced by chronic exposure usually bear little or no relationship to those produced by acute exposures. Often different organ
systems or biomechanical mechanisms are involved. For example, acute
effects of overexposure to chlorinated solvents, such as chloroform
or methylene chloride, are primarily those of the central nervous system, such as dizziness, narcosis, and headache, whereas chronic exposures often produce liver damage.

Just as the symptoms of acute and chronic toxicity of a given chemical are hardly related, so too are their relative potencies. A chemical
that is highly toxic acutely is not necessarily highly toxic chronically,
and a chemical that is of a low order of toxicity acutely is not necessarily low in toxicity chronically. Examples of such differences in
toxicities include vitamin D, sodium fluoride, and table salt, all of
which can be acutely toxic in high doses, but are nontoxic and, in
fact, essential in chronic, low dose amounts. The converse is illustrated
by metallic mercury. One-time ingestion of a large amount of metallic
mercury is not likely to cause death or illness. The mercury will pass
through and be eliminated in the feces, and the amount absorbed from
the intestinal tract will be very small. However, if a small amount
were swallowed daily, the amount absorbed each day could give rise
to chronic mercury poisoning.

Another effect, known as cumulative action, is often associated
with the heavy metals, such as mercury, lead, arsenic, etc. and pesticides such as DDT. Over a period of time the chemical is absorbed
into the body and only partially excreted. This results in increasing
amounts, which accumulate in storage areas of the body. Eventually,
the quantity in the body is great enough to cause a toxic effect.

Thresholds. Paracelsus, a sixteenth-century physician and
alchemist, practiced medicine in a way that was, in many aspects, the
precursor of modern toxicology. In defense of his introduction and
use of mercury for the treatment of the then lethal disease syphilis,
he put forth the concept that all things are poisons and that "the dose
alone makes a poison." This principle is the keystone of toxicology
and industrial hygiene.

Most toxicologists have accepted the dictum of Paracelsus that for
all poisons a dose can be found that produces no toxic effects (3).

In contrast, recent years have seen the rise of a "new" toxicology, which proclaims that for certain effects no dose other than zero is safe, and no dose other than zero can be tolerated for substances that produce irreversible effects such as cancer. It is this latter philosophy that was incorporated into the Delaney Clause of the Food Additives Amendment of 1958, which prohibited the Food and Drug Administration from setting any tolerance other than zero for any substance found to induce cancer when ingested by man or other animal. More recent events, such as the implication of saccharin and methylene chloride as carcinogens, are causing legislative and regulatory officials to reconsider such totally exclusionary laws (4). There is, however, a body of respectable scientific opinion which holds that thresholds exist even for carcinogens. Although a single molecule of a carcinogen has the potential to induce a malignant transformation in a single molecule of DNA, there is a long step between a molecule and an intact organism. The organism possesses effective defense mechanisms. Carcinogens may be detoxified before they reach target cells. Genetic damage can be repaired at the molecular level. If a cell becomes malignant, it may be dealt with by the body's immunologic defenses (5). Nevertheless, the debate over the existence of thresholds contines and forms one of the more interesting contemporary problem areas of toxicology.

Toxicity versus Hazard

A special distinction that needs to be made is the difference between toxicity and *hazard*. While hazard has come into common use as a synonym for toxicity, it is a much more complex concept, in that it includes conditions of use of a substance.

There are two components to the hazard presented by chemicals. First, there is the inherent ability of a chemical to do harm by virtue of its properties, such as explosiveness, flammability, toxicity, etc.; second, there are environmental conditions under which the chemical exists, such as its use in a closed system, protection of the skin from contact with a liquid, etc. In other words, the amount of a substance needed to produce a toxic effect and how likely it is that this amount of material will enter the body where it can actually produce this effect are the primary factors in hazard evaluation. It is these conditions that determine the potential for exposure to a harmful substance. Every chemical has some set of exposure conditions in which it is toxic and some set of conditions in which it is nontoxic. It is also these conditions that can be managed to reduce the overall or specific risk due to chemical exposures in laboratories.

The amount of a chemical that actually enters the body is known as the dose. And the health hazard of a particular chemical is related to the dose received by the body *and* the toxicity of the substance.

Since there are no easy methods to determine the actual dose of most
chemicals in the human body at a given time, the dose is estimated
by determining the "exposure" of an individual to the chemical in question. Exposure is determined by measuring the concentration of the
substance in the breathing zone of the environment or the amount in
contact with the skin. These measurements are then used to estimate
the dose of the compound that might be absorbed in the body.

Factors that Influence Levels of Exposure

A number of factors affect the amount of exposure and subsequent
toxicity to a specific chemical. The primary factor affect-ing the
health hazard due to a chemical is the amount of the substance actually
taken into the body in relation to the toxicity of that chemical. A
very toxic material, such as cyanide gas, requires only small amounts
to cause a serious health effect, while significant amounts of sodium
chloride would be required to cause ill health effects. Until they reach
and produce effects inside the body, toxic substances are not harmful.
    An understanding of the major factors that influence the levels
of exposure of an individual will aid in evaluating the overall risk to
a person. Three factors of importance when discussing level of exposure are the chemical and physical properties of the material, the
laboratory procedures used, and the environmental control methods
employed in the facility.

    Chemical and physical properties. The chemical and physical prop-
erties of a substance that are most important when evaluating the potential for personnel exposure are those that have an impact on the
ability of the material to enter a person's body. The major routes of
entry are through inhalation and penetration through the skin.
    Inhalation hazards exist when the chemicals are able to become
airborne. In the case of absorption through the respiratory tract,
evaporation of the chemical into the workplace air largely determines
whether a potential for exposure exists for the worker. Evaporation
depends on the vapor pressure of the substance and on its temperature.
Substances with higher vapor pressures, such as ethyl ether, benzene,
and xylenes, etc., tend to evaporate more readily than substances
such as petroleum ether and other low vapor pressure solvents. Also,
the higher the temperature, the higher the rate of evaporation.
    For solids and aerosols, the physical properties of major concern
are the particle size (aerodynamic diameter) and solubility. The aerodynamic diameter is equal to the diameter of spherical particles of unit
density, which have the same falling velocity in air as the particle
in question (6). Generally, only particles below 10 μm in diameter are
regarded as of concern as a health hazard, as these particles can penetrate into the deep reaches of the lung and be deposited where the
mechanisms for removal are very slow. Particles larger than this are

normally deposited in the upper respiratory system and removed by mucociliary flow to the mouth, where they may enter the gastrointestinal system.

Solubility of the chemical in water is important in evaluating inhalation exposure. Particles enter the lungs during inhalation. If they are soluble in water, then upon contact with lung tissue they will go into solution (solubilize) and enter or diffuse through the lung tissue and enter the circulatory system.

The solubility of solids is also important to the understanding of dermal exposure. As a general rules, compounds that readily penetrate the skin are designated fat- or lipid-soluble and water-insoluble. This means that they are soluble in many common organic solvents such as alcohol, ether, and chloroform. Benzidine, for example, is easily solubilized and readily penetrates the skin. In fact, for benzidine, skin exposure is a primary concern. Other substances that are highly skin-permeable include aniline, nitrobenzene, and dimethyl sulfoxide (DMSO).

Laboratory procedures. The procedures used by an individual in a laboratory contributes directly to the level of exposure of all the occupants of the laboratory. As indicated previously, inhalation and skin contact are the two major exposure routes. Any procedure that increases the amount of chemical that can become airborne or come into contact with skin will increase the risk of exposure to the substance. Allowing substances with high volatility, such as organic solvents, to stand in open beakers or uncapped bottles in the room increases the exposure potential. Any procedure that causes agitation of the chemical, such as stirring, pouring, mixing, centrifuging, heating, etc., causes development of aerosols and an increased rate of volatilization.

The amount of skin contact is also increased with certain procedures. Grinding, blending, pouring, and mixing of powdered material generate aerosols and extensive surface contamination. Failure to clean up after a procedure may increase the risk of exposure to others in the laboratory. Not all contamination is visible, and someone may assume that the bench area on which they are working is clean and result in skin contamination. The use of appropriate gloves is an effective method to protect skin; however, when the gloves are not removed when answering the telephone, opening the door, or handling noncontaminated items, the exposure potential to other personnel in the laboratory who subsequently handle these items is increased. Proper procedures are extremely important factors to consider when assessing chemical health risk in a laboratory environment.

Environmental control methods. Environmental control methods used in the laboratory have the greatest impact on exposure reduction. The more extensive the controls, the lower the levels of exposure.

Section II includes details of methods of environmental control that
play an important role in the laboratory.

Routes of Exposure

Before a chemical can cause a toxicological response in an organ system
or to the body, it must gain entry into the body. There are four main
routes of entry to consider. These are skin and eye contact, inhalation, ingestion, and injection. In actual practice, it is difficult to
have an exposure to a toxic substance solely by a single route. Thus,
when route of exposure is specified, it means that exposure is *primarily* by that route.

   Contact with the skin and eyes. The skin is normally a relatively
effective protective barrier. If a chemical cannot penetrate the skin,
it cannot exert a toxic effect by the dermal route. If, however, a
chemical can penetrate the skin, its toxicity depends upon the degree
of absorption that takes place. As a general rules, inorganic chemicals
are not absorbed through intact skin. Organic chemicals may or may
not be absorbed, depending on a number of conditions. Usually, a
substance in dry powder form is absorbed to a slower degree than
the same substance in an aqueous solution, suspension, or paste. An
oily or organic solvent solution often permits greater absorption than
aqueous solutions. Certain solvents, such as dimethyl sulfoxide
(DMSO), aniline, xylenes, benzene, and others, greatly enhance the
skin absorption of a wide variety of compounds. The main parameters
that determine the dose received via skin absorption, in addition to
the permeability of the chemical itself, is the amount or concentration
of the toxic chemical and the amount of time it is in contact with the
skin. This is especially important when an accidental spill occurs onto
the laboratory worker's clothing. The contaminated clothing should
be removed immediately to reduce the amount of chemical in contact
with the skin and the time of contact. Laboratory workers should keep
a separate change of clothing readily available for emergency use (hope-
fully they will never have to be used). Past experience has indicated
that persons are reluctant to remove contaminated clothing because
they have nothing else to put on. Skin contact frequently results
in localized irritation, whereas recurrent contact may result in dermatitis. The histology laboratory, where alcohols and other organic sol-
vent-clearing agents are used in fairly significant amounts, typifies
the type of biomedical environment where there is the potential for
skin exposure. Due to the delicate nature of the work, appropriate
protective gloves cannot be used to prevent contact. As a result,
many of these workers end up with extremely dry skin (especially on
the fingers), which, if not treated appropriately, will crack and develop
open wounds, thereby creating an open pathway for chemicals to the

bloodstream. Once in the blood, the chemical can be carried to its target organ (site of toxic action).

Due to their sensitivity to irritants, eye exposure to chemicals is of particular concern. The eye has a rich vascular supply and is connected directly to the brain. Therefore, few substances are innocuous when in contact with eyes. A number of substances cause pain and irritation, and a considerable number are capable of causing burns and loss of vision. Alkaline materials, phenols, and strong acids are particularly corrosive to the eyes and can cause permanent loss of vision. Probably one of the most ignored procedures in biomedical research laboratories is the use of protective eyewear.

Inhalation. Inhalation is the most common route by which toxic materials can enter the body. Absorption of toxic chemical vapors, mists, gases, or dusts through the mucous membranes of the mouth, throat, and lungs can cause damage to these tissues. Inhaled gases or vapors may pass rapidly into the capillaries of the lungs and be transported to the circulatory system. Absorption through the lungs can be very rapid due to the relatively large surface area. It has been estimated that the average adult has about 20 ft^2 of skin surface area and about 750 ft^2 of lung surface (7). The type and severity of the action of toxic substances depends upon the nature of the substance, the amounts absorbed, the rate of absorption, individual susceptibility, and a number of other host factors.

Chemicals that cannot become airborne cannot enter the lungs and consequently do not represent an inhalation hazard. Chemicals can become airborne in two ways, either as very small particles or as gases or vapors. The difference between a gas and a vapor is determined by its physical form at room temperature. A gas is a chemical that is in the gaseous state at normal room temperature and pressure (e.g., carbon monoxide, hydrogen sulfide, and nitrogen dioxide. A vapor is the gaseous form of a chemical that is a liquid or solid at normal room temperature and pressure, such as benzene, chloroform, etc. An aerosol is a dispersion of a particulate in air. Aerosols are considered to be small liquid or solid particles, often generated by many normal laboratory operations, such as stirring, blending, centrifuging, pouring, etc.

Ingestion. Exposure by ingestion is much less common in the workplace than inhalation or contact exposure to toxic chemicals. If the substances are not absorbed, they cannot cause an effect on the body's systems. The toxic contaminants in the foods we eat and liquids we drink can enter our bodies and can be absorbed from the gastro-intestinal tract.

In the laboratory, intentional ingestion of chemicals is not usual. However, unintentional ingestion of toxic materials, although not prevalent, can occur. For example, laboratory personnel who handle toxic substances with latex gloves and then proceed to perform other functions, such as answer the phone, write down data, open the door, etc., without removing the contaminated gloves can transfer the contamination to those items. Personnel who drink and eat in the laboratory run a risk of contamination of the materials they are ingesting. Although this is not a likely source of exposure, it is possible, and personnel should be aware of it.

Inhaling a toxic aerosol can also result in exposure through ingestion as particles are brought up by mucociliary action and swallowed with saliva.

Injection/inoculation. Injection of toxic chemicals into the body is not a common source of exposure. However, accidental injections have been known to occur, either from a syringe, or from a sharp object, such as contaminated broken glass, piercing the skin. Although exposure via this method is rare, one of the aspects of such exposure is that the total amount of contaminant that enters the body is often very small.

One of the main differences between accidental injection with a chemical contaminant and biological contaminant is the fate of the material in the exposed individual. When biological materials enter the body, they can replicate and increase their concentrations to the point where they may cause illness. With chemical contamination, the amount will not increase over the initial dose and will most likely decrease due to metabolism and excretion. Therefore, although exposure via this route is not without risk, it usually does not carry the conditions of risk associated with a biological exposure.

By knowing the most likely route of exposure along with the properties of the chemicals, appropriate controls can then be developed to minimize the amount of chemical to which an individual or environment is exposed. The following seciton discusses some methods of environmental control that are important in minimizing the chemical health risk in laboratories.

III. CONTROL OF THE CHEMICAL HEALTH RISK

As indicated at the outset, there is no practical method available to completely eliminate the risks from chemicals used in research. These risks can, however, be controlled in a manner that helps to minimize the associated risk. The basic principles for controlling chemical risks in the research environment include substitution, isolation, and ventilation. Also of significant importance is proper administrative control of the laboratory, that is, assuring that personnel are following proper procedures.

A. Substitution

Usually, one thinks automatically of controlling a hazard by adding something to provide the control. For example, an inhalation hazard from a chemical such as benzene is often more likely to be controlled by using the chemical in a fume hood than by substituting a less hazardous material for the benzene. Substitution of acetone for benzene or EDTA for dichromate as a glassware-cleaning solvent is seeing wider acceptance. Yet, substitution with a less hazardous material or processing equipment, or even use of a less hazardous process may be the least expensive as well as the best method to control the chemical health risk.

Chemical Substitution

Substitution is not a technique easily taught, especially in the research laboratory. Many research investigations utilize "recipes" of earlier investigators, often with minor variations. The original investigators had little information available regarding the potential health risks due to the chemicals they were using in their research work. Unfortunately, many researchers today use these same formulas for their work, without considering the additional toxicological information now available. As in the case of benzene, there are other chemicals now available that work as effectively and, in some instances, more effectively than the hazardous chemical in question. Information of appropriate substitutes can often be obtained from the technical support staff of the major biomedical chemical supply companies as well as from more recent literature searches of the experiment under consideration. The individual researchers should be made aware of the possibility of availing themselves of appropriate substitutes that will lower the associated health risk of the research work. Then, in conjunction with other resource personnel, they should attempt to determine if a substitute exists for the high hazard chemical that will not interfere adversely with the desired research outcome.

Material substitution can be the least expensive and the most beneficial of the control mechanisms for reducing risks. However, as with the earlier researchers, the attendant hazards of new or substitute products must also be investigated.

Process and Equipment

Choosing a substitute process is not always difficult. For example, dipping an object into a container of paint almost always creates much less of an inhalation problem than does the process of spraying the object. Mechanical stirring generates less aerosol than does sparging. A basic principal that emerges regarding processes is that the more closely the process approaches being continuous (verses intermittent), the less hazardous that process is likely to be. Research environments, due to their very nature, tend to use intermittent processes. Awareness of the potential of a particular process for creating an environmental hazard is most important. Once recognition of the process hazard potential is made, substitution can then be considered.

When a material or process change to reduce the risk is not possible, the necessary control might be achieved through equipment substitution. Equipment is frequently changed or substituted more often than either processes or materials in the research laboratory. This is due to development of newer and more sophisticated equipment as well as the dynamic nature of the research itself. It is important to keep in perspective that much of the attendant risk occurs not during the analytical process itself, but in the preparation of the materials for processing. One of the main parameters to consider when purchasing new equipment should be the health and safety features of the equipment. Too often new equipment is purchased with no provision for offgassing of vapors, odors, etc., and facilities must then be renovated at considerable expense to provide the appropriate controls for the equipment. For example, a cell sorter used for sorting blood cells from an AIDS patient or hepatitis patient should have ventilation control to prevent escape of aerosols. Or, if a large-scale fermenter is installed, provision should be made for odor control.

B. Isolation

Isolation refers to interposing a barrier between a hazard and those who are at risk. The barrier can be physical, such as using a totally enclosed system, or distance may provide the isolation considered necessary by locating a higher risk operation in a secluded area away from higher populations. The latter option is often used when working with chemicals that have explosive potential. For substances considered high health risks, such as arsine or phosphine, establishment of a regulated area within the facility limits the number of persons with access to the area and decreases the overall risk. In the area of biological research, isolation is a well-documented control factor for containment of pathogenic organisms (8).

Stored Chemicals

Stored chemicals rarely pose an overt hazard, and therefore those concerned are likely to take it for granted and to assume that it poses no threat. This assumption can be dangerous.

Due to uncertainty in the integrity of chemicals, there is an unwillingness on the part of research investigators to work from a common chemicals storeroom. This results in both the extensive duplication of chemicals purchased as well as the dispersal of these chemicals throughout the research facility. This in itself is a practice that leads to an increase in personnel and property at risk. Aside from the inherent safety risks of fire, explosions, and violent reactions of accidently mixed incompatible chemicals, inhalation hazards due to the storage of materials must also be considered.

When the principal hazard of a liquid is due to inhalation rather than a safety hazard such as fire, the imposition of a physical barrier becomes much more difficult than simply putting the material in a storage cabinet. When the quantities are relatively small, the best storage technique uses both isolation and ventilation, such as the ventilated storage cabinets used in some laboratories. Such cabinets are usually made of fire-resistant material, and air is drawn through them constantly by means of a fan that discharges out-of-doors. This type of arrangement interposes both a physical and ventilation barrier between the contents of storage vessels and the laboratory environment and, in addition, may free much valuable hood space for other than storage use. A number of vendors now manufacture chemical fume hoods with the base cabinet vented for the storage of flammable materials as an option.

The best option for any facility is to maintain a centralized chemical stores. This allows for better control options to be installed and effectively isolates the laboratory and other personnel from an increased risk of chemical hazards.

Equipment and Process

Most equipment in research laboratories, if used properly, is designed to be safe. There are times, however, when this is far from true. Equipment that is operated under very high pressure, for instance, may well pose a severe hazard even when operated properly. In such cases, the proper action to take is to isolate the equipment from the laboratory envrionment. Usually physical barriers are used, and the barriers may range from very simple to formidable ones, depending upon the hazard potential.

Inhalation hazards can often be reduced by equipment isolation. The placement of equipment into chemical fume hoods is one example of isolation, combined with ventilation. The use of fume hood glove boxes is another. Proper planning should include the use of isolation as a control factor in the laboratory.

The Worker

From the time when Pliny the Elder wrote about the use of pigs' bladders by miners to reduce the amount of dust inhaled until now, isolation of employees from their occupational environment has, and perhaps will always be, an important control strategy (9).

The use of personal protective equipment exemplifies this principle.
Such equipment should be designed for emergency or temporary use,
but this does not always hold true. The use of eye protection in research labs, for example, is always stressed, because the loss of vision
is such an extreme penalty to pay for a moment's inattention.

It is difficult to isolate anyone from any environment for extended
periods of time. We have successfully moved people through the vacuum
of space to the moon, and we have submerged them in the depths of
the ocean, completely protected from the extreme external environments.
While such complete protection is possible, it is rarely used in the normal
work environment.

Because the complete isolation of workers form the environment
is difficult and expensive, partial isolation is normally used. Even
partial isolation can result in discomfort to the worker. Consider wearing a respirator all day, for example. In such cases other techniques
of controlling the environment should be considered.

The primary method for the protection of laboratory personnel
from airborne contaminants should be to minimize the amount of such
material entering the laboratory air. However, when effective engineering controls are not possible, suitable personal protective equipment can be substituted. By its very nature, personal protective equipment is designed for use at times when all of the hazards are not readily
delineated, where, in fact, the actual hazards may never be known.
Such is often the case in the research laboratory. For those instances
where other control factors may not be practical, such as a short-
term or one-time use of a process of high-inhalation hazard material,
the use of respiratory protection might be appropriate. However,
do not forget that most research laboratories are only a small part of
a total facility, and the potential for environmental contamination of
other areas of the facility must be considered when choosing appropriate controls.

C. Laboratory Ventilation

Ventilation is used in laboratory facilities for two primary purposes;
to ensure thermal comfort of the building inhabitants and to help maintain the exposure of laboratory employees to biological, radiological,
and chemical agents at safe levels. Ventilation for comfort control involves the heating or cooling of outside air, which is circulated throughout the laboratory facility. Both heating and cooling involve energy
consumption resulting in economic cost to a facility. The more heating
or cooling requires, the higher the cost. To minimize the costs associated with heating and cooling large facilities, part of the air from
laboratory room environments is often recirculated. The amount of
room air recirculated can range from 0 to 100%. Recirculation of 100%
of the room air means that little or no outside air is being brought into

the building or laboratories. This can lead to serious prol ems for
the building inhabitants as a minimum amount of "fresh" ou.side air
is necessary to provide for the removal of normal contaminants of respi-
ration and building product offgassing in nonlaboratory settings. In
laboratories, the supply of outside air is critical as the amount of low-
level contaminants given off by the various research activities, as well
as chemicals in storage, is more than in a nonlaboratory use facility.
Exhaust air from a laboratory should not be recirculated within the
building. Laboratory exhaust air should be discharged directly out
of the facility.

The increase in the cost of energy in recent years has resulted in
a conflict between the desire to minimize the costs of heating or cooling and dehumidifying laboratory air and the need to provide laboratory workers with improved ventilation as a means of protection from
toxic gases, vapors, aerosols, and dusts. However, regardless of
the substantial costs associated with tempering the supply air for laboratories, cost considerations should never take precedence over ensuring that laboratories have adequate ventilation systems to protect workers from hazardous concentrations of airborne toxic substances. An
inadequate ventilation system can be worse than none, because it is
likely to give laboratory workers an unwarranted sense of security
that they are protected from airborne toxic substances when in reality
they are not.

Local Exhaust Ventilation

Local exhaust ventilation refers to the removal of a toxic contaminant
at the source of emission by directing air movement away from the
worker.

There are two main principles governing the correct use of local
exhaust ventilation to control airborne hazards. The first is to enclose the process, reaction, or equipment physically as much as possible.
The second is to withdraw air from some physical enclosure, for example, a chemical fume hood, at a rate sufficient to assure that the
direction of air movement at all openings is always into the physical
enclosure. If these principles are followed, no airborne material will
escape as long as the enclosure is intact and the ventilation system is
operating and used properly.

There are times when enclosure is impossible and where control
of airborne hazards must be accomplished simply by the direction and
velocity of air movement into an enclosure or hood. Too often the alternative types of local exhaust control mechanisms are overlooked
for application in laboratories (10).

Three main problems associated with local exhaust systems in laboratories have been identified as poor design, inadequate exhaust power,
and inadequate makeup air (11).

Many ventilation systems appear to have been designed by persons who seem to have no knowledge of how to handle air properly. These types of systems are easily identified by the abrupt expansions and contractions, right angle entries, and overuse of dampers in duct systems to alleviate problems.

The problem of inadequate exhaust power is typified by continual addition to the exhaust system, until nothing associated with the system functions adequately. All systems, old as well as new, must be well engineered.

Inadequate makeup air supply is a common problem in laboratories. Additional hoods are often added to a system with little or no thought given to the provision of adequate makeup air. Too often the exhaust system is already overburdened and the supply is already inadequate. The addition of more exhaust requirement burdens the system even more and eventually creates a useless ventilation system.

In the past, it was thought that makeup air would be provided by building leakage and infiltration. However, with the energy crisis, these older, leaky buildings have been insulated and tightened so that infiltration cannot provide adequate makup air supply to the already existing system, much less to additional exhaust. Adequate supply air must be provided mechanically for proper functioning of the exhaust systems. Such air can be filtered (cleaned), tempered, and delivered to the right location so as to minimize interference with the proper operation of the local exhaust hoods.

Another significant problem is the improper use of chemical fume hoods by the laboratory workers themselves. All too often hoods are used to store any number of chemicals and laboratory workers expect the protection factors of an unobstructed hood to be retained. Just as with any laboratory procedure or piece of equipment, the chemical fume hood will perform properly if and only if it is used properly. Proper use includes:

1. Assuring the hood is functioning before using it. Some continuous monitoring device should be present and checked before using the hood.
2. The hood sash should be kept closed except when performing adjustments of manual manipulations in the hood, in which case the opening should be the minimum necessary to complete the task.
3. Hoods are not intended primarily for storage of chemicals. Materials placed in them should be kept to a minimum and should not interfere with the airflow pattern or block the vents.
4. Training of personnel in the proper use of the chemical fume hood.

Although many laboratory workers regard chemical fume hoods strictly as local ventilation devices to be used to prevent vapors from entering the general laboratory environment, hoods offer two other

significant types of protection. The use of the hood with the sash
door closed places a physical restriction between the worker and the
chemical reaction which can provide protection from chemical splashes
or sprays, fires, and minor explosions. Also, the hood can be an effective containment device for accidental spills of materials.

General Ventilation

General laboratory ventilation is intended to provide a source of noncontaminated air for the laboratory workers to breathe and as a source
of makeup air for the local ventilation systems. The overall ventilation
system should ensure that the laboratory air is continuously being
replaced so that concentrations of odorous or toxic substances do not
increase during the working day. The air pressure in laboratories
should always be negative with respect to the rest of the building,
and all air from laboratories should be exhausted outdoors and not
recycled.

 General ventilation provides only modest protection from toxic gases,
vapors, aerosols, and dusts, when these materials are released into
the laboratory in any quantity. The primary rule when working with
toxic materials is that the quantities of vapors and dust that might
produce adverse effects are to be prevented from coming into contact
with the skin and are to be prevented from entering the general laboratory environment. Any operation that deals with toxic chemicals should
be performed in a hood or other containment device.

Maintenance of Ventilation Systems

Ventilation is the most important mechanical control device used in
the research laboratory for the control of toxic health hazards. As
such, assurance of its proper functioning plays a vital role in the overall hazard control program.

 Routine inspection and maintenance of all ventilation systems is
one of the most important aspects of assuring the systems are performing as they were designed. Blocked or plugged air intakes and exhausts may change the performance of the entire ventilation system.
Loose belts, unlubricated bearings, motor breakdown, duct corrosion,
and other minor failures can affect ventilation performance. It is important that anyone who works on ventilation systems must protect
themselves accordingly.

IV. EDUCATION AND TRAINING

Regardless of the number and type of control factors in place, unless
the worker is aware of the need for and proper use of the methods

to reduce risk to themselves, their fellow workers, and the facility, the total institutional risk will remain higher than is necessary.

Two main goals that should be a part of every facility's health and safety program are to increase the employees' awareness of the factors that contribute to increased health risks and to make available to them the technical resources for gaining further information.

All people working in the laboratory share responsibility for controlling the laboratory environment. Too often, laboratory personnel fall into the habit of disregarding safety precautions, even those that are viewed as common sense. Reinforcement of correct techniques as well as proper procedures is necessary to maintain minimum risk levels.

In addition, no engineering control is effective unless it is used correctly. Therefore, laboratory personnel must be informed of the correct use of and limitations to any control equipment they might be using. Without such understanding, overconfidence in the ability of the control equipment can lead to excessive exposures and adverse outcomes.

Ancillary and support services personnel also need to be informed of the potential risks involved with the type of work being conducted within the facility, as well as the controls that are in place to minimize their personal risks. Once they have an awareness of the risks and an understanding of the need for appropriate control devices, they are more likely to follow proper procedures in carrying out their responsibilities. In addition, a general awareness on their part can alleviate some of the anxiety that inevitably builds in persons not familiar with the properties and characteristics of hazardous chemicals. The availability of technical resources in the area of chemical health hazards is also very important. When personnel have a question regarding health and safety of working with a specific chemical, there should be some resource for them to turn to to gain further information. This resource may vary depending upon the personnel and their ability to understand the information in the format it is presented. Laboratory research personnel may need only good chemical health reference materials to be able to gain the information necessary. On the other hand, maintenance, housekeeping, and other personnel may require personal communication or dialogue with a technically competent person, who can also effectively communicate with the employees. The method of communicating this information is very dependent upon the abilities and cognitive levels of the employees.

REFERENCES

1. National Research Council. Committee on Hazardous Materials in the Laboratory. (1981). *Prudent Practices for Handling Chemicals in Laboratories*. National Academy Press, Washington, D.C.

2. Ottoboni, A. M. (1983). *The Dose Makes the Poison.* Vincente Books, Berkeley, Calif., p. 32.

3. Zapp, J. A. (1981). In *Patty's Industrial Hygiene and Toxicology.* G. Clayton and F. Clayton (Eds.). Wiley and Sons, New York.

4. *Chemical and Engineering News,* Specific issue to be cited.

5. Old, L. J. (1977). Cancer immunology. *Scientific American* 236(5):62.

6. Roach, S. A. (1973). Sampling air for particulates. In *The Industrial Environment—Its Evaluation and Control,* U.S. Department of Health, Educaiton and Welfare. NIOSH. U.S. Government Printing Office, Washington, D.C.

7. Olishifski, J. B. (1979). In *Fundamentals of Industrial Hygiene,* Olishifski, J. B. (Ed.). National Safety Council, Chicago, Ill.

8. U.S. Department of Health and Human Servcies. (1984). *Biosafety in Microbiological and Biomedical Laboratories.* Public Health Service. Centers for Disease Control-National Institutes of Health. U.S. Government Printing Office, Washington, D.C.

9. C. Plinius Secundus, *Historias Naturalis Lib.,* 33, Sec. 11.

10. American Conference of Governmental Industrial Hygienists. (1982). *Industrial Ventilation,* 17th ed. Committee on Industrial Ventilation, Lansing, Mich.

11. Peterson, J. E. (1973). Principles for controlling the occupational environment. In *The Industrial Environment—Its Evaluation and Control.* National Institute for Occupational Safety and Health, U.S. Government Printing Office, Washington, D.C.

III. Regulatory Agency Consideration

16

Controlling Laboratory and Environmental Releases of Microorganisms

HARLEE S. STRAUSS   *H. Strauss Associates, Inc., Natick, Massachusetts and Massachusetts Institute of Technology, Cambridge, Massachusetts*

I. INTRODUCTION

One of the great uncertainties and potential risks of using either pathogenic or not fully characterized genetically altered microorganisms is that they may be released into the environment, proliferate, and cause harm in a manner that is difficult (if not impossible) to control. Thus, it is incumbent upon an investigator using such microorganisms to reduce the potential for "environmental release" to an appropriate extent. The appropriate extent should be a reflection of either the potential hazard of or the level of uncertainty about the microbe.

In this chapter, we will examine the routes by which microorganisms can be released from the laboratory, growth chamber, greenhouse, or field trial site in which they are initially placed. Knowledge of the relative importance of the routes of environmental release facilitates the development of procedures and facilities that will most effectively reduce (control) these releases. However, it is also important to monitor the extent of the release to ensure that the control procedures are effective.

There are several alternate definitions of the "environmental release" that could be useful for government regulatory purposes (1), although no formal definition has yet been adopted (2). Currently, the Environmental Protection Agency (EPA), one of the federal agencies that regulates the deliberate environmental release of microorganisms, focuses on when an organism is contained rather than when it is released. The EPA considers a microorganism to be contained if it is subject to physical containment, such as the four walls and roof of a laboratory or other structure.

In this chapter, we will use a slightly different notion of environmental release: A microorganism is released into the environment when it is available for dispersal. We use the term "environment" in a general sense to include the laboratory spaces outside of the intended container as well as the surroundings beyond the laboratory walls. In order for a microorganism to be dispersed, or to be available for dispersal, three requirements must be met: (1) some event has to displace the microbe from its intended location, (2) the microbe must have access to modes of transport such as air, water, surfaces, or a host organism that will move beyond the spatial boundaries of its initial location, and (3) it must be able to survive the displacement and transport processes. Unless otherwise stated, in this chapter we will assume that the microorganism has the potential to survive the displacement and transport processes.

Several international committees of experts in pathogenic microorganisms and U.S. governmental organizations have produced guidelines and standards for reducing the release of pathogenic and/or genetically engineered microorganisms from research and clinical laboratories and for minimizing the exposure of the investigators (3-7). These guidelines establish three or four categories of organisms based on their potential for causing serious disease in individuals and communities. The procedures and equipment recommended for working with these categories of organisms become more elaborate as the potential hazard increases. Although no widely adopted guidelines or standards yet exist for growth chambers, greenhouses, and field trials, the same principle should apply. Specifically, the higher the potential risk or uncertainty of environmental behavior posed by an organism, the more elaborate the dispersal reduction procedures.

II. RELEASE OF MICROORGANISMS
FROM LABORATORIES

As stated previously, in order for a microorganism to escape from a laboratory into the outside environment, it must be displaced from its intended location, be provided with a means of transport, and survive the displacement and transport processes.

There are many scenarios by which displacement and transport of a microorganism result in its release from the laboratory. Almost all common laboratory procedures will displace microorganisms from their intended location through the production of aerosols, because microbes stick to the surfaces of containers, or because microbes can be carried by water. For example, manipulations such as the pipetting, shaking, opening of sealed test tubes or flasks, and centrifuging create microbial aerosols, which then allow the microbe to be transported on air currents around the laboratory or perhaps even further. Sometimes the aerosolized microorganism may settle out onto a surface, such as a laboratory coat or paper towel, that is then transported out of the laboratory. If the microorganism is a human pathogen, it may be inhaled, transported out of the laboratory in the infected cells of the human investigator, and then shed by any of several mechanisms (see below). Alternately, it may settle onto food or coffee, be ingested by the investigator and subsequently deposited in the sewage system. Secondary transfer can also occur. For example, an aerosolized microbe may settle out onto a laboratory bench or beaker and be transferred to an investigator's hands. Ingestion may result if the investigator then touches his or her hands to the mouth or eats prior to handwashing.

There are several bodies of literature that contain information about the dispersal of microorganisms. In the laboratory, the effects of the dispersal of highly pathogenic organisms have been monitored by surveying laboratory-acquired infections (8-10). Modes of dispersal may be inferred from these studies by examining the likely incidents leading to the infection. Researchers who study infectious diseases commonly monitor the manner in which pathogenic organisms are shed from humans. Several investigators have measured the number of viable particles aerosolized following common laboratory operations and accidents in research and clinical laboratories. Finally, a computer simulation has recently been developed to examine the quantities of microorganisms likely to be released from a laboratory and their escape routes (11,12). In the remainder of this section, we will briefly summarize dispersal information obtained from each of these bodies of literature.

A. Laboratory-Acquired Infections

Sulkin and Pike (8) conducted a survey of laboratory-acquired infections. A relatively recent update (9,10) of this survey reveals that there have been at least 4,079 laboratory-acquired infections, which have resulted in 168 deaths. Pike (10) reviewed and categorized the probable events leading to the laboratory-acquired infections. He found that known accidents accounted for 18% of the infections. Other causes of infection were cited as due to working with the infectious agent (21%), working animals or ectoparasites infected with the agent (17%), or exposure to aerosols (13%). Twenty percent of the infections were listed as having unknown sources. Within the accident category, Pike found that accidents involving a needle and syringe accounted for one quarter of the infections and accidents involving spills and sprays accounted for another quarter. Other common accidents were injury with broken glass or other sharp object, aspiration through a pipette, and a bite or scratch of an animal or ectoparasite.

The route of transmission, and by inference the means of microbial dispersal for many of the accidents, is apparent. The needle and syringe accidents are largely self-inoculations, although earlier accidents also included spray from the needle or syringe when they came apart under pressure (10). Injuries involving cuts and scratches are likely to lead to infection through direct contact. Aspiration accidents result in ingestion of the infectious agent. Accidents with spills and sprays may cause transmission by any of several routes including inhalation of aerosols, ingestion of something to be eaten (or smoked) after touching with contaminated hands, ingestion of a substance after the aerosol lands it, or through direct contact with small cuts in the skin. The 38% of the infections caused by working with cultures of the infectious agent or with infected animals could also have been transmitted by any of these routes.

In the laboratory, pathogens may invade individuals by routes, such as aerosols, that are unusual portals of entry for the microorganism when causing disease in the community at large. For example, aerosol transmission has accounted for at least one case of fatal typhoid fever, which is more commonly transmitted by ingestion of contaminated water or food. In addition, encephalomyelitis and typhus fever, which are normally transmitted by arthropod vectors, have repeatedly caused laboratory-acquired infections due to inhalation of the infectious agent (13).

B. Release from Humans and Laboratory Animals

Humans and laboratory animals can contribute to the dispersal of microorganisms in three distinct ways:

Microorganisms can adhere to and subsequently shed from surfaces
    such as hair, skin, fur, and clothes
Individuals may become infected by the microbe and shed the microbe
    during (and subsequent to) the infection
Individuals may ingest the microorganism through food, drinks, or
    smoking and release the microbe in urine or feces

Shed from Surfaces

Microorganisms adhere to or are nonspecifically associated with many types of surfaces including skin, hair, fur, shoes, and clothes. In fact, there is an almost universal association of microorganisms with surfaces or with each other (14). Microorganisms can be shed from humans or animals when skin, hair, or fur are sloughed off. They can also be shed if the surface is disturbed in such a way as to cause an aerosol, if washed off by water, or if they attach to another surface that comes into direct contact with it.

Shed during Infection

Pathogens can be shed from infected individuals in bodily fluids and secretions such as blood, mucus, pus, vomitus, sweat, urine, and feces. Not all microorganisms will be present in all fluids and secretions; where the microorganism is likely to be found is a characteristic of the individual microorganism.

The presence of microorganisms in the body fluids and secretions can be detected either analytically or because they result in further infections. For example, the potential of blood containing the infectious agent for AIDS to transmit the disease is well publicized at present. Other examples abound. According to Wedum et al. (15): "There have been some dramatic examples of laboratory infections from accidentally aerosolized dried urine and feces during research on the hemorrhagic fever viruses, in psittacosis-ornithosis, and in Q fever." Pike (10) cites the example of a physiologist who died of yellow fever, which he contracted while examining the black vomitus of a yellow fever patient. Chronic enteric carriers of *Salmonella typhi*, the agent which causes typhoid fever, often excrete 10^6 or more viable bacilli per gram of feces. The typhoid bacillus can survive for weeks in water, ice, dust, and dried sewage (16).

Shed after Ingestion

Following ingestion, some microorganisms can be recovered from urine and feces (both human and animal); others cannot. It is not necessary that organisms cause disease for them to survive enteric passage. For example, in an experiment to test the safety of recombinant DNA experiments, 15 volunteers ingested 5×10^{10} colony forming units of a marked, nonpathogenic strain of *E. coli* (17). The colons of all 15 persons became heavily colonized with this strain and all 15 excreted both the marked strain and indigenous coliforms during the 10-day test period. The levels excreted ranged between 10^6-10^8 bacteria per gram of stool. In some cases, the organism ingested may be pathogenic, but not cause disease in the particular carrier individual. An example of this is the previously cited typhoid carriers who excrete *S. typhi* in their feces.

C. Aerosols

Several investigators have investigated the number of bacteria in aerosols following common laboratory procedures or accidents (11,12,18-21).

In early studies at the U.S. Army's Fort Detrick Laboratories, Reitman and Wedum (18) and Wedum (19) used easily identifiable indicator organisms (*Serratia indica*, *Bacillus subtilis* var. *niger* and coliphage T_3) in simulations of many common microbiological procedures such as pipetting, centrifuging, inoculating and lyophilizing cultures,

and autopsy of animals. They measured the formation of bacterial
aerosols by surrounding the work area with sieve-type air samplers.
Each sampler drew air at the rate of 1 cubic foot per minute through
340 small openings, allowing the bacteria-containing aerosols to impinge
on the surface of a petri dish agar plate 1-2 mm below the openings.
The number of colonies on the petri dish were counted after a suitable
incubation period. Each viable particle gave rise to one colony, although more than one bacterium could have been associated with each
viable particle or aerosol droplet. For example, the average number
of staphylococci per viable particle is four and ranges from two to
eight (22). Other studies have shown that an airborne viable particle
can contain up to 200 bacteria (11).

  Reitman and Wedum (18) found wide variations in the number of
bacteria aerosolized, depending in part on small differences in technique used by each individual monitored. Some of their results are
summarized in Table 1. It can be seen that nearly every operation,
including seemingly gentle ones such as using an inoculating loop,
will aerosolize at least some bacteria.

  Kenny and Sabel (20) measured both the concentration and the
particle size distribution of bacterial aerosols created during common
laboratory procedures or accidents. Their results are summarized
in Table 2. The particle size is important, especially for human pathogens, because it determines the site of deposition of the aerosol in
the respiratory system. Particles with diameters between 1-5 μm are
most likely to be deposited deep within the lungs. The particle size
is also an important determinant of the distance an aerosol will travel
on an air current.

  Kenny and Sabel (20) used *Serratia marcescens* as their indicator
organism. They monitored for bacterial aerosols using two Andersen
air samplers, one on each end (but outside) of the biosafety cabinet
in which the simulations were conducted. The probes were approximately 0.4 meters from the source of the aerosol. These investigators
found that removing the top of a Waring blender generates a highly
concentrated bacterial aerosol (1,500 viable particles/cu. ft. of air)
with a small median particle diameter (1.7 ± 0.5 μm), which would allow
deposition in the lung aveoli. Dropping a lyophilized or flask culture
also generated concentrated aerosols, but the median particle diameter
is larger, somewhat limiting their penetration of the lung and aerial
transport around the laboratory.

  Rutter and Evans (21) determined the concentration of bacterial
aerosols generated by some common clinical laboratory apparatus and
blood sample tubes. Their simulated experiments utilized *S. marcescens*
mixed with heparinized blood to give a final concentration of 10^9 par-
ticles/ml. Aerosolized bacteria were sampled using a Bourdillon slit
sampler with a sampling rate of 1 cu. ft. (0.028 m^3) per minute.

TABLE 1 Aerosols Produced by Common Bacteriological Operations

Operation	Average # colonies on sampler plates
Agglutination, slide drop technique	0.3
Animal injections	
1. 10 shaved guinea pigs injected i.p. with 0.5 ml culture, no disinfectant	15
2. Same as above, but with disinfectant	0
Autopsy, guinea pig	
1. Immediately after 1 ml culture injected i.p.	4.5
2. Immediately after 10 ml culture injected i.c.	3
3. Grinding tissue 2 min with mortar and pestle	
Liver after i.p. injection	1.8
Heart after i.c. injection	19.5
Centrifuging	
1. Pipetting 30 ml culture into 50 ml tube	1.2
2. Removal of one cotton plug after centrifuging	2.3
3. Removal of one rubber cap after centrifuging	0.25
4. Decanting supernatant into flask	17.6
5. Adding 30 ml saline to one tube of packed cells and resuspending by sucking and blowing a pipette	4.5
6. One 50-ml tube breaking, but all 30 ml staying in trunnion cup	4
7. As in (6), but culture splashing on side of centrifuge	1183
Inoculating loop	
1. Streaking 1 agar plate with 1 loopful broth culture	0.6
2. Same as (1), but rough agar plate	25.1
3. Streaking 1 agar plate with 1 loopful agar culture	0.26
4. Loopful of broth culture striking edge of tube	0.6
5. Inserting 1 hot loop into 100 ml culture in a 250-ml Erlenmeyer flask	8.7
6. Same as (5), but loop cold	0.8
Lyophilization	
1. Breaking ampule containing 2 ml lyophilized culture by dropping on floor	
first 10 minutes	2029
50-60 minutes after breakage	741

TABLE 1 (continued)

Operation	Average # colonies on sampler plates
2. Opening ampule by filing and breaking tip	86
3. Same as (2), but wrapped in ethanol-soaked cotton pledget	0.08
Petri dish plates	
1. Pipetting 1 ml inoculum on a pour plate without blowing, adding melted agar and mixing	2.6
2. Streaking smooth agar plate with 0.1 ml culture; spread with glass rod	0.06
Plug, stopper, or cap removed from culture container	
1. Escher rubber stopper removed immediately after shaking up and down	5.0
2. Same as (1), stopper removed after 30-sec wait	2.5
3. Plastic screw cap removed from 8-oz prescription bottle immediately after shaking	4.0
4. Cotton plug removed from 250-ml flask immediately after shaking, dry plug	5.0
5. Same as (4), but plug wet	10.2
High speed blendor, culture mixed 2 min	
1. Screw capped, no rubber gasket	8.7
2. Screw capped, rubber gasket, worn bearing	61.0
3. Removing tight fitting cover immediately after mixing	too many to count
4. Same as (3), but 1 hour after mixing	8.2
One drop culture falling 7.5 cm onto	
1. Steel surface	1.3
2. Painted wood	0.3
3. Dry paper towel	0.11

i.p., intraperitoneal; i.c., intracardial

Air samples near a microhematocrit centrifuge were taken at a height of 1.68 meters before, during, and after spinning blood samples either with clean filled tubes or when three tubes leaked. While no detectable aerosols were generated prior to the centrifugation or while the capillaries were filled, 3-10 particles were detected during the 3-minute

TABLE 2 Bacterial Aerosol Concentrations and Particle Sizes
Generated by Common Laboratory Procedures and Accidents

Operation	Mean viable particles/ cu. ft. air sampled	Median diameter
Harvesting infected egg[a]	22.0	3.5 ± 1.6
Mixing culture with pipette	6.6	2.3 ± 1.0
Use of blender		
top on blender in operation	119.6	1.9 ± 0.7
top removed after operation	1500.1	1.7 ± 0.5
Mixing culture/mechanical mixer		
for 15 sec	0.0[b]	0.0
overflow	9.4	4.8 ± 1.9
Use of centrifuge		
no spilled material	0.0	0.0
material spilled on rotor	1.9	4.0 ± 1.8
Use of sonic oscillator	6.3	4.8 ± 1.6
Opening lyophilized cultures		
suspended in 3% lactose	134.7	10.0 ± 4.3
suspended in mother liquor	32.6	8.0 ± 3.4
Dropping lyophized culture		
suspended in 3% lactose	4838.7	10.0 ± 4.8
Dropping infected egg[c]	85.2	3.0 ± 1.8
Dropping flask culture	1551.0	3.5 ± 2.0
Spilling culture from pipette	2.7	4.9 ± 2.6

[a]Approximately 1.4×10^{11} viable cells/egg.
[b]A total of 4 viable particles sampled in 10 trials.
[c]Approximately 9.8×10^{10} viable cells/egg.
Source: Kenny and Sabel (20).

spin with clean filled tubes. When three tubes leaked, there were too
many bacteria to count aerosolized above the centrifuge. Five minutes
after this spin, there were also numerous bacteria detected in a sample
taken on the other side of the room. Small numbers of aerosolized
bacterial particles (0-15) were detected due to several operations using
an AutoAnalyzer stirrer and a blood microtonometer. Finally, Rutter
and Evans (21) sampled aerosols generated by opening plug cap and
cone-shaped screw top test tubes. They detected 1-42 viable particles

per operation if the inner surfaces of the caps were wet for both types
of tubes.

Knowledge of the sources of aerosols lead to a better understanding of how a microorganism can escape from a laboratory or infect
laboratory personnel. It also points the way to effective prevention
or reduction of the escape and protection of laboratory personnel.
For example, aerosolized microorganisms can escape from the laboratory:

on air currents through faulty filters, air vents, or open windows
on surfaces after deposition onto diverse places including waste mate-
      rials (that are then incompletely sterilized), laboratory coats, and
      people's hands and shoes
in wastewater after deposition into water or cleanup of a contaminated
      area
in infected organisms including humans, laboratory animals, and insects

Measures taken to reduce the dispersal of microorganisms into the
environment should be directed at all of the escape routes. The amount
of effort devoted to shutting off the escape routes depends upon an
assessment of the hazard of the microorganism. No single measure
is likely to completely halt the dispersal of microorganisms; it will only
decrease the fraction of microorganisms initially present that are dispersed.

Probably the most effective method to reduce the dispersal of aerosolized microorganisms is to perform experiments in a biological safety
cabinet. Instead of dispersing through the room air on air currents
and settling on some unknown object, the organisms should be trapped
on the HEPA filters through which all the cabinet air should pass.
To be effective, of course, the safety cabinet must be in good working
condition, and operated using appropriate rates of airflow and good
technique (23). Closing the windows should reduce the escape of the
aerosolized microorganisms into the outside environment.

D. Integrated Models of Escape

To the author's knowledge, no thorough sampling has been performed
on multiple modes of release of microorganisms from laboratories at
the same time. In addition, even if such measurements were available,
their applicability to all laboratory situations is limited because the
number of microorganisms released is highly dependent upon the procedures actually being performed, the safety equipment provided in
the laboratory (e.g., safety cabinets), the design of the laboratory,
and the skills and caution of the technicians performing the procedures.
However, Lincoln and Fisher and their co-workers (11,12) have developed a computer simulation of the release to the environment and
exposure to technicians of microorganisms being manipulated according

to two common protocols used in applied genetics laboratories. Protocol A stimulates the insertion of a recombinant DNA molecule into a microorganism and subsequent colony selection for incorporated DNA. Protocol B simulates the growing and harvesting of 10 liters of each of three selected colonies. Each of these procedures was estimated to take 1.5 days to complete. Preliminary calculations of the releases during the operation of a 250-liter fermenter were also made.

Lincoln, Fisher, and co-workers (11,12) considered the following routes of release of microorganisms from laboratories: aerosols, technicians' hands, technicians' clothes, inhalation and ingestion by the technicians, liquid waste, and solid waste. The model was run to simulate five different physical containment levels and three microorganisms with different aerobiological stabilities. We shall only consider their results from simulations under BL1 conditions with organisms that have complete aerobiological stability (designated "no-die" by the authors) but are somewhat sensitive to the decontamination procedures used. The results of this simulation are summarized in Table 3.

The dominant routes of release are via aerosols when large cultures are grown (the release actually occurs during sampling of the culture) and via the liquid wastes from these cultures. The number of viable microorganisms released via liquid waste is extremely sensitive to the effectiveness of the decontamination procedures used. Lincoln et al. assumed 99.99% decontamination, which corresponds to decreasing the initial number of viable microorganisms by a factor of 10^4 prior to disposal down the drain. More effective decontamination procedures will, of course, decrease the number of microorganisms released due to liquid waste disposal. The level of decontamination of microbes is likely to be highly variable from laboratory to laboratory and from time to time.*

III. RELEASE OF MICROORGANISMS FROM GROWTH CHAMBERS AND GREENHOUSES

Growth chambers and greenhouses are coming into more common use for experimentation with microorganisms because of the potential for using genetically altered microorganisms in the environment. Green-

---

*
If commercially available hospital disinfectants are used for decontamination, the level of disinfection may be quite poor indeed. It has been estimated that 20-30% of commercially available products do not meet the appropriate standards based on the testing of several state laboratories that monitor disinfectants. The U.S. EPA, which had a national program to test the efficacy of hospital disinfectants, no longer does so (24).

TABLE 3 Environmental Releases of Microorganisms from a BL1 Laboratory[a]

Route of release	Number of microorganisms released		
	Protocol A[b]	Protocol B[c]	250-1 fermenter
Aerosol to outside air	$2\text{-}3 \times 10^4$	1×10^8	$2\text{-}3 \times 10^6$
Liquid waste[d]	3×10^6	$4 \times 10^7\text{-}3 \times 10^{9f}$	n.a.[g]
Solid waste[e]	none	none	n.a.
Technician's clothes	140	6×10^5	$2\text{-}3 \times 10^4$
Technician's hands	none	none	none
Technician inhalation	350	1×10^6	3×10^4

[a]From Strauss (1) with the numbers derived from the computer simulation of Lincoln et al. (11,12).
[b]Protocol A corresponds to screening for a particular trait incorporated into a microorganism.
[c]Protocol B corresponds to growing and harvesting 3 batches of 10 liters of cells.
[d]This assumes 99.99% effectiveness of the decontamination process.
[e]Lincoln et al. assumed solid waste is only generated from contaminated lab coats and gloves. Since these are not required for BL1 containment, they assumed no solid waste. However, there are other sources of solid waste including paper towels, glassware, plasticware, pipet tips, etc. These are supposed to be decontaminated before disposal, but the effectiveness of the procedures are rarely verified. Thus, in reality, there would be some release of microbes on solid waste, but it is likely to be small compared to other routes of escape.
[f]The lower estimate is taken from Lincoln et al. (11); the upper estimate is based on my recalculation assuming the cells grow to a density of 10^9 cells/ml.
[g]not available

house experiments are useful for preliminary testing of both the efficacy and the safety of a microorganism prior to small-scale field trials and large-scale commercial use. Growth chambers may be used as facilities with containment properties between those of laboratories and greenhouses. At the present time, there are no federal regulations about containment structures, procedures, or equipment for greenhouses or growth chambers.

Growth chambers are essentially incubators that provide a physically enclosed artificial environment for the growth of organisms. The

potential modes of microbial release from growth chambers are the same
as from an incubator or a similar piece of laboratory equipment. The
most likely sources of microbial release are on surfaces such as the
investigator's hands, aerosols that may be generated during any operation performed in the growth chamber, and insufficient decontamination
of solid and liquid wastes stemming from the experiment.

Because microorganisms, especially microorganisms other than
plant pathogens, have been seldom studied in greenhouses, most greenhouses have not been designed to reduce the dispersal of microorganisms at all. On the contrary, greenhouses are designed to exchange
air rapidly with the outside air, and many are operated with their windows open without screens to prevent access to birds and insects.
Often, greenhouses have no facilities for decontaminating drainage
water or contaminated soil (25). However, greenhouses can be (and
have been) designed to contain microorganisms, at least to the level
released from BL1 laboratories. Changing operating procedures and
some retrofitting can reduce the release of microorganisms from currently existing greenhouses (although whether the level of reduction
is sufficient depends on the risks posed by the microorganism and
relevant governmental regulations).

Microorganisms are transported from greenhouses by the same
media that transport microbes out of laboratories; namely by air and
water, on surfaces (including solid wastes and investigators' clothing),
and in contaminated material. However, microbes may be displaced
from their initial site and reach the transport media in slightly different ways because the routine operations in laboratories and greenhouses
differ.

Microbes may be displaced into the air as either particulate dusts
or as aerosols. Viable dust particles may become airborne following
soil cultivation. Aerosols may be generated intentionally during the
application of microorganism onto the plants or unintentionally such
as from the splash of water onto a microorganism-occupied surface
during overhead irrigation. Microbial aerosols can also be generated
through the interaction of wind currents with leaves and soil (26).

Microbes may be displaced into drainage water during irrigation
and then transported from the greenhouse. Microbes can adhere to
and be transported on many surfaces including solid wastes; the hands,
clothing, and/or shoes of the human investigator; and on insects and
birds passing through unscreened open windows. Finally, microbes
may be transported from the greenhouse in infected plant material or
contaminated soil that is to be disposed of upon termination of the
experiment.

As discussed previously in relation to laboratories, identifying
the routes of release of microorganisms leads directly to designing
ways to reduce (but not fully prevent) the release. And as before,
the elaborateness of the procedures taken to reduce the dispersal of

the microbe should depend upon its hazard. Some procedures might
include, for example, the decontamination of solid or liquid waste prior
to disposal and putting screens on open windows to reduce the entry
of birds and insects. Cultivation, irrigation, and perhaps microbial
application systems could be used that minimize aerosols. Technicians
could incorporate the same hygiene measures, such as laboratory coats
and handwashing, that are used in laboratories.

To the author's knowledge, no data directly measuring the release
of microorganisms from greenhouses are available. However, Harrison
and Hattis (25) have calculated the expected number of microorganisms
released from a greenhouse under various assumptions about greenhouse construction and experimental procedures. Major alternative
assumptions included whether or not the outgoing air was filtered and
whether or not the soil was sanitized (or decontaminated) following
the experiment. Harrison and Hattis considered the following modes
of release in their calculations: airborne, waterborne, on solid wastes,
and on technicians' clothes and skin. They considered airborne releases
due to aerosols generated by foliar application of the microorganisms,
splash from watering, and the interaction of air currents with the surfaces of leaves and the soil. The waterborne release was due to drainage water and the solid waste was due to the disposal of plant and
soil following termination of the experiment.

Harrison and Hattis (25) based some of their calculations on an
experiment in which 10^{12} bacteria were applied to plants using a foliar
spray in a greenhouse with 1,700 sq. ft. of plants. The number and
method of application were based on their review of the patent of S.
Lindow of the University of California-Berkeley for using "ice minus"
strains of *Pseudomonas syringae* (27). A summary of their results
for this case is provided in Table 4.

The single largest source of emissions was due to the drainage
water regardless of whether a conventional or a BL1 type greenhouse
(i.e., the water subject to some kind of decontamination procedure,
the windows screened and closed, and the exiting air passed through
particle filters and a fan) was considered. The disposal of solid wastes
such as soil and plant material, in the absence of decontamination,
was the second major emission source. Aerosols resulting from microbial
application using a foliar spray also constituted a major emission mode
if the outgoing air was not passed through a HEPA filter. Under conditions commonly used in conventional greenhouses, Harrison and Hattis
estimated all the microorganisms originally placed in the greenhouse
could be released (assuming no increase or decrease in microbe numbers during its residence in the greenhouse). Further, they suggested
that from 1 to 10% (in this example, 10^{10}-10^{11}) may be released as an
aerosol immediately following each application of a foliar spray due
to the rapid air exchange in greenhouses (approximately 1 change/min-
ute, higher on hot days). If a BL1-type greenhouse were used, that

TABLE 4 Estimated Emissions from Greenhouses[a]

Source of emission	Average daily emission:[b] conventional greenhouse (microbes/day)	Average daily emission: BL1-type greenhouse (microbes/day)
Aerosol from:		
watering	3×10^7	HEPA[c]
foliar spray	3×10^8-3×10^9	HEPA
leaves and soil	7×10^8-7×10^9	HEPA
Solid wastes	3×10^{10}	n.a.[d]
Drainage water	10^{11}	10^{7e}
Clothes/skin	700	negligible

[a]From Strauss (1) and adapted from the calculations of Harrison and
Hattis (25). Their calculations assumed that 1×10^{12} microorganisms
were applied by a foliar spray and that this number of microorganisms
was maintained in the greenhouse throughout the 30-day experiment.
The number of microorganisms were maintained, despite some amount
of daily emissions, by either reproduction or reapplication of appropriate levels.
[b]Emissions that occurred only once during the experiment, such as
the disposal of solid wastes, were divided by 30 (the assumed duration
of the test) to obtain average daily emissions. In practice, however,
daily emissions may vary greatly from day to day, depending upon
the operations being performed.
[c]Harrison and Hattis could not estimate the effectiveness of HEPA filtration without data on the rate of bypassing the filter system and
particle diameter. If all air passed through the filter, bacterial emissions would be reduced at least 99.99% (a factor of 10^4).
[d]Not available. Harrison and Hattis did not have data on the effectiveness of autoclaving solid wastes.
[e]This assumes 99.99% decontamination of wastewater.

is, one in which outgoing air had to pass through a HEPA filter, and
liquid and solid wastes were subject to decontamination, then emissions
would be severely decreased. The major source of emissions would be
drainage water and the level of emissions highly dependent upon the
efficacy of the decontamination procedures. If drainage water is not
decontaminated, then emissions from drainage water could be as high
as 10^8-10^{10} organisms/day. If a disinfeciton efficiency of 99.99% is
assumed, then the drainage water emissions would be reduced to 10^4-
10^6 organisms/day.

IV. RELEASE DURING FIELD TRIALS

The testing of a microorganism in the open environment, or field trials, is an important step in its development path from laboratory to commercial use in the environment. Field trials are conducted out-of-doors in order to realistically evaluate the efficacy and safety of the test microorganisms while they are subject to complex environmental vaiables. Field trials can vary in size from inoculating small quantities of test microorganisms onto a few square meters of soil to spraying large quantities of the test microorganisms onto fields of many acres. They can also vary in time from a few days to a year or more. Some field trials may require that microorganisms be inoculated into water, such as ponds or rivers, or injected into deep wells.

In most cases, microorganisms will be dispersed from the field trial site during, and perhaps after, a field trial. However, the magnitude of the dispersal can be decreased if the field test site is carefully chosen and appropriate control measures are instituted. In general, with a given microorganism, the magnitude of the dispersal will increase as the number of viable organisms in the inoculum increases. Thus, it may be appropriate to keep initial field trials small while safety tests are being conducted. The scale of the field trials can be increased as experience is gained with the test organism.

The modes of transport of microorganisms from field trial sites are the same as those we have discussed earlier: air, water, surfaces, and infected materials. Some of the means by which microorganisms are displaced from their initial location are similar to those we have discussed in relation to greenhouses, although there are other ones to consider in the open environment as well. In addition, the frequency with which the microorganisms may be displaced is likely to increase in field trials compared to greenhouse studies. Several recent reports have discussed the dispersal of microorganisms resulting from field trials (28-30).

A. Displacement of Microorganisms from Field Trial Sites

Displacement Caused by Rainfall

Rainfall on plants and soil could release microorganisms from field trial sites by generating aerosols, by splashing organism-containing water or soil onto other locations, by washing organisms further into the ground where contact with groundwater is possible, or by carrying organisms into surface runoff.

Graham et al. (31) have examined the generation of aerosols by the impact of raindrops. In a closed chamber, these investigators measured the release of *Erwinia carotovora*, a tuberborne pathogen, from the stems of potato plants. Stem cuttings infected with measured quantities of the bacteria were hit by simulated raindrops 3 mm and 5 mm in diameter while a windspeed of 23 m/min was also simulated. The aerosol particles released, of which over 90% were between 1 and 5 µm in diameter, remained airborne for at least 60 to 90 minutes. The number of organisms released during more than 30 minutes of simulated rain was roughly proportional to the number of organisms present in the stem cutting. Cuttings, which contained 10^8 to 10^9 organisms per gram, released approximately 0.003% of the organisms (i.e., 3×10^3 - 3×10^4 organisms per gram of stem cutting). These findings suggest that aerosol generation during normal rainfall or overhead irrigation may be a significant potential source of release.

The other mechanisms by which rainfall displaces microorganisms are dependent upon local conditions. For example, in general, coliforms will not move through the soil. However, if the soil is saturated, fecal coliforms can move significant distances and reach groundwater or surface waters and be transported long distances by these transport modes (29).

Displacement Caused by Wind

Wind can aerosolize microorganisms found in the soil or on plant surfaces, thus displacing them from their initial sites and providing access to a mode of transport. For example, Lindemann et al. (26) measured aerosol bacteria concentrations at two locations above a plant canopy during relatively calm wind conditions. They then calculated the vertical flux of organisms from the field. Fluxes of between 57 and 543 particles/m^2-s were found for crops with surface bacteria concentrations of 10^5 to 10^8 particles/g during calm wind conditions of 1.3 to 3.5 m/s. A vertical flux of 499 particles/m^2-s was calculated for a field of beans which supported $10^{7.72}$ bacteria per gram. If we assume a crop yield of 1.8 tons/acre for beans (32), we can estimate that there are 2×10^{10} surface bacteria per square meter of crop. The daily release based on the reported flux would be 4×10^7 particles/day, or about 0.1% of the total bacteria. This suggests that the potential for release from plant surfaces during normal, calm weather conditions should not be ignored. Organisms that are found primarily in the soil and on plant leaves will probably have differing potentials to displacement by wind.

Displacement by Mechanical Forces

The mechanical disruption of land being used for agriculture can displace enormous quantities of microorganisms into the air. For example, about 40% of the total aerial bacterial loading in the spring in the Willamette Valley in Oregon has been estimated to come from plowing (32). Disturbance of plants during the growing season or harvest may dis-

place microorganisms living on the surface of plants as well as in the
soil. Plowing the field after harvest will also result in displacement if
the microbes remain viable.

Microorganisms can also be displaced into the air by mechanical
forces during application. For example, microorganisms may be applied
to an area using a foliar spray, which is intended to produce a fine
mist of organism-containing solution. This presents an obvious means
of generating aerosol droplets. Plant leaves dusted with a powder
that contains the organisms are also likely to cause microbial aerosols.

*Displacement through Contact with Surfaces
and Animal Vectors*

As previously discussed in relation to laboratories and greenhouses,
microorganisms can be displaced by direct contact with surfaces and
dispersed when the surfaces are transported elsewhere. Such surfaces
include animal fur and feet, equipment such as plows and shovels,
and the clothes, shoes, hair, and skin of personnel working in the
environment. Animal vectors can also displace microorganisms in their
digestive tracts following ingestion.

Most microorganisms will come into contact with some type of animal.
For example, microbes on the top of soil, water, or plant leaves are
accessible to insects and birds, while those deeper in the soil may come
into contact with earthworms, soil arthropods, and burrowing mammals
(29). Bees and other insects can carry 10^5 cells of *Erwinia amylovora*
per insect (34).

B. Transport

As discussed previously, air, water, and surfaces can transport microorganisms. In field trials, once microorganisms have reached these
modes of transport, there are no artificial physical barriers, such as
walls and ceilings, to impede transport and thus dispersal. The number of viable microorganisms that are dispersed depends upon the number of microorganisms that are displaced, the percentage of microorganisms that survive displacement, and the percentage of microorganisms
that survive transport.

C. Methods to Reduce Microbial Release
   from Field Trial Sites

Efforts to reduce microbial dispersal from field trial sites can be directed toward:

reducing the frequency of activities that will displace microorganisms
reducing access to modes of transport

reducing the percentage of microorganisms that survive displacement
and/or transport

There are numerous methods available to accomplish these reductions, although no single method, or combination of methods, is likely
to reduce the dispersal of microorganisms to zero. Many properties
of the field trial site itself and the experimental protocols will have a
large impact on the amount of dispersal. The following discussion is
intended to point out some of the features of field sites and experimental protocols that influence microbial dispersal and some means of
reducing this dispersal. It is mainly directed at agricultural field trials,
although many of the principles will be the same for other field trails.

Field Trial Site Considerations

When evaluating a potential field trial site, attention should be given
to aspects that may affect the release and subsequent transport of
microorganisms away from that site. The geography, climate, hydro-
geology, and the flora and fauna of a site are among the important
considerations in this regard.
　　Geographical features such as the close proximity of surface water
or heavy animal or human traffic present obvious means for the transport of microorganisms from the site. Climatic factors such as the
amount of wind or rain will determine the magnitude and frequency
of aerosol generation or the number of microorganisms dispersed by
runoff or by percolation through saturated soil. Other site characteristics such as soil permeability, slope, potential for leaching and chan-
neling, and the depth to ground water will also affect dispersal. Large
populations of birds and insects increase the opportunity for dispersal
of microorganisms through surface contact. If the test microorganism
is a pathogen, the presence of susceptible plant or animal hosts will
also increase the opportunity for dispersal.
　　Field sites can be modified to reduce the dispersal of microorganisms
(35). For example, wind screens can be erected that reduce the potential for aerial dispersal. Fences, traps, and buffer zones can reduce
land animal traffic. Netting over the area can reduce the access of
birds and insects. Grass buffers and perimeter barriers can reduce
surface runoff.

Experimental Protocols

Aspects of the experimental protocol of a field trial, including the method of microbial application, subsequent operations at the site, and
experimental termination protocols will affect the amount and likely
mode of dispersal of the microorganisms. For agricultural field trials,
subsequent operations may include cultivation, irrigation methods,
and application of chemical pesticides to control insects.

There are several methods of applying microorganisms for agricultural field trials, including seed inoculation, soil inoculation, dusting plant leaves, spraying of leaves with an aqueous solution, and dipping the roots of seedlings in an aqueous solution prior to replanting (27). Microbial applications incorporating the dusting or spraying of plant leaves are more likely to initially generate aerosols than the other modes. Since dusting and spraying will also lead to a predominance of microorganisms on plant leaves and on top of soil, these application methods will also increase the likelihood of that subsequent aerosols or microbe-containing fugitive dust will be generated by wind. Dispersal during dusting and spraying operations may be reduced by performing these operations only during low-wind conditions.

For agricultural field trials, cultivation and irrigation practices will affect the amount of dispersal. For example, ridging or no-till practices will lead to far less microbial aerosolization than traditional methods. Similarly, drip irrigation will lead to less microbial aerosolization than overhead methods. Proper management of the amount of irrigation can reduce dispersal due to runoff. Site modifications such as fences and netting can reduce dispersal by animal and insect vectors. Dispersal by insects could also be reduced by the application of chemical insecticides (35).

Experiment termination protocols can be important determinants of microbial dispersal. Once a trial is completed, the site can be subjected to physical or chemical procedures that reduce the population of the test microorganism, although procedures that extensively disrupt the soil (or water) should be avoided as these are likely to lead to heavy microbial aerosols. It may be appropriate to choose or engineer the test microorganism to be sensitive to some chemical compounds or environmental factors to facilitate their elimination at the end of the experiment. For example, an organism used in agriculture in the summer could be sensitive to (killed by) the cold temperatures of the winter.

If viable microorganisms remain at the field test site after the experiment, monitoring (see next section) should be continued for an appropriate length of time after the microbe is no longer detectible. For agricultural field trials, an appropriate length of time might be the next growing season.

Reducing the Fraction of Microorganisms that Survive Dispersal

It may be possible to decrease microbial dispersal by choosing or engineering microorganisms that have limited ability to survive displacement and transport. For example, in order for a microorganism to survive aerial transport, it must be resistant to dessication and uv light. A microorganism sensitive to these factors is less likely to be

dispersed by aerosolization. Some microorganisms have special morpho-
logical forms, such as spores, that make them resistant to these environmental stresses. Thus, the elimination of the ability to sporulate
is likely to reduce dispersal.

V. MONITORING THE RELEASE OF MICROORGANISMS

Monitoring of microorganisms, like monitoring toxic chemicals, is a multi-
step process. There are two major approaches to monitoring microorganisms: (1) monitoring the *presence* of the microorganism and (2) monitoring the *effects* of the microorganism. In addition, in the case of
some recombinant microorganisms in which DNA has been inserted (especially if it has been inserted on a mobile genetic element), it may
be necessary to monitor for the inserted DNA itself. The inserted
DNA may reside either in the original microbial strain or in other microorganisms to which the DNA may have been transferred.

The important steps for monitoring the presence of the microorganism or inserted DNA can be categorized as follows:

sampling strategy, including the place, time, and frequency of sampling
    and the appropriate statistical methods
sampling procedure, including methodology for collecting and process-
    ing the samples
detection methods, including its sensitivity, specificity, reliability,
    and practicality

In some cases, enumeration (quantification) and confirmatory identification methods may also be required. For the monitoring of effects,
sampling strategy and detection methods must be considered. In some
cases, the detection method is simple observation of the health of the
"sentinal" plant or animal. Thus, sampling procedures would not be
necessary. The remainder of this section provides a brief overview
of important considerations in designing monitoring protocols.

A. Sampling Strategy

An appropriate sampling strategy is a key feature in any monitoring
plan. It helps assure that useful, believable, and statistically valid
results have been generated. Important elements in any sampling strategy include designating the exact location of the sampling, the time
of the sampling, and the frequency with which samples are taken. Also
important are the number of samples taken, including replicate samples,
and the statistical method(s) employed to interpret the raw data.

In developing a sampling strategy to monitor the microbial release
from any site, be it a laboratory, greenhouse, or open field, it is first

necessary to determine the most likely locations of the microbe and
the escape routes open to it. For example, if the microorganisms to
be monitored are located on plant leaves (as determined by knowledge
of its natural habitat and because the application protocol placed it
there), then escape routes from plant leaves should receive the most
intense monitoring. This would include sampling air above the plant
canopy, insects and birds leaving the field trial site, and runoff. The
timing of the sampling is also important. Microorganisms are most likely
to be aerosolized from plant leaves on windy days. Thus, sampling
should take place during windy periods, not during dead calms. Runoff should be sampled immediately after irrigation or a rainstorm.

Decisions about the frequency, the number of samples, and the
number of replicates in a monitoring strategy depend on why the monitoring is being performed and on how rapidly the parameters that influence the release are likely to change. If monitoring is being performed to detect any release from a highly contained laboratory, then
sampling should be frequent and comprehensive, especially while experiments are taking place. The number of replicates depends, in
part, upon the sensitivity and reliability of the deteciton method (see
below) and the certainty of detection required. If the monitoring is
routine surveillance of a field trial site testing a "low-risk" microorganism, then less frequent sampling would suffice, especially after a
baseline is established.

B. Sampling Procedure

The sampling procedure includes the means of collecting and processing samples in order to prepare them for the detection assay. In most
cases, the sampling procedure will depend upon the detection assay
chosen.

When collecting the sample, care should be taken not to contaminate the sample with other microorganisms. In general, the sample
should be isolated from contaminating sources immediately. Care must
also be taken when storing the samples for processing. The microorganisms of concern should not be subject to conditions that may kill them
during sampling (if the detection assay requires viable microorganisms).
For example, anaerobic microbes should be maintained under anaerobic
conditions, whereas aerobic microbes should not be subject to anaerobic conditions.

The type and amount of processing is dependent upon the sample
matrix (e.g., soil, water, fur, feathers) and the detection assay selected. For example, if the microorganism is to be monitored in soil samples
using a fluorescence assay, then the microorganisms should be sufficiently separated from the soil to reduce the soil fluorescence to acceptable background levels. When separating microorganisms from
their sample matrix, a measure of the efficiency of the separation should
be obtained.

C. Detection Methods

A number of methods are available for detecting the presence of microorganisms in environmental samples. Methods for detecting the effects
of microorganisms may use the principle of a "sentinel" organism—one
that is particularly sensitive to the effects of microorganism. This
is the same principle as the use of canaries in mines to detect asphyx-
iating gases. In this section, we will discuss four criteria with which
to judge the suitability of a detection technique for use in monitoring
the release of microorganisms into the environment. These criteria
are sensitivity (limit of detection), specificity, reliability, and practicality.

Sensitivity

A method should be sufficiently sensitive to detect the quantity of
microorganisms that may cause harm. Ideally, the method should be
able to detect a single microbe since, in principle, a single microbe
can multiply to a virtually unlimited extent. However, this degree of
sensitivity may prove impossible to attain, at least in the near future.
Instead, the level of detection should reflect the degree of hazard and
the number of microorganisms necessary to cause that hazard. For
example, if the hazard being considered is infection, the number of
microorganisms required for infection must be determined. This quantity will not be a single number (all or nothing effect) but a probability
of response per dose. Furthermore, the number required for successful infection will be different for each microorganism, and if the microbe has a broad host range, it may be different for each type of
host (36).

Specificity

In most cases, a method of detection should be specific for the microbe
being monitored (although sometimes indicator organisms will suffice;
in which case confirmatory tests should be available). There will be
a variety of "native" microorganisms already established in the environment into which the microorganism of concern will be released. A detection method must be able to distinguish between the newly introduced
microbe and those already present in the environment.

Reliability

A detection method should be reliable; if it indicates the microorganism
is there, it should really be there. The frequency of false positives
that indicate the microbe is present when in fact it is not should be
low. Similarly, the frequency of false negatives that indicate the
microbe is absent when in fact it is present should also be low. In
addition, the test should be reproducible.

Practicality

Detection tests must be performed on many samples, generally by many
different people, and sometimes in field situations. In addition, the
samples will not be pure, uniform cultures of a microorganism, but
will also contain various extraneous materials such as soil. These considerations require that a method of detection be low in cost and fairly
simple to perform. It must be somewhat insensitive to varying background contaminations. If the procedure is to be performed outside
of the laboratory, the reagents should be safe to handle and show good
stability.

VI. REGULATIONS REGARDING THE RELEASE OF
MICROORGANISMS IN THE ENVIRONMENT

Federal regulations regarding the environmental release of microorganisms have recently been developed in response to the proposed environmental uses of genetically engineered microorganisms. Prior to this
time, release of microorganisms from laboratory and manufacturing
facilities were regulated under the pollution laws, and commercial products were regulated under the appropriate product laws. Small-scale
releases for research and development and commercial uses that did
not fall under existing product permit laws were not regulated (except
for some pathogens considered extremely dangerous).

In June 1986, the United States Office of Science and Technology
Policy published a new U.S. policy, "A Coordinated Framework for
Regulation of Biotechnology," in the *Federal Register* (2). In this
notice, the regulatory responsibilities of several governmental agencies
were delineated and their new regulatory procedures were set forth.
In general, the intended use of the microorganism remains the determinant for which agency has jurisdiction and which statute will be applied.
However, under the new framework, the appropriate federal agency
must be notified prior to nearly all environmental releases of genetically altered microorganisms, including those small-scale experiments
conducted only for research purposes.

Federal agencies have been assigned jurisdiction over the environmental release of microorganisms as follows. The EPA Office of Pesticide Products will have primary jurisdiction over microbial pesticides
under the authority of the Federal Insecticide, Fungicide and Rodenticide Act (FIFRA). The EPA Office of Toxic Substances will regulate
"new" microorganisms intended for general commercial and environmental applications (including metal leaching, pollutant degradation,
enhanced oil recovery, and energy production) under the authority
of the Toxic Substances Control Act. All microorganisms in which
at least one of the source organisms is a pathogen will be reviewed by

the U.S. Department of Agriculture (USDA). Microorganisms that by intended use would be regulated by EPA, and in which at least one source organism is a pathogen, will be reviewed by both EPA and USDA. Microorganisms used in food, drugs and cosmetics will be regulated by the Food and Drug Administration (FDA). Research laboratories are expected to comply with the NIH Recombinant DNA Guidelines (5). All facilities must also comply with federal antipollution laws, such as the Clean Water Act, the Marine Protection, Research, and Sanctuaries Act, and the Resource Conservation and Recovery Act. Federally funded research and development must also comply with the National Environmental Policy Act and the Endangered Species Act. State and local regulations may also be applicable.

A. Environmental Proteciton Agency Regulations

As discussed in the introduction, the EPA has not formally defined environmental release. Instead, it considers an organism to be released if it is not contained. Functionally, the EPA defines a microorganism to be environmentally contained if it is used:

in a laboratory that complies with NIH guidelines (4)
in a contained greenhouse, fermenter, or other contained structure

A contained structure is a building or structure that has a roof and walls. It should also have a ventilation system to minimize microbial release to the outdoors, a system for sterilizing water runoff and wastes, and a system for restricting insects. A contained greenhouse is similar to the BL1-type greenhouse described in Section III.

Toxic Substances Control Act

Under TSCA, EPA has the authority to regulate new chemical substances or mixtures.* EPA has asserted that microorganisms and their DNA molecules are "chemical substances" and has published an operational definition of "new microorganisms" in the *Federal Register* (2). Briefly, a "new microorganism" is one that is formed by deliberate combinations of genetic material from organisms of different taxonomic genera. The exception to this is that a well-characterized, noncoding regulatory region may be inserted into any microorganism without the resulting microorganism being deemed to be "new."

As of June 26, 1986, "new microorganisms" are subject to TSCA review before any environmental release, including small-scale field testing and other environmental research and development. Compliance

---

*However, EPA will not regulate research and development for noncommercial purposes, including academic R&D.

includes notification of the EPA Office of Toxic Substances at least
90 days before such a release takes place and submission of identification, health, environmental, and other data about the microorganism.

EPA also intends to use TSCA to regulate environmental releases
of microorganisms not used for agricultural purposes in which at least
one of the parents is a pathogen (except if the only contribution of
the pathogen is a well-characterized, noncoding regulatory region),
even if it is not "new." Although compliance with this regulation is
not mandatory until EPA has gone through formal regulatory procedures,
EPA expects voluntary compliance with this rule after June 26, 1986.
For nonengineered pathogens, EPA expects to be notified of an environmental release intended to cover more than 10 acres of land or some
equivalent, but as yet undetermined, standard for nonland uses (e.g.,
in water or subterranean). Again, EPA expects voluntary compliance
with these proposals even though they must go through formal rule-
making procedures to make them binding.

Federal Insecticide, Fungicide, and Rodenticide Act

Under FIFRA, the EPA has the authority to regulate the distribution,
sale, and use of pesticide products, including microbial (bacteria,
blue-green algae, fungi, viruses, and protozoa) pesticides. In the
past, this regulation has been accomplished by requiring that experimental use permits (EUPs) be obtained prior to large-scale field experiments (terrestrial studies that involve greater than 10 acres of land
and aquatic studies that involve more than 1 surface acre of water)
and that products be registered before the pesticide is sold. Small-
scale field trials (i.e., those smaller than large-scale field trials) could
proceed without any government notification.

Under EPA's new policy, EPA must be notified prior to any field
trial of a microbial pesticide that has been genetically engineered or
that has not been engineered but is nonindigenous (2). As with chemical pesticides, EUPs will be required for large-scale field trials. For
small-scale field trials of microbial pesticides, EPA has established
two levels of notification: level I for those microorganisms that are
less likely to pose a risk, and level II for microorganisms that may be
riskier. EPA will use the information obtained from each notification
to determine whether an EUP will be required for the small-scale field
trial. The information required in each level of notification is outlined
in the *Federal Register* notice and will not be reviewed here.

The following categories of microorganisms must undergo level II
review prior to any field experimentation: (1) genetically engineered
microorganisms formed by deliberate combinations of genetic material
from organisms of different genera ("new" microorganisms as defined
under TSCA); (2) microorganisms formed by genetic engineering where
one of the source organisms is a pathogen (even if the sources are

from the same genus); and (3) microorganisms that are nonindigenous pathogens, even if no genetic engineering has been performed. Categories of microbial pesticides that must undergo level I review are: (1) nonpathogenic microorganisms that have been genetically engineered, but are not intergeneric; and (2) nonpathogenic microorganisms that have not been genetically engineered, but are nonindigenous.

For microbial pesticides requiring level I reporting, the EPA must be notified at least 30 days prior to any field test. Limited amounts of information about the microorganism and the field trial must be included in the notification. The field trial is presumed to be approved unless EPA notifies the applicant otherwise within 30 days.

For microbial pesticides requiring level II notification, adequate information to evaluate the small-scale field trial must be submitted to the EPA. The Agency then has 90 days to review the notification and determine whether an EUP is required. If an EUP is not required, the applicant may conduct the field test. If an EUP is required, the applicant must then apply for a permit.

EPA is expected to publish proposed rules regarding microbial pesticides and other microbial products of biotechnology in the Federal Register in the Spring of 1989. The proposed rules will further define the 1986 policy guidelines for FIFRA and TSCA. The proposed TSCA rule is expected to have substantial differences from the proposals outlined in the 1986 Federal Register notics.

Other EPA-Administered Environmental Laws

The release of microorganisms from laboratories and other facilities are subject to several federal (and in some cases state) environmental pollution laws. These include the Clean Water Act; the Marine Protection, Research, and Sanctuaries Act (Ocean Dumping); and the Resource Conservation and Recovery Act (RCRA). No new regulations have been proposed under these general pollution laws for genetically engineered microorganisms.

B. USDA Regulations

The USDA administers several laws that give it jurisdiction over the release of microorganisms in the environment. These include the Virus-Serum-Toxin Act (VSTA) of 1913, the Federal Plant Pest Act, and the Plant Quarantine Act.

Under the VSTA, the USDA regulates the shipment (interstate, intrastate, import, and export) of veterinary biologics, a category of materials that includes "all viruses, serums, toxins, and analogous products of natural or synthetic origin, such as diagnostics, antitoxins, vaccines, live microorganisms, killed microorganisms, and the antigenic or immunizing components of microorganisms intended for use in the diagnosis, treatment, or prevention of diseases of animals"

(2). USDA authorizes the shipment of unlicensed animal biologics for the purpose of treating limited numbers of domestic animals if USDA determines that the conditions under which the experiment is to be conducted are adequate to prevent the spread of disease.

Under the Plant Pest Act and the Plant Quarantine Act, "USDA has regulatory authority over the movement into or within and through the United States of plants, plant products, plant pests, and any produce or article which may contain a plant pest at the time of movement" (2). Plant pests include microorganisms such as bacteria, fungi, viruses, and protozoa. USDA has proposed new regulations that are designed to prevent the release into the environment of genetically engineered organisms that are plant pests or for which there is reason to believe are plant pests. In particular, USDA proposes to regulate any genetically engineered microorganism in which the donor or recipient organism or the vector used in the transfer is a pathogen. As with the EPA rules, well-characterized noncoding regulatory regions are exempt from this requirement.

C. Speculation about Future Regulations

The regulation of the environmental release of microorganisms is in its early stages and will probably evolve rapidly over the next few years. Whether future regulations will be more lax or more stringent than those described here will depend upon the successes and perceived safety of the early field trials and commercial products.

If no problems arise from field experiments and early commercial uses of genetically engineered microorganisms, the EPA will probably develop categories of microorganisms that are exempt from regulation at the research and development phase. This will be similar to the evolution of the NIH guidelines for research laboratories where exempt categories were developed as experience was gained with genetically engineered microorganisms. The NIH has already proposed exempting genetically engineered microorganisms that are deletion mutants from review prior to small-scale field trials (37).

If an environmental or health hazard does occur during a field trial, regulation of environmental releases of microorganisms are likely to become more stringent. At the EPA, this may take the form of requiring permits, rather than just notification under TSCA (this would require an amendment to the current law or a new law entirely); reduction of categories of expedited review; and perhaps increasing the coverage of the laws to nongenetically engineered microorganisms. At the USDA, this may take the form of more extensive review of the permit applications. Additional state legislation and local ordinances would also almost certainly occur.

ACKNOWLEDGMENT

Dr. Dale Hattis of the MIT Center for Technology, Policy and Industrial Development participated in the early phases of this chapter.

REFERENCES

1.   Strauss, H. S. (1986). Defining an "environmental release" of microorganisms: Consequences of tying the definition to emissions from a "typical" BL1 laboratory. Report #CTPID 86-6 available from the Center for Technology, Policy and Industrial Development, MIT, Cambridge, MA.
2.   Office of Science and Technology Policy. (1986). Part II: Coordinated framework for regulation of biotechnology. *Federal Register 51*(123):23302-23350.
3.   Collins, C. H. (1984). Safety in microbiology: a review. *Biotech. and Gen. Eng. Rev. 1*:141-165.
4.   National Institutes of Health. (1976). Recombinant DNA research: Guidelines. *Federal Register 41*(131) Part II.
5.   National Institutes of Health. (1986). Guidelines for research involving recombinant DNA molecules. Part III. *Federal Register 51*(88):16958-16985.
6.   Centers for Disease Control. (1983). *Biosafety in Microbiological and Biomedical Laboratories*, Atlanta.
7.   World Health Organization. (1983). *Laboratory Biosafety Manual.* WHO, Geneva.
8.   Sulkin, S. E., and Pike, R. M. (1951). Survey of laboratory-acquired infections. *Am. J. Public Health 41*:769-781.
9.   Pike, R. M. (1978). Past and present hazards of working with infectious agents. *Arch. Pathol. Lab. Med. 102*:333-336.
10. Pike, R. M. (1979). Laboratory-associated infections: Incidence, fatalities, causes, and prevention. *Ann. Rev. Microbiol. 33*:41-66.
11. Lincoln, D., Fisher, E., Lambert, D., and Chatigny, M. (1985). Release and containment of microorganisms from applied genetics activities. *Enzyme and Microbial Tech. 7*:314-321.
12. Fisher, E., and Lincoln, D. R. (1984). Assessing physical containment in recombinant DNA facilities. *Rec. DNA Tech. Bull. 7*:1-7.
13. Pike, R. M., and Sulkin, S. E. (1952). Occupational hazards in microbiology. *Scientific Monthly*, October, 222.
14. Marshall, K. C. (1984). *Microbial Adhesion and Aggregation.* Springer-Verlag, New York, pp. 1-3.
15. Wedum, A. G., Barkley, W. E., and Hellman, A. (1972). Handling of infectious agents. *JAVMA 161*:1557-1567.

16. Benenson, A. B. (1975). Typhoid fever. In Beeson, P., and
     McDermott, W. (Eds.), *Textbook of Medicine*, 14th ed. W. B.
     Saunders Co., Philadelphia.
17. Levine, M. M., Kaper, J. B., Lockman, H., Black, R. E.,
     Clements, M. L., and Falkow, S. (1983). Recombinant DNA
     risk assessment studies in humans: Efficacy of poorly mobilizable
     plasmids in biologic containment. *J. Infect. Dis.* 148:399-709.
18. Reitman, M., and Wedum, A. G. (1956). Microbiological safety.
     Public Health Reports 71:659-665.
19. Wedum, A. G. (1964). Laboratory safety in research with in-
     fectious aerosols. *Public Health Reports* 79:619-633.
20. Kenny, M., and Sabel, F. (1968). Particle size distribution
     of *Serratia marcescens* aerosols created during common laboratory
     procedures and simulated laboratory accidents. *Appl. Microbiol.*
     16:1146-1150.
21. Rutter, D., and Evans, C. (1972). Aerosol hazards from some
     clinical laboratory apparatus. *British Med. J.* 1:594-597.
22. Mackel, D., and Forney, J. (1986). Overview of the epidemiology
     of laboratory-acquired infections. In Miller, B. (Ed.), *Labora-
     tory Safety: Principles and Practices*. American Society for Micro-
     biology, Washington, D.C.
23. Macher, J. M., and First, M. W. (1984). Effects of airflow rates
     and operator activity on containment of bacterial aerosols in a
     class II safety cabinet. *Appl. Env. Microbiol.* 48:481-485.
24. Pear, R. *New York Times*, Sunday, June 22, 1986, p. 16.
25. Harrison, K., and Hattis, D. (1985). Containment of genetically-
     engineered microorganisms: A comparison of expected releases
     during greenhouse trials with releases in ordinary research and
     development. Report #CTPID 85-2. Center for Technology, Policy
     and Industrial Development, Massachusetts Institute of Tech-
     nology, Cambridge, MA.
26. Lindemann, J., Constantinidous, H. A., Barchet, W. R., and
     Upper, C. D. (1982). Plants as sources of airborne bacteria,
     including ice nucleation-active bacteria. *Appl. and Environ.
     Microbiol.* 44:1059-1063.
27. Lindow, S. E. (1983). U.S. Patent 4482160.
28. Gillett, J. W. (Ed.). (1986). *Prospects for Physical and Bio-
     logical Containment of Genetically Engineered Organisms*. Pro-
     ceedings of the Shackelton Point Workshop on Biotechnology Im-
     pact Assessment. Ecosystems Research Center Report No. 114,
     Cornell University, Ithaca, NY.
29. Andow, D. A. (1986). Dispersal of Microorganisms with Empha-
     sis on Bacteria. *Environmental Management* 10:470-487.
30. Strauss, H. S., Hattis, D., Page, G., Harrison, K., Vogel, S.,
     and Caldart, C. (1985). Direct release of genetically-engineered
     microorganisms: A preliminary framework for risk evaluation

under TSCA. Publication #CTPID 85-3. Center for Technology, Policy and Industrial Development, Massachusetts Institute of Technology, Cambridge, MA.

31. Graham, D., Quinn, C., and Bradley, L. (1977). Quantitative studies on the generation of aerosols of *Erwinia carotovora* var. *atrosptica* by simulated raindrop impaction on blackleg-infected potato stems. *J. Appl. Bact. 43*:413-424.

32. Golden, J., Ouellette, R., Saari, S., and Cheremisinoff, P. (1980). *Environmental Impact Data Book*. Ann Arbor Science Publishers, Ann Arbor, MI.

33. Lighthart, B. (1984). Microbial aerosols: Estimated contribution of combine harvesting to an airshed. *Appl. and Environ. Microbiol. 47*:430-432.

34. Miller, T. D., and Schroth, M. N. (1972). Monitoring the epiphytic population of *Erwinia amylovora* on pear with a selective medium. *Phytopathology 62*:1175-1182.

35. Strauss et al. (1986). Controlling the dispersal of genetically engineered bacteria and fungi during field trials. In J. W. Gillett (Ed.), *Prospects for Physical and Biological Containment of Genetically Engineered Organisms*. Proceedings of the Shackelton Point Workshop on Biotechnology Imapct Assessment. Ecosystems Research Center Report No. 114, Cornell University, Ithaca, NY.

36. Stanier, R., Douderoff, M., and Adelberg, E. A. (1976). *The Microbial World*. Prentice-Hall, Inc., Englewood Cliffs, NJ.

37. National Institutes of Health. (1986). Recombinant DNA research; Proposed actions under guidelines. *Federal Register 51*(158): 29423.

17

The OSHA Hazards Communication Standard as It Applies to Laboratories

STANLEY F. BIELICKI *National Hazards Control Institute, Easton, Pennsylvania*

I. INTRODUCTION

A. What Prompted This Standard?

To fully appreciate what affected organizations must do to comply with OSHA's Hazards Communication Standard, we should reflect on what may have been a major reason leading to its enactment. The problem was the failure of American enterprise and public institutions alike to effectively train employees about the potential hazards presented by many chemicals they use when improperly handled or when involved in an emergency. Historically the only training a new employee received concerning his or her job was a brief initial assignment working alongside a fellow employee. Certainly, not every organization has failed to effectively train its employees, yet many organizations, even some of our country's major industrial leaders, can be included in this group.

B. What Is the Intent?

In the late 1960s American workers became aware of the possible connection between some illnesses and chemical exposure. It was natural for workers to question whether or not they were being adequately trained to play an active role in the protection of their own safety and health. As awareness of these problems by workers and workers' rights activists grew, the pressure on our federal and states' governments to develop regulations mounted. Pressure increased until the federal and many states' governments acted in the early 1980s.

As is often the case, when many levels of government act simultaneously, their actions overlap, resulting in problems of jurisdiction
and costly duplication of effort. Many states acted to establish their
own right-to-know laws prior to implementation (November 25, 1985)
of the Federal Standard (1). The Hazards Communication Standard,
however, preempts these laws. The Third Circuit Court of Appeals
upheld the federal government's right to preempt states' laws in this
matter (2). As a result, confusion exists as to how the federal and
states' governments will ultimately divide responsibility for workplace
hazards communication programs. Despite these problems, organizations should proceed with the steps necessary to meet the requirements
of OSHA's Hazards Communication Standard. In doing so, most of
the major requirements of the various states' right-to-know laws will
also be met.

The current OSHA standard differs from many of the state laws
in its scope. The OSHA standard only applies to exposed employees
in manufacturing industries with Standard Industrial Classification
(SIC) codes (20-39). Many believe the scope of the current OSHA
standard is too limited. It is, therefore, in the best interest of any
regulated organization to build flexibility into its overall programs
so that hazards communication information can be made available to a
broader range of individuals or other organizations that could benefit
from having it. Incorporating flexibility now will make it easier to
comply with the possible future enactment of broader revised individual
states' right-to-know laws or possible expansion of the scope of the
OSHA Standard to currently unregulated industries and/or to external
organizations. For example, the state of Pennsylvania has recently
announced its intent to regulate nonmanufacturing businesses that
use hazardous chemicals under a revised state right-to-know law (3).

C. What Will Be Its Benefits?

It is important to keep in mind that effective training as required under
this standard will provide benefits in many areas of business. Training and information are important parts of the communication program.
Surveys usually show that better informed and better trained employees
are more productive. A great deal of effort and money are and will
continue to be spent in areas related to producing higher quality American products so that American industry can compete in the world market.
Training and information, as required under this standard, will provide
many of the same benefits derived from the use of "Quality Circles,"
meetings involving management and labor discussing product quality
objectives and how they can be met or exceeded. By conducting regular
training sessions, there is greater opportunity for constructive management employee interchange. When workers are made to feel involved

in their work and believe management is interested in them as human beings, they are usually more productive.

II. MAJOR COMPONENTS OF THE HAZARDS COMMUNICATION STANDARD

A. Laboratory Requirements

Why is a chapter on the OSHA Hazards Communication Standard being included in a biohazards management book? At present, many biotechnology companies are still operating only at the research level. The Hazards Communication Standard places laboratories in a special category. It requires the following, and exempts them from all other provisions of the standard (4):

1. Employers shall ensure that labels on incoming containers of hazardous chemicals are not removed or defaced.
2. Employers shall maintain any material safety data sheets that are received with incoming shipments of hazardous chemicals and ensure that they are readily accessible to laboratory employees.
3. Employers shall ensure that laboratory employees are apprised of the hazards of the chemicals in their workplace in accordance with section (h) of the standard. Section (h) states the following:

> Employers shall provide employees with information and training on hazardous chemicals in their work area at the time of their initial assignment and whenever a new hazard is introduced into their work area (5).

It is important to realize that modern research is a blend of many disciplines, therefore, as biotechnology research begins to produce commercial products, these same companies will move into large-scale manufacturing operations and, therefore, will be required to have a hazards communication program. Often biomedical researchers use a variety of extremely hazardous chemicals and radioactive isotopes as tools in their research. Many of these commonly used chemicals are thought to have potentially serious chronic effects on humans if accepted exposure limits are exceeded. It has been my experience as a trainer that research personnel, including those with advanced degrees, are anxious to learn better ways of protecting themselves from these insidious health hazards.

Every employee who handles hazardous chemicals, regardless of the quantities and despite their previous education or job title, should receive some training tailored to the nature of their work. The extent of the training required may differ by activity, but a formal, regular program for all areas should exist.

B. Implementation of Your Program

Several things must be considered at this point. Although organizational implementation of this Standard is likely to fall on the shoulders of the safety and health professional or staff, these individuals will need the assistance of a number of others within the organization. Upper management must be made aware of the Standard's requirements. The financial requirements must be defined, and a budget sufficient to implement the program must be prepared and submitted to management for approval. Middle management must be involved in helping identify hazardous materials being used, the types of operations where they are used, the level and amount of training previously received by their employees, as well as helping plan the additional training that is necessary.

Of major concern is who will actually conduct the training. Many organizations plan to use their first-line supervisors. This approach can be used, but two points must be kept in mind. In the past, supervisors were given the responsibility for employee training, but when time conflicts arose, priorities were shifted from training to other tasks. Second, many supervisors are inadequately trained to provide the type of information required by this standard; so either the instruction will be poorly executed or a great many hours of supervisory training will be required before supervisors will be capable of conducting the training. Organizations that lack a budgeted safety and health training program, even though they have people capable of providing the needed training, will find the problem just discussed to be of major proportions.

The remainder of this chapter describes actions required to comply with the standard. We will assume that the organization has upper/middle management support, a budget, and sufficient time and staff to complete the tasks described.

Activities required for compliance with OSHA's Hazards Communication Standard can be divided into six major areas: (1) the development and presentation of a written program describing all of the following elements; (2) a list of all hazardous chemicals at the site; (3) employee training about chemical labeling/warning systems and improvement of in-house labeling systems if used; (4) development of a Material Safety Data Sheet (MSDS) program and employee training about the program, including their right to information contained on these sheets; (5) the development and presentation of an employee hazards information program for both new and current employees; and (6) the development of a program to inform contractors working on-site of any potential hazards they may encounter and what steps to take to protect themselves (6).

C. Your Written Hazards Communication Training Plan

The importance of the written program should not be overlooked. Its complexity will vary from organization to organization, but regardless

of the organization's size or business, the written program must accurately reflect the nature of the operation and steps actually being taken
to comply. OSHA inspectors have the right to review this document,
then compare its statement of program elements to the actual implementation of those same elements. An organization's written hazards communication program must include provisions for container labeling,
material safety data sheets, and an employee training program. It
must also contain a list of hazardous chemicals in each work area, the
means the employer will use to inform the employees of the hazards
of nonroutine tasks, the hazards associated with chemicals in unlabeled
pipes, and the way the employer will inform contractors in manufacturing facilities of the hazards to which their employees may be exposed.
Employers may rely on existing programs in total or in part to comply
with the above requirements. The written program must be made available upon request to employees, their designated representative, the
Assistant Secretary for Occupational Safety and Health, and the Director of the National Institute for Occupational Safety and Health (NIOSH)
(7). The following is a checklist of things to consider when developing a written program. Numbers or sections marked with an asterisk
are suggested additions to your written program if they can be implemented.

Written Program Checklist

1. Title of the individual responsible for implementation of your hazards communication program.
2. * Introductory statement of the organization's overall safety and
   health policy, signed by the CEO.
3. Description of the procedure to be followed by an employee to obtain access to the written program.
4. Statement, if a chemical user, indicating that the organization is
   relying on information supplied by the chemical manufacturer to
   make the necessary hazards determinations.
5. Description of any in-house labeling/posting systems in use, as
   well as a description of the training program to familiarize employees
   with labeling systems used on chemicals purchased by the organization.
6. Description of the system for receiving, circulating, using, and
   filing Material Safety Data Sheets. This section should include
   the title of the person responsible for this activity if different
   from the person with overall program responsibility.
7. Description of the hazards information training program for new
   employees before they start to work.
8. Description of the hazards information training program for current employees and how ongoing training will be conducted.

9. Description of the system for supplying hazards information re-
    quired to safely perform nonroutine tasks. Two major areas among
    those to be considered are (1) nonroutine maintenance and sani-
    tation of equipment having contained or containing hazardous
    chemicals and (2) entry into confined space containing or having
    contained hazardous chemicals.
10. Description of the system for informing contractors of the poten-
    tial chemical hazards to be found in the operation. *This system
    should also include the means to review the potential hazards,
    if any, of chemicals that contractors bring to the job site.
11. A list of all hazardous chemicals used in the workplace.
12.* A description of the training program for laboratory personnel,
    if it is to be of a more limited nature than the overall program
    because of the laboratory exclusion.
13.* Description of the organization's program of employee preemploy-
    ment physicals and ongoing medical surveillance if required.
14.* Description or sample of the record-keeping system to be used
    for training documentation.

D. Material Safety Data Sheet Program

What is needed in the organization's Material Safety Data Sheet (MSDS)
program? The OSHA standard requires manufacturers of hazardous
chemicals to supply copies of MSDSs with the initial shipment of their
products. Importers and distributors also must provide these sheets
for products they supply. This requirement went into effect on
November 25, 1985.

The first step in the MSDS program should be to incorporate a
statement into the organization's purchase orders, that a MSDS is required with the initial shipment of all hazardous chemicals as per the
standard, and failure to provide them in a timely manner will void the
order. Next, a system to gather the MSDSs as they arrive that includes purchasing, receiving, and safety and health should be developed. It is wise to furnish suppliers with the office or title of the
individual to whom these sheets should be sent.Without this simple
step, sheets will arrive at any one of a number of different locations
creating a logistical nightmare. A follow-up system for contacting
suppliers that have not sent the required MSDSs is also needed. Always
document the requests for missing sheets. Enlist the aid of your local
OSHA office if a supplier doesn't respond to your requests. Chemical
users are responsible for initiating a follow-up effort for missing MSDSs.
Once the program to gather these sheets has been thought through,
the next step is to decide where and in what form these sheets will
be stored for easy access by employees. The standard does not specify where they must be stored or in what form, only that they be
readily available and, if sotred via electronic means, that individuals

be available to access the system when information is required. Table 1 lists possible locations where these sheets may have to be stored:

TABLE 1 Possible MSDS Storage Locations

Materials handling office	Shops
Sanitation office	Production office(s)
Safety and health office	Medical office
Laboratory and/or R&D library	Personnel office
Pilot plant	

While Material Safety Data Sheets can take a variety of forms, activity is under way to develop a standardized format to be used nation wide. The following information areas must be present on all MSDSs regardless of their physical style. Remember, no section is permitted to be left blank. If data is not available for any section or does not apply, it should be noted accordingly.

General MSDS Information Sections

1. Manufacturer's information
2. Hazardous properties of mixtures or hazardous properties of ingredients of a mixture if the properties of the mixture are unknown
3. Physical data
4. Fire and explosion hazard data
5. Reactivity data
6. Health hazard data
7. Control measures
8. Special precautions
9. Recommended first aid information
10. Revision date

Often because of the preoccupation with accumulating MSDSs, the matter of how the information is to be used, which is a matter of major importance, is overlooked. Each sheet must be carefully reviewed by health and safety personnel, the facilities engineer, and produciton manager.

The purpose of the MSDS program is not simply the accumulation of the sheets but the use of information they contain to make the workplace safer and more healthful. Information on these sheets should be incorporated into work practices, emergency procedures, facilities

design requirements, storage requirements, medical surveillance programs, employee training programs, as well as being used for the
proper selection of personal protective equipment, etc.

   Every employee including new hires must be informed of the Material Safety Data Sheet program and how their request for this information will be processed. Employees should be made aware that information furnished by suppliers of hazardous chemicals is reviewed and
used daily in the development and implementation of chemical handling,
storage, emergency, and disposal procedures. As an example, the
type of personnel protective equipment to be worn when handling a
chemical can be found on these sheets, therefore, the MSDS should
be used by the organization purchasing the chemical not only to inform
all the employees that may come in contact with it, about the need to
wear specific protective equipment, but also to be used by purchasing
so that the correct type of protective equipment is ordered and is available when needed.

E. Labeling and Workplace Warnings

Many organizations buy chemicals in bulk and transfer them into storage tanks at their facility, or they may buy multiple units for cost
savings and make these chemicals available to users. In a sense they
have become a second supplier of chemicals to their employees. Often,
chemical storage tanks or chemical storage areas are not identified
as to the hazards posed by their contents. The same is often true
in plants where process vessels are routinely not labeled. Other problems, such as the failure to label small portable containers of hazardous chemicals, may also result in avoidable exposures.

   The standard requires organizations storing or using hazardous
chemicals in vessels to label them in a proper fashion (8). Vessels
in storage areas may be individually identified or, if it is clear that
several vessels in an area contain the same chemical or chemicals with
the same potential hazard(s), they may be labeled as a group rather
than individually. An example might be posting an entire tank farm
containing flammable liquids with warnings such as these:

"Authorized Personnel Only"
"Flammable Liquids Stored Here"
"Potential Fire Hazard"
"No Smoking"
"No Use of Open Flames or Other Ignition Sources"

   To avoid uncertainty, the employee training program should define who is "authorized" and any limits associated with such authoriza-
tion. If someone needs to ask if they are authorized, they probably

are not, and should not enter any limited access areas until they have obtained permission from the supervisor in charge.

There is an exemption in the labeling requirement of the Standard (9). This exemption deals with small portable containers to be used by a single person for a single shift and kept under that person's direct control. It is important to keep in mind that the unexpected can happen and take a person from the job site without warning. Thus, we could have a small container holding an extremely hazardous chemical of unknown nature left behind for the next person to guess its contents and hazards. It is always better to require a proper label on any container regardless of its size, since one simply does not know when a "small" quantity of chemical in an unlabeled or improperly labeled container could cause a problem. Little oversights can destroy your safety record. It is important to take a more thorough approach. Insist that all containers, large or small, that are to be used to store hazardous chemicals be completely identified using chemical name and the physical and/or health hazard it presents. In the long run it will save time and money and avoid unnecessary suffering.

The Standard does not require the use of a specific labeling system. It does require that the system used be taught to employees who, in the foreseeable course of their work, will come in contact with hazardous chemicals and, therefore, need to be familiar with the labeling system. Three systems are currently in widespread use in the United States. Where and when each system or system combination is used depends to a great extent on how the chemical is being handled. For example, is it being shipped, used, or stored, and what quantities and types of containers are involved? An organization must use a that explains in sufficient detail what the health and physical hazards of the chemical are and what the likely results of improper handling will be. For this reason, the "ANSI Hazardous Chemical Precautionary Labeling System" (10) or a modification of it will most completely meet this requirement. Keep in mind that, regardless of the system used, it must clearly identify the physical or health hazards presented. Remember, there are three systems currently in use in the United States; therefore, make certain that employees who may encounter them are familiar with them.

The three systems currently in use in the United States are: The Department of Transportation (DOT) system (11), (Figure 17-1), required during the transportation of hazardous chemicals and hazardous chemical waste; The American National Standards Institute (ANSI) Hazardous Chemical Precautionary Labeling System (Figure 17-2), used on most chemical packages including small bulk packages such as 55 gallon drums; and The National Fire Protection Association System (NFPA) 704 (12) (Figure 17-3), used on many bulk storage vessels and chemical warehouses. It should be pointed out that some chemical manufacturers are also using parts of the ANSI and NFPA systems to

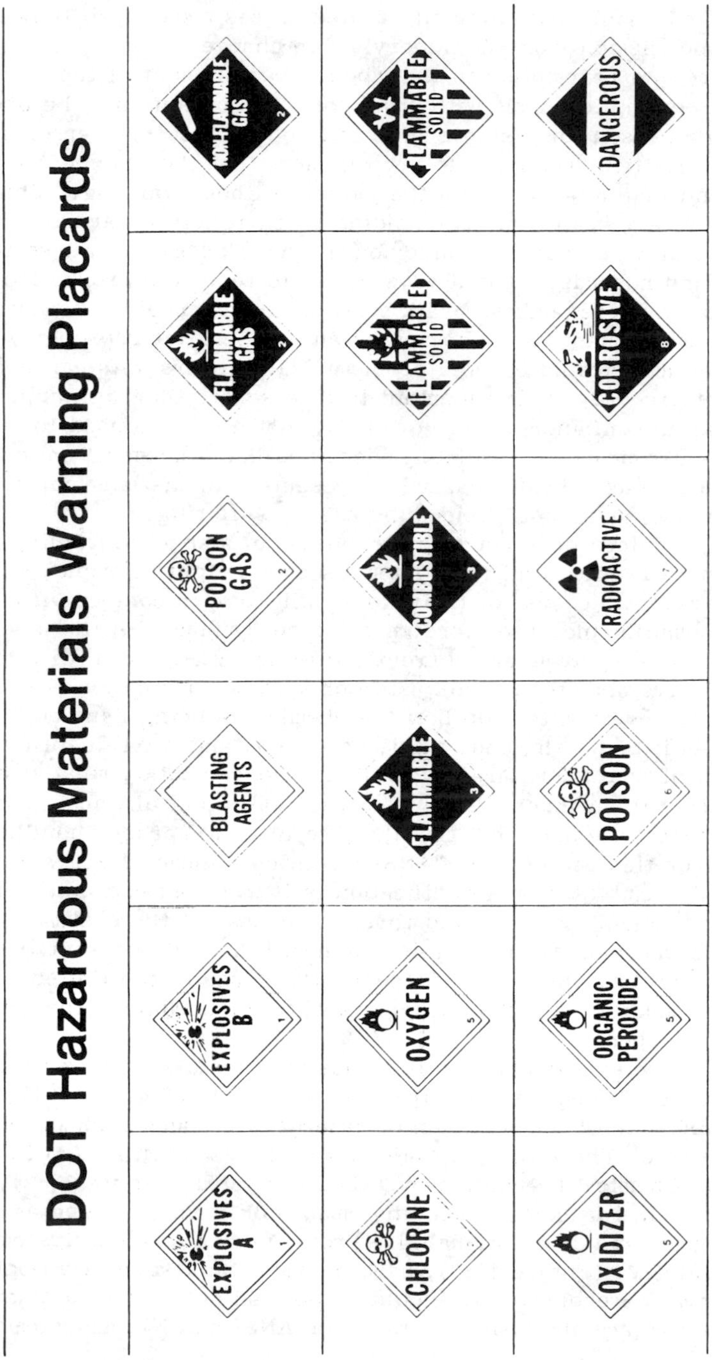

DOT Hazardous Materials Warning Placards

FIGURE 17-1 The Department of Transportation labeling system.

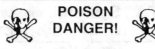

**POISON
DANGER!**

LIQUID AND VAPOR CAUSES
SEVERE BURNS
HARMFUL IF INHALED AND MAY CAUSE
DELAYED LUNG INJURY
SPILLAGE MAY CAUSE FIRE OR
LIBERATE DANGEROUS GAS
MAY BE FATAL IF SWALLOWED

Do not breathe vapor. Do not get in eyes, on skin,
on clothing. Keep container tightly closed and in a
cool area. Use with adequate ventilation. Wash
thoroughly after handling. In case of fire, soak with
water. In case of spill or leak, keep upward and flush
away by flooding with water applied quickly to entire spill.

CALL A PHYSICIAN

FIRST AID RECOMMENDATIONS

FIGURE 17-2 The American National Standards Institute Hazardous
Chemical Precautionary Labeling System.

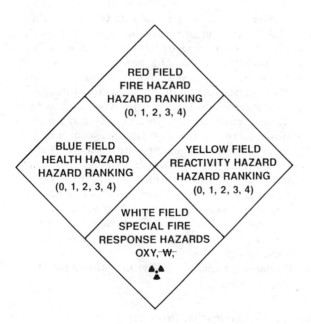

FIGURE 17-3 The National Fire Protection Association system of labeling.

FIGURE 17-4 J. T. Baker Chemical Company's SAF-T-DATA system.

make a composite system. Figure 17-4 is an example of the J. T. Baker
Chemical Company's SAF-T-DATA system used on their laboratory
and small bulk packages (13). If you intend to develop your own system you should start with the ANSI system as the base.

   Independent of the system selected, the standard requires you
to train your employees so they understand how to read and interpret
the system being used. It is important to inform employees of the existence of all three systems. It is a waste of time to limit their training
to just one system to save time under the usually false assumption that
it is unlikely that a gorup of employees will not encounter more than
one system during the course of their work, even if you develop your
own system. Exposing employees to all the systems also helps them
to gain a better understanding of the rationale used to label chemicals.
If you decide to conduct a general three-system training program,
Table 2 is a breakdown by system and job type that may help you
concentrate greater emphasis in your training on the most often encountered system by that work group.

   A number of good audiovisual programs describing these systems
are available (10-12).

F. Your Contractor Hazards Communication Program

The standard requires manufacturers and users of hazardous chemicals to inform contractors about the hazards of chemicals they may be

TABLE 2 Labeling Systems Classified by Job

Job description	Labeling systems		
	Primary	Secondary	Tertiary
Receiving	DOT	ANSI	NFPA
Shipping	DOT	ANSI	NFPA
Materials handling	DOT	ANSI	NFPA
Production	ANSI	DOT	NFPA
Sanitation	ANSI	DOT	NFPA
Shops	ANSI	DOT	NFPA
Pilot plant	DOT	ANSI	NFPA
Laboratory	ANSI	Mod. NFPA	NFPA

exposed to while working on-site. Most organizations have been advised by their legal counsel, however, not to take the responsibility of actually performing the training, because they may be assuming additional liability for these nonemployees. The correct approach is to involve your engineering, safety, and contracting staff members in this process. The following items should be included as part of any agreement with an outside contractor. Prior to beginning work, and within a specified lead time, one of the above-mentioned job titles will send to the contractor's safety/training officer all the MSDSs for the hazardous chemicals in the area where the contractor will be working. In turn, the contractor, prior to beginning work, will send all the MSDSs for hazardous chemicals to be used on the job, if any, to your safety/engineering department. As part of the contract terms, each organizaiton will assume total responsibility for communication of any necessary safety and health information about the chemicals involved to their affected employees. Also, as part of the terms of the contract, contractors will be informed as to what areas are off-limits to their employees and how these areas are identified.

G. Your Hazardous Chemicals Information Training Program

As discussed above, the Standard requires that new employee training be conducted before these employees begin work, and that existing employees receive training in the law's requirements, in the chemical labeling systems used in your operations, in your MSDS program, and in any specific technical areas such as corrosive chemical or toxic chemical safety, unless such training has been previously provided and this has been documented. It is important to realize that training means more than having been assigned to spend time watching or helping a fellow employee. It means instruction by a person considered to be knowledgeable in the subject. Such an individual should be a member of the management staff and/or a qualified third party instructor.

Let's look at those topics that must be presented to all employees working with or around hazardous chemicals. The training program must explain what rights all employees have under the Hazards Communication Standard. This includes their right to review the organization's written hazards communication program, their right to chemical hazards avoidance training before they begin working with or around the chemical(s), and their right to see MSDSs for the chemicals used in their work or work area.

Another element of the training program should explain what procedure will be used to prevent a new employee from beginning work prior to having received the required training. One method that might be used is to limit the days that new employees can begin working to those days when someone is available to conduct the required training. This could be accomplished by having the employee's rights section of the program presented by personnel or a member of the safety staff before the individual reports to his or her work area. Then, through careful scheduling, new employees would only be released to their supervisor if the hazards training had been provided or if the supervisor was prepared to provide the training because his or her schedule was purposely kept free for this training. Once the general training concerning employee rights has been completed, the following information, which will depend on the nature of the business and its operations, must be provided to all employees not covered by the laboratory exclusion. The following is a suggested outline that can be used to help develop the needed material for your presentation.

Principles of Chemical Safety:

A. Hazards Identification
    1. Signs/postings
    2. Labeling
    3. Material Safety Data Sheets
B. Hazards Versus Risk
    1. Physical hazards
    2. Health hazards
C. Types of Hazards to Be Found in the Area
D. Physical Hazards
    1. Flammable/combustible liquids
    2. Corrosives

   3. Reactives
   4. Compressed gases
   5. Explosives
E. Health Hazards/Toxic Chemicals and Toxicity
   1. Terms and definitions
   2. Exposure limits
   3. Acute/chronic exposure effects
   4. Toxic actions
F. Potential Health Hazards Control
   1. Preemployment physicals
   2. Hazards determinations
   3. Engineering controls
   4. Administrative controls
   5. Personal protective gear
G. Discussion of Unique Hazards, If Any, for the Chemicals Used at Your Site
   1. Supertoxics
   2. Product intermeidates
H. Administrative Controls and Safety Practices for Nonroutine Tasks
   1. Maintenance and sanitation of equipment that contains or has contained hazardous chemicals
   2. Entrance of confined spaces that contain or have contained hazardous chemicals

III. RESEARCH LABORATORY EXEMPTION

A. Why Such an Exemption for Laboratories?

Some assumptions always have to be made when drafting a regulation as sweeping in scope as the Hazards Communication Standard. Several of these assumptions possibly led to the Research Laboratory exemption. The first of the two most important assumptions is that, because of their education laboratory personnel, have a greater awareness of the potential hazards of the chemicals they use or are more aware of the resources available for obtaining such information. The second important assumption is that laboratories usually use small quantities of hazardous chemicals, thereby limiting the exposure and subsequent potential for risk. This is supported by the fact that pilot plants are regulated under the Standard, whereas laboratories are not. The areas of required compliance deal with container labeling, Material Safety Data Sheets, and employee training prior to a new assignment (5).

B. What Elements of the Standard Should Still Be Put Into Effect?

Two important points should be considered before we answer the above question. First, in many laboratories the individuals working at the bench are not the directing scientists. The individuals working at the bench often have only a high school education. In this situation, the directing scientist is usually responsible for reviewing safe practices for using hazardous chemicals with his or her research staff. Second, in organizations where bench work is performed by individuals with college level training, another problem can exist. Most academic institutions share the same problem as many industries, in that they often lack programs to teach students pursuing degrees in the sciences the proper procedures (good laboratory practices) for safely handling hazardous chemicals. Therefore, many graduates with strong academic credentials often have poor safety habits.

Laboratory personnel should be informed, as part of the overall organizational information and training program, about the organization's written hazards communication program, the organization's MSDS program, the organization's preassignment training for new hires program, and how all employees will be kept informed of new health and physical hazards information for the chemicals they use. If these steps are taken, then the organization is in compliance.

The problem, as mentioned earlier, however, is that many laboratory professionals are uninformed about current health hazards information for many of the chemicals they use routinely. This problem is also complicated by the fact that many of the potential health hazards being reported are chronic effects involving chemicals that have been used in laboratories for years. For example, chloroform is now considered to be a carcinogen (14). Since many of these discoveries concern basic laboratory reagents, our educational problem is not identifying the hazards but rather determining how to handle all laboratory chemicals so that overall exposure is limited, thus reducing the risk of developing these chronic effects.

The Third Circuit Court of Appeals in its May 1985 ruling also ordered OSHA to broaden the scope of the standard (2). This may mean that in the future OSHA will remove its laboratory exemption. A good laboratory practices training program should be instituted along with a brief manual summarizing accepted standard procedures for laboratory-scale operations. The topics to be included in both the manual and the training depend on the nature of the research being conducted. The following list should be considered:

1. Laboratory chemical labeling systems
2. Glassware handling tips
3. Flammable liquid safety
4. Corrosive chemical safety
5. Compressed gas safety
6. Toxic chemical handling
7. Proper use of fume hoods and biological safety cabinets
8. Reactive and unstable chemicals handling

9. Biohazards
10. Radioisotopes
11. Chemical storage practices
12. Handling and disposal of laboratory wastes (chemical, biological, and radioactive)
13. Miscellaneous laboratory practices (good housekeeping)

It can not be overemphasized that the management of most organizations assume that college graduates have been properly trained in the above 13 areas. Those of us in the field, however, know that this is rarely the case.

C. Incorporating MSDS Information into Existing Information Programs/Systems

The standard requires organizations to provide employees with access to Material Safety Data Sheets. Certainly, this is an important use of the sheets, but information provided on these sheets should be used in a number of other ways. For example, information provided on MSDS's can be used to help design a facility, to store chemical properly, to develop emergency response procedures, or to correct a, heretofore, unrecognized need for the use of special personal protective equipment, etc. Material Safety Data Sheets should be reviewed by personnel in a number of functional areas. A partial list of these areas and the section(s) of MSDSs which are important to them is shown in Table 3.

The fire and explosion data section on a MSDS is very useful. The flash point, when appropriate, allows the proper ranking of a chemical by its potential as a fire hazard. This in turn is important in the development of proper storage, handling, disposal, and emergency procedures. Recommended extinguishing media and unusual fire or explosion hazards information also found in this section is useful in developing fire control and emergency response procedures.

The reactivity data seciton on a MSDS can be used to develop proper storage procedures, because it provides information about chemical incompatibility and stability, such as limited shelf life. Hazardous products of decomposition are usually also found in this section. This information is useful in fire control and emergency response procedures. It can be used to select the proper personal protective gear for individuals assigned responsibility for reacting to these emergency situations.

The health data and control measures sections have numerous applications within organizational practices and procedures. For example, exposure limits are useful in designing ventilation systems or in developing handling procedures. Routes of entry information is useful in selecting the necessary means to protect employees that must handle chemicals, for example, if the chemical poses an inhalation hazard selection and use of containment or respiratory protection devices, or if

TABLE 3 Functional Areas with a Need to Review MSDS

Functional area	MSDS section
Purchasing	Manufacturers information Physical/chemical characteristics
Materials handling (receiving/shipping)	Physical/chemical characteristics Physical/health hazards Special handling and use information Control measures
Engineering/process development	All sections
Safety and health	All sections
Quality Assurance (Q/A)	Physical/chemical characteristics Physical/health hazards Control measures Special handling and use information First aid information
Medical	Health hazards data First aid information
Maintenance and sanitation	Health hazards data Control measures First aid information
Production	All sections
Pilot plant	All sections
Laboratory	All sections

the chemical presents an absorption hazard proper selection of gloves and glove material. The signs and symptoms of exposure information can be used to train employees to recognize that there is an exposure problem or to help medical personnel to evaluate whether an exposure problem exists and, if so, to what extent. Also in this section are suggested work practices. This information is useful for developing the precautions section of several procedures, such as Batch Production Records (BPRs) and Quality Assurance (Q/A) raw material sampling procedures. First aid information is an important MSDS item. It can be incorporated into the organization's medical program and used in special training for first aiders or EMTs. The documents and/or procedures where MSDS information can be useful is summarized in Table 4.

TABLE 4 MSDS Information Classified by Documents/Procedures Where it Should Be Incorporated

MSDS section	Organization document/procedure
Manufacturer's information	Purchasing vendor records Medical emergency contact documents Quality Assurance (Q/A) sampling procedures
Hazardous ingredients of mixtures	Medical emergency procedures Industrial Hygiene (IH) documents Batch Production Records (BPRs) Q/A sampling procedures Emergency response procedures Environmental documents Maintenance Procedures Facility design documents
Physical data	BPRs IH documents Q/A sampling procedures Emergency response procedures Environmental documents
Fire and explosion data	Maintenance nonroutine task permits Emergency procedures Facility design documents BPRs Q/A sampling procedures Materials handling procedures Environmental documents Facility design documents
Health hazard data and control measures	Maintenance nonroutine task permits Medical documents IH documents BPRs Q/A sampling procedures Materials handling documents Facility design documents Environmental documents
Special precautions	Maintenance nonroutein task permits BPRs Materials handling documents Facility design documents

TABLE 4 (continued)

MSDS section	Organization document/procedure
	Q/A sampling procedures
	Environmental documents
Recommended first-aid	Medical documents
procedures	Q/A sampling procedures
	BPRs
	Materials handling procedures
	Maintenance procedures

In conclusion, MSDS information should be reviewed as an aid in developing or improving the following general types of organizational documents listed in Table 5.

D. Use of Internal Programs that Already Exist to Comply with the Standard

The Hazards Communication Standard is a performance standard. This means each regulated organization will be evaluated for compliance or noncompliance based on the results of its efforts, not its specific efforts. For example, employee training is a requirement of the standard, but each organization is free to develop its own style of training

TABLE 5 Documents that Should Include MSDS Information

Document	Document section
1. Medical documents	esp. first aid procedures
2. BPRs	esp. precautions section
3. Q/A sampling procedures	esp. precautions section
4. Nonroutine maintenance permits	esp. precautions section
5. Engineering designs	esp. ventilation
6. IH documents	esp. exposure levels and sampling procedures
7. Job Safety Analysis (JSA)	esp. chemical handling
8. Materials handling procedures	esp. precautions section
9. Environmental procedures	chemical handling
10. Organizational safety manual	chemical handling
11. Laboratory manual (good laboratory practices)	chemical handling

program. In this case the measure of compliance or noncompliance would be how effectively, not how, the employees were trained. Obviously, though, failure to develop an adequate training program will usually produce less than satisfactory training results.

Organizations can use programs already in existence to comply with the standard. This means that an evaluation of existing efforts to provide safety and health information should be made before drafting the organization's written hazards communication program. Some organizations will find that many existing programs meet the requirements of the standard with little or no modification. Some of the more common programs that can be used for compliance are as follows:

(1) Development and dissemination of an organizational or area safety and health manual. To meet the Standard's performance requirement, a manual should also be used as part of the training program, instead of just being issued to each employee or supervisor. (2) Many manufacturing organizations, as part of their BPRs, have a precautions section prior to the actual work directions. If this section is present and is reviewed with new employees before they start an operation, then this could be included in the written program. (3) More and more organizations have developed a new employee safety and health orientation program. It is usually presented by a member of the Safety and Health staff to new hires before they report to their respective job site. If such a program exists it should also be included in the organization's written program. (4) Properly conducted, effective safety meetings and an effective safety committee can also be included in the organization's written program, if they actually provide an effective means for new safety and health information to reach employees as it becomes available. (5) Along these same lines, any internal publications that are sent or given to employees that contain safety and health information should also be included. For example, some organizations have been publishing a safety and health newsletter for years. (6) A limited number of organizations have also developed their own internal hazardous chemical labeling system. If this system provides the needed information as described in the standard, then it should also be included as part of the written program. (7) A program to assure proper confined space entry, which includes a permit system and preentry review, instruction as to the potential hazards, and the means to avoid them, should be included.

These are just a few of the activities or programs that should be considered for inclusion in the organization's written hazards communication program. The key question asked should be, does their use with or without modification increase employee safety and health awareness of the physical and or health hazards of the chemicals they use, as well as their awareness of how to avoid these hazards?

IV. GRAY AREAS, COMMON QUESTIONS, AND MISCONCEPTIONS

A. Trade Secrets

A manufacturer of a chemical may withhold formula information from the MSDS by claiming trade secret protection. If such protection is claimed, it must be stated on the MSDS. This claim usually appears in the manufacturer's or hazardous ingredients of mixtures section of the MSDS. The Third Circuit Court of Appeals in its May 1985 ruling dealt with the issue of trade secrets, but despite this ruling it is still a controversial issue (2). In the event of a medical emergency, a chemical manufacturer must supply the treating physician or nurse with any chemical identity information necessary for proper first aid. The manufacturer, after the emergency, may require the persons involved to sign a confidentiality agreement. In nonemergency situations, health professionals—defined as physicians, industrial hygienists, toxicologists, or epidemiologists—who are providing medical or other occupationally related care to employees may request information about the identity of trade secret chemicals. Again, the signing of a confidentiality agreement may be required, and the organization claiming trade secret proteciton does not have to disclose process or percentage information considered to be nondiscoverable by reverse engineering methods.

B. States' Right-to-Know Laws Versus OSHA's Hazards Communication Standard

There are a number of differences between the various states' right-to-know laws and the Federal Standard. In this section, we will highlight the major areas of difference, but readers should make their own comparison for the state(s) in which their organizations operate.

The most striking difference between the Federal Standard and various states' laws is which industries are regulated and which are not. The Federal Standard limits its coverage to employees exposed to hazardous chemicals in manufacturing industries in SIC codes (20-39), while most states include employees in nonmanufacturing industries and nonindustrial organizations, such as fire departments, as well as emergency response groups. Many state laws also include residents living in the general area of a manufacturing facility using hazardous chemicals.

Another area of difference is the chemicals that are covered under the Federal Standard versus those covered under the vairous state laws. The Federal law has a floor of approximately 2,300 chemicals known to be hazardous, and, therefore, covered, while many states include more chemicals. Some states even include radioactive materials on their list. While most states list the chemicals covered, the Federal

Standard, since it is a performance standard, lists those criteria that make a chemical a potential physical hazard and/or a health hazard.

The definition of an exposed individual also differs widely between the Federal Standard and the states' laws, even between the various states. The Federal Standard covers those employees routinely exposed during the course of their work or anyone who could be exposed during a foreseeable emergency. Some states define exposure in terms of the number of days and the minimum airborne concentration that must be present before a person is considered to be exposed.

While the Federal Standard addresses labeling using performance criteria, many states require specific types of information on container labels. In a few states, the National Fire Proteciton Association (NFPA) 704 system is required on containers or product labels. States also differ widely regarding the size of containers that are required to have hazards warning labels.

The Federal Standard and the various states' laws do not differ significantly concerning the requirement for chemical users to obtain MSDSs and to develop a system so that MSDSs are available to employees.

Generally, the Federal Standard and the state laws all provide some form of trade secret protection, except in the case of medical emergency or if there exists a legitimate need to know by a health care professional.

Individuals responsible for the organization's "hazards communication/right-to-know" program should become very familiar with the similarities and differences between the Federal Standard and the laws of the state in which the organization is located. Do not overlook the fact that even though preemption was upheld by the courts, it also ordered OSHA to expand the coverage of the Standard. Preemption does not prevent states form passing more stringent laws or from passing laws with broader coverage, such as regulating nonmanufacturing industries using hazardous chemicals.

C. Areas Subject to Review and Revision

Throughtout the previous sections we have indicated that changes in the current Standard are likely. The Third Circuit Court of Appeals ruling is likely to be a major factor in bringing about these changes. Under the court's May 1985 ruling, OSHA must expand coverage under the standard to all workers that come in contact with hazardous chemicals. If this occurs, all employees coming in contact with hazardous chemicals, regardless of the industry in which they work or their occupation (laboratory staff included), will be covered by all sections of the standard. In addition, states are likely to retain or enact laws that will require industry to better inform public agencies of the number and nature of the hazardous chemicals they have on-site. Pressure already exists for the development and use of a standardized format

for Material Safety Data Sheets. There will also likely be clarification
of what constitutes an exposure. Parameters are needed to remove
possible ambiguities between workers that should receive training as
specified under the standard because they are at risk and those where
such formal training is not necessary because, for all intents and purposes, they are not at risk. This is not to suggest that employees
should not be made aware of their right to this information, but rather
that the degree of training and the amount of information furnished
is proportional to the actual need as dictated by the nature of the job.

D. Conclusion

Sound information about the physical and health hazards of the chemicals used by society is critically important. The actions discussed
in this manuscript as required by either the Hazards Communication
Standard or right-to-know laws are steps in the right direction. We
must keep in mind that our information about the health hazards of
many chemicals is still limited. We must develop programs to respond
to the needs of the employee and do it in such a way that the results
of our efforts lead to corrective, protective actions that do not become
hung up in the courts or delayed by differences among factions within
the scientific and regulatory community. Once the information is available, we must develop practical approaches to assure that even small
organizaitons can afford to provide the necessary training to their
employees at possible risk. Perhaps then we can begin to meet the
objective of a sound Occupational Health and Safety program by providing workers who use/handle chemicals a safe and healthful workplace.

NOTE ADDED IN PROOF

Two important developments have taken place since the writing of this
chapter. The first is the expansion of the Hazards Communication Standard to include all employees exposed to hazardous chemicals, regardless
of the industry in which they work. And second, OSHA has published
for comment, a proposed rule specifically: Occupational Exposures to
Toxic Substances in Laboratories (15).

REFERENCES

1. *Federal Register* 48(228), Friday, November 25, 1983. Rules
   and Regulations, U.S. Government Printing Office, Washington,
   D.C., pp. 53333-53348.
2. *United Steelworkers of America* v. *Auchter*, U.S. Court of Appeals
   for the Third Circuit, May, 1985.
3. Laws of Pennsylvania, Worker and Community Right to Know Act,
   No. 1984-159.

4. *Federal Register 48*(228), Friday, November 25, 1983. Rules and Regulations, U.S. Government Printing Office, Washington, D.C., pp. 53340, (b)(3)(i-iii).
5. *Federal Register 48*(228), Friday, November 25, 1983. Rules and Regulations, U.S. Government Printing Office, Washington, D.C., pp. 53344, (h).
6. *Federal Register 48*(228), Friday, November 25, 1983. Rules and Regulations, U.S. Government Printing Office, Washington, D.C., pp. 53343-44, (e-h).
7. *Federal Register 48*(228), Friday, November 25, 1983. Rules and Regulations, U.S. Government Printing Office, Washington, D.C., pp. 53343, (e)(3).
8. *Federal Register 48*(228), Friday, November 25, 1983. Rules and Regulations, U.S. Government Printing Office, Washington, D.C., pp. 53343, (f)(1-9).
9. *Federal Register 48*(228), Friday, November 25, 1983. Rules and Regulations, U.S. Government Printing Office, Washington, D.C., pp. 53343, (f)(6).
10. American National Standard for hazardous industrial chemicals—precautionary labeling, American National Standard Institute, New York, NY, Z129.1-1982.
11. CFR Title 49, Sections 100-199. U.S. Government Printing Office, Washington, D.C., Cur. Ed.
12. Recommended System for the Identification of the Fire Hazards of Materials, 704. National Fire Protection Association, Quincy, MA, Cur. Ed.
13. Baker SAF-T-DATA Gudie. J. T. Baker Chemical Co., Phillipsburg, NJ, 1982.
14. Third Annual Report on Carcinogens Summary, September 1983. National Technical Information Service, Springfield, VA, PB-83-135855, pp. 40-41.
15. *Federal Register 51*(142), Thursday, July 24, 1986. Proposed Rule, U.S. Government Printing Office, Washington, D.C., pp. 26660-684.

18
United States OSHA Laboratory Standard: "Regulation of Toxic Substances in Laboratories"

ALAN M. DUCATMAN *Massachusetts Institute of Technology, Cambridge, Massachusetts*

I. INTRODUCTION

Few regulations have as far reaching-implications for scientists as the proposed OSHA rule CFR Part 1910 (Docket H-150), "Occupational Exposures to Toxic Substances in Laboratories" (5). Laboratories in the United States may well be faced with toxic chemical regulation specifically intended for laboratory use. The components of this proposed law, while far from perfect, are sufficiently realistic that they will be enforceable in the laboratory environment.

The proposal recognizes that laboratories, with their many highly trained workers, small quantities of numerous toxins, and nonrepetitive tasks, require different regulation from production facilities. The "Hazard Communication" (4) (or "Right to Know") regulation has not been enforceable in laboratories. It is even regarded contemptuously by laboratory workers, whose health it might have originally intended to protect. OSHA has recognized the shortcomings of the Hazard Communication or "Right to Know" regulation in the laboratory environment and has met the challenge with what it terms a "performance-based" standard.

By "performance-based," OSHA means that laboratory workers are best protected by assignment of specific safety responsibility, by facility and engineering requirements designed to minimize exposures, and by written operating plans for the handling of certain chemicals. The proposed standard requires that each laboratory create a chemical hygiene plan. This plan can be formulated only with the cooperation of laboratory workers, and therefore it addresses their perceived needs. The plan is also accessible to regulators. The appropriateness of its components and the degree to which they actually become laboratory practice will undoubtedly become the basis for regulatory inspections and decisions concerning whether safe "performance" has been attained. Although the standard was clearly written for chemistry laboratories, it will apply to biomedical laboratories as well.

II. RELATIONSHIP TO OTHER REGULATION

The proposed regulation supersedes previous regulations that deal with toxic chemicals. Other regulations that deal with physical safety hazards (fire, explosion, electrical) and employee rights (OSHA 1910.20– access to records) will still apply to laboratories. The proposed regulation does not mention microbiological hazards, so future regulations pertaining to these (sharps, waste handling) remain unaffected as well.

Some elements of the Hazard Communication Act are reproduced in this laboratory standard. Material Safety Data Sheets (MSDSs) and reference materials must still be available, for instance. The modification is that laboratories no longer face the logistical (and safety) nightmare of providing these materials in each laboratory space. Laboratories will escape from many of the labeling requirements of production facilities, but they are still prohibited from defacing or removing preexisting labels. Most important, laboratories within production facilities should recognize that they must still comply with the training and education requirements of the Hazard Communication Standard. Thus, laboratories in production facilities will live under two very different kinds of regulations.

A major change will be displacement of portions of Code of Federal Regulations (CFR) 1910. Within CFR 1910, Permissable Exposure Limits (PELs, from the 1910.1000 subpart Z tables) will remain as a toxic exposure standard for laboratories (7). The remainder of subpart Z and the single substance standards (CFR 1910.1001-1910.1047) are superseded for laboratories. An interesting outcome of this change is that medical surveillance action levels might no longer form the legal basis of surveillance decisions for laboratory personnel. The proposed standard's description of an "exposure evaluation" does "not require monitoring of concentrations." This may even suggest to some readers that specific industrial hygiene measurements will no longer be required. This interpretation should be viewed with caution for any routine procedure, as neither health outcomes nor legal liability for inappropriate exposures will be affected by the introductory language of this or any other regulation. Laboratories using chemicals should certainly get on with the business of employing industrial hygiene techniques to document the effectiveness (or ineffectiveness) of their controls upon routine procedures.

Finally, the 25 states and territories with their own OSHA plans (see Appendix I) may promulgate comparable superseding regulations.

III. QUALIFICATION AS A LABORATORY

Deciding if one's facility is or is not a laboratory is a critical decision, with important implications for how health and safety resources are to be allocated. Academic and hospital laboratories are included; office-based medical, dental, and veterinary labs are not. OSHA considers that only "laboratory scale" operations should qualify. "Laboratory scale" is defined as operations that can be conducted by a single individual without the aid of mechanical transfer devices. A single type of transfer equipment is permitted in a "laboratory scale" operation; dollies and other similar equipment may be used to move gas cylinders and other similar supplies to and from the laboratory.

The presence of other transfer equipment implies that a facility is really a pilot plant to be operated under the General Industry Hazard Communication Standard. This distinction, if left in the regulation, will be critical for laboratories involved in large-scale culture of organisms. Such activities require transfer equipment; thus, laboratories with large bioreactors may not meet a strict interpretation of "laboratory scale." Their laboratory chemicals could be regulated under the general industry standard even though they have only biological hazards in any quantity. Interpretation of this problem will require an opinion from OSHA.

IV. CENTRAL DEFINITIONS

The most important definition is whether or not a facility is a laboratory. Six additional definitions provide guidelines for comprehending the proposed standard. The terms "carcinogen," "toxic substance," "overexposure," "regulated area," "chemical hygiene plan," and "chemical hygiene officer" are intended to have specific regulatory meaning for laboratories.

A. Carcinogens

The proposed regulation makes no distinctions among strong carcinogens, weak carcinogens, and potential carcinogens. Instead, OSHA proposes that any substance already regulated as an OSHA carcinogen, or which appears in either the most recent editions of the International Agency for Research on Cancer (IARC) (3) or the U.S. National Toxicology Program (U.S. Department of Health and Human Services) tests of known or potential carcinogens (8), be regulated as carcinogenic.

There are two interesting outcomes for this proposed definition, and
both clearly trouble OSHA. First, strong, weak, and possible carcinogens may all receive the same regulatory treatment. And agencies
without regulatory context may suddenly find their published (albeit
not commonly distributed) evaluations elevated to regulatory status.
While both of these outcomes have the virtue of simplicity, some discussion is required to determine whether this is appropriate for regulatory
purposes.

B. Toxic Substances

The definition of a "toxic substance" resembles the definition of a carcinogen. Again, there are three lists. *Two of them are the IARC
and NTP carcinogen lists*! The only noncarcinogenic materials to be
regulated as toxic substances are chemicals whose PELs are listed in
OSHA's Z tables. To a large degree, OSHA's proposed definition of a
"toxic substance" is therefore a real or potential carcinogen.

C. Overexposure and Exposure Evaluation

"Overexposure," under the proposed regulation, means exposure in
excess of OSHA-PELs. Overexposure triggers a medical consultation.
Thus, exposures to chemicals from the NTP and IARC carcinogen lists
have both procedural and regulatory meaning but do not trigger medical surveillance or consultation because they are not OSHA PELs. An
"exposure evaluation" is an assessment of whether overexposure has
occurred. Air monitoring, while clearly valuable for this purpose,
is not always required by the draft standard and should not constitute
the whole of any exposure evaluation. Employees may request and
receive "exposure evaluations." Also, any reason for suspecting possible overexposure should result in such an evaluation.

D. Regulated Area

This is a poorly defined concept and may involve the most difficult
administrative decisions for compliance with this law. A regulated
area is intended as a demarcated work area, where "toxic substances"
including carcinogens are handled. Access to these areas must be
limited to qualified personnel. The draft regulation states that the
"regulated area" does not have to be an entire laboratory but could
be a single hood or other separable work station. A number of points
of uncertainty exist. It is unclear whether the regulated area could
be as small as a labeled part of a bench top, how best to post signs
designating "regulated areas," or what mechanisms adequately restrict
access to qualified personnel trained to handle toxic substances.
Realistic solutions to questions concerning "regulated areas" may turn

out to be expensive and require considerable sign-posting. The use
of physical lock-out mechanisms, not required in the draft standard
but worth considering in some circumstances, will increase procedural
and equipment costs. While potentially attractive, the mere declara-
tion that an entire institution's laboratories are all "restricted" at the
entrance from the parking lot is short-sighted and does not meet the
spirit of the regulation (adds no safety). A more rational solution is
to plan adequate facilities to handle all operations involving "toxic substances."

E. The Chemical Hygiene Plan and Officer

The Chemical Hygiene Plan and the Chemical Hygiene Officer who administers the plan represent the centerpiece of the proposed regulation.
The plan is further defined in terms of seven essential elements. These
are:

1. Standard operating procedures for work with toxic chemicals.
2. Criteria for the use of control measures.
3. Enumeration of unusually dangerous or unconventional processes
   that require special permissions from responsible authorities.
4. Provisions for expert consultation and medical evaluations.
5. Selection of regulated or restricted areas for working with toxic
   substances.
6. Descriptions of training programs, emergency procedure training,
   and available reference and training materials pertaining to toxic
   chemicals.
7. The Chemical Hygiene Officer's role, which is to assist with ad-
   ministraiton of the plan. This is discussed in the following section.

V. ADMINISTERING THE OSHA LABORATORY STANDARD

A. Line of Responsibilities

The standard defines the responsibility for chemical hygiene lists,
requires specific facility and engineering capability to prevent overexposure and aid in emergency situations, and creates a chemical hygiene plan for all processes and procedures involving toxic chemicals.
The responsibility for implementation belongs successively to the Chief
Executive Officer, department heads, and laboratory or project supervisors. Penalties for failure to implement the law would presumably
also be their successive responsibility. At this time, the nature of
penalties for noncompliance have not been stated.

It is worth noting that the Chemical Hygiene Officer can carry
on administrative functions without specific credentials. The Officer

need not be an industrial hygienist, occupational physician, or other specific type of safety professional, as long as there is access to industrial hygiene and occupational medicine expertise. Institutions may create a "Chemical Hygiene Committee" whose members can have additional expertise. Such a committee can be used to increase expert consultation and, if desired, to serve as an administrative check upon the activities of the Hygiene Officer.

B. Administrative Duties of the CHO

The Chemical Hygiene Officer (CHO) is an administrator and consultant without direct line responsibility. His first role should be to assist laboratories by providing a format for laboratory chemical hygiene plans. This format must meet the policy and procedural requirements of the law, include helpful suggestions, and remind responsible individuals of related facility modification and record-keeping issues. While the format and suggestions should come from the Officer, the plan must be created by the responsible authority in each laboratory. Once the plan is in motion, the officer should audit both inspection reports and periodic resubmissions of the plan from regulated facilities. It may also prove useful to monitor the distribution, use, and disposal of hazardous chemicals. As an aside, extending these audits to chemical stocks can serve an additional function of meeting anticipated Environmental Protection Agency directives for reporting the presence of highly toxic chemicals in inventory (2). Finally, the audit function serves to remind line authorities of record-keeping requirements for exposure evaluations and medical evaluations. The proposed standard requires that the Hygiene Officer personally maintain reference materials such as copies of Hygiene Plans, the Laboratory Standard, up-to-date toxicology data, and required supplementary information such as the American Conference of Government Industrial Hygienists (ACGIH) Threshold Limit Values (TLVs) (1). The CHO need not personally maintain evaluation records but ought to assure that they are maintained and available to workers. Medical evaluations should obviously be secured in the medical department.

The officer cannot have primary responsibility for providing required training, facilities, policies, or inspection. The Officer merely assists the laboratory personnel primarily responsible for these functions.

Facility recommendations and requirements are designed to prevent overexposure and enhance emergency response. Fume hoods, maintained to meet individual laboratory ventilation specifications, are clearly required to prevent overexposure. The plan also requires designation of "regulated areas" for work with "toxic substances," including closed systems such as glove boxes or fume hoods. These need to be segregated from other work stations, and protective equipment appropriate to anticipated exposures need to be available. Respirators are prominently mentioned in this regard and need to comply with existing federal regulation (29 CFR 1910.134) (6).

Creation and administration of the chemical hygiene plan is the backbone of this proposed OSHA law. Ten elements that OSHA expects to be present in each laboratory plan are outlined in Appendix II. There is little doubt that laboratories will be greatly aided by a written outline, which discusses required components and puts them in a logical framework. The Chemical Hygiene Officer should provide such an outline. Laboratories should be careful to avoid commitments to processes, procedures, or programs that may not actually be carried out; OSHA inspectors will expect plans to reflect real practice. Appendix II is a general outline, which could be modified for use by laboratory personnel. It makes a distinction between what is apparently recommended and what is required with the proposed law. Plans created by laboratory personnel on the basis of an outline should be reviewed for completeness and appropriateness by the Chemical Hygiene Officer. Laboratories may further wish to create a mechanism for automatically informing the Officer of the dates and findings of required inspections, maintenance, and training, as well as any accident reports. The Hygiene Officer can provide more reliable quality assurance if there is such a formal mechanism for providing documentation of these periodic requirements of the plan.

VI. DISCUSSION

There are problems inherent in any proposed regulation. This proposed OSHA laboratory standard will be problematic for employer and employee interest groups. For instance, laboratories that exist within production facilities are placed in an unfortunate position. The proposed regulation will require that they still come under the very provisions of the Hazard Communication Standard, whose unwieldiness in laboratories prompted the search for a performance standard. Rather than try to exist under two contradictory standards, industry-based laboratories may decide to simply limit their regulatory liability by declaring themselves to be "pilot plants." The largest group of potential beneficiaries of the law, namely workers and safety personnel in industrial laboratories, may continue to live under less effective "Hazard Communication" regulation.

The exclusion of medical, dental, and veterinary office-based laboratories is equally unfortunate. Numerous hazards exist within these presently unregulated environments. The health of office-based medical laboratory workers would have benefitted enormously from performance-based requirements that include expert consultation and hygiene plans.

Elements of the proposed regulation will also be worrisome for organized labor. OSHA feels that the enormous procedural, training, and engineering requirements of this standard can somehow be implemented for an average of $85 per laboratory and $6 per worker ("Summary of Industry Profile and Costs" and Table 1 of the proposed standard). This absurdly low estimate has nothing to do with the nature or implementation of the regulation. Yet it will certainly alienate organized labor and thereby forfeit a potential ally of performance-based regulation. A more substantial concern may also affect the organized labor perspective of this regulation. Employers appear to have been given fairly wide latitude regarding the use of methods for determining "overexposure" as well as deciding upon the need for subsequent medical consultation. This will undoubtedly provoke anxiety about compliance with the intent of the Laboratory Standard. The proposed law also blurs some distinction between legal requirements and prudent recommendations. Some of what appears to be merely recommended for safe facility operation should be considered absolute requirements. This wide latitude of interpretation should rightly concern labor interests.

Chemical Hygiene Officers should assist laboratories to ensure that appropriate respirators are available for all anticipated procedures. General ventilation systems with air intakes and exhausts, sinks, eyewash fountains and drench showers, and even arrangements for waste disposal could also be regarded as merely "recommended" under a narrow interpretation of the proposed law. Store rooms are "recommended." A legal interpretation that these essential elements of safe laboratory practice are somehow optional should be resisted on common-sense and economic liability grounds. Conversely, a suggestion that every laboratory have two exits is unpractical for some kinds of clean rooms and should be interpreted as a recommendation. Finally, a legal facility requirement that vacuum pumps and lines be protected from contact with contaminants is a nice but unenforceable idea. The technology for monitoring for such contamination is not accessible.

Employers and employees will be disturbed with the rather circular OSHA definitions of toxic substances and carcinogens. The emphasis on chemial carcinogenesis as the featured enviornmental hazard shows OSHA to have inadequately considered the types of hazards present in, say, microelectronics or biotechnology laboratories. These environments are at least as likely to contain acute respiratory, dermal, and chronic neurologic, hematologic, or reproductive hazards as they are to be carcinogenic. The law has the potential for creating a misleading emphasis upon carcinogenic hazards by featuring long lists of weak or potential carcinogens. Laboratory directors and Hygiene Officers should not focus upon these lists to the exclusion of more significant and immediate laboratory hazards. These categorical lists were not intended for regulatory purposes. If they are not modified or removed, the proposed laboratory standard's central engineering and procedural

thrusts run the risk of being ignored. This is reminiscent of the list
approach taken by the Hazard Communication Standard, which has
been ineffective in laboratories.

In conclusion, several problems in the proposed laboratory standard may leave it without political allies in industry or labor. It will
be interesting to see if it can survive, as this regulation offers significant protection of workers in unpredictable environments such as laboratories. The strength of the regulation is easy to understand. One
strength is that the regulation cleverly circumvents traditional friction
between line management and staff safety functions in laboratories.
Under this regulation, laboratory managers have no option to give
away safety responsibility. Unsatisfactory plans or failure to carry
out written plans will clearly be the problem of laboratory managers.
They will have every motivation to avail themselves of consulting expertise in areas such as industrial hygiene and occupational medicine.
Whether these consultations are done or merely administered by the
Chemical Hygiene Officer becomes a matter of little consequence. Decisions to ignore consulting advise now falls directly where it should,
on responsible line management. The proposed standard also describes
some specific ventilation, engineering, and safety facility requirements.
These are not now present in every laboratory. The proposed standard will support their installation, maintenance, and use. The importance of this improvement should not be underestimated. In addition,
creation of regulated areas and segregation of hazardous tasks will
decrease exposures and send an important safety message to laboratory
workers and supervisors. Implementation of these changes alone would
represent a significant boost to the protection of workers in the complex
laboratory environment.

APPENDIX I: U.S. STATES AND TERRITORIES WITH OSHA-APPROVED PLANS THAT MAY ADOPT LABORATORY REGULATION

Alaska	Michigan	South Carolina
Arizona	Minnesota	Tennessee
California	Nevada	Utah
Connecticut*	New Mexico	Vermont
Hawaii	New York*	Virginia
Indiana	North Carolina	Virgin Islands
Iowa	Oregon	Washington
Kentucky	Puerto Rico	Wyoming
Maryland		

---

*Plans cover only state and local government employees.

APPENDIX II: CHEMICAL HYGIENE PLAN OUTLINE

A. Required Designation of Responsible Personnel
1. Authority within the laboratory
2. Institutional authority
3. Institutional administrator–The Chemical Hygiene Officer (and, if desired, the Chemical Hygiene Committee)

B. Facility Requirements and Recommendations
1. Fume hoods required for meeting toxic substance OSHA PELs, recommended for other exposures.
2. Laboratory recommended to have general ventilation with air intake and exhaust, giving $\geqslant 4$ room air changes/hr.
3. Recommended designated stock and storerooms.
4. Recommended sinks, eyewash fountains, drench showers, fire extinguishers (required by other regulation), telephone access.
5. Recommends two exits for each laboratory (not appropriate for all laboratories).
6. (*Author's note:*) Laboratories should prudently incorporate above recommendations and consider them requirements.
7. Recommends an alarm system for isolated areas such as cold rooms.
8. Sign posting:
    A. Emergency telephone numbers recommended.
    b. Identity labels of substances recommended.
    c. Location signs of pertinent equipment and hazards recommended.
    d. Designated restricted areas required.

C. Recommended Procurement, Distribution, and Storage of Chemicals
1. Recommends central receiving for bulk chemicals.
2. Recommends use of ventilated store and stock rooms versus storage on bench tops and in hoods.
3. Recommends that chemicals be carried within containers or buckets during distribution.
4. Recommends periodic inventories and reduction of excess laboratory stores.

D. Standard Operating Procedures for Primary Handling of Toxic Substances and Carcinogens
1. Procedures that require establishment and use of restricted areas, including specially designated hoods, spaces, and enclosures. Requires designation of such restricted areas.
2. Procedures that require the use of respirators in accordance with 29 CFR 1910.134.
3. Procedures that require other protective equipment and a list of such apparel.

4. Requires personal hygiene practices within and upon exit of a restricted area, including skin washing.
5. "Recommended" prohibitions (*Note*: These should be required by prudent laboratory directors):
   a. No horse play.
   b. Avoid tasting or smelling chemicals.
   c. Prohibition of eating, smoking, drinking, gum or tobacco chewing, cosmetic application, food or utensil storage in laboratories.
   d. No mouth pipetting.
   e. No bare feet, sandals, perforated shoes.
   f. Avoid contact lenses (some authorities approve with safety goggles).
   g. Use gloves, respirators, lab coats, and other pertinent personal protection when working with toxic chemicals.
6. Requires procedure to protect vacuum pumps and lines from contamination.
7. Requires enumeration of other control measures for general reduction of exposure:
   a. Engineering, including general ventilation and hood performance criteria.
   b. Personal protective equipment to be worn in regulated areas.
   c. Other hygiene practices, including spill and accident procedures

E. Maintenance and Housekeeping
   1. Requires maintenance of fume hoods, eyewash fountains, showers, and other safety engineering equipment. Recommends quarterly to biannual periodicity for such maintenance.
   2. Requires maintenance of personal protective equipment. For respirators, maintenance must comply with 29 CFR 1910.134.
   3. Recommends regular cleaning such as floor washing and maintenance of clear passageways.
   4. Requires arrangement for regular disposal of contaminated waste (but allows discretion for defining what is contaminated and for establishing periodicity of pick-up):
      a. Recommends weekly pick-up.
      b. Recommends incineration as disposal method.
      c. Urges that waste chemicals not be poured down drains. (*Note*: EPA regulations increasingly require this. Nuclear Regulatory Commission also has interlocking regulation for radioactive chemicals.)
      d. Requires that waste removal procedures be described.

F. Training and Information Aspects
   1. Requires that employees know the location and availability

of the Chemical Hygiene Plan and OSHA Laboratory Standard (including appendices).

2. Requires that employees know the location and availability of supporting reference materials. (*Note*: It is advantageous that the Chemical Hygiene Officer maintain a reference library; material need not be in each laboratory.)
   a. MSDS required.
   b. Relevant OSHA PELs required.
   c. ACGIH-TLVs required.
   d. Additional standard toxicology and medical texts recommended.
3. Requires that employees be instructed in the selection, use, and maintenance of routine protective equipment.
4. Requires that employees be instructed in spill and accident procedures including location and use of emergency equipment, and pertinent phone numbers. Recommends a written emergency plan (EPA regulations require a written plan).
5. Recommends instruction that hoods containing toxic substances be left on when not in use unless adequate ventilation can otherwise be assured.
6. Recommends that receiving and stock room clerks be included in laboratory training.
7. Recommends that periodicity of training be more frequent than annually.

G. Requires provisions for evaluations and consultations. These may be provided, but need not be provided, by the Chemical Hygiene Officer
   1. Exposure evaluations, conducted when there is a possibility of overexposure or when workers request such evaluation, must be performed by qualified persons (such as industrial hygienists) and must be without charge to workers.
   2. Requires medical consultation if overexposure is deemed likely:
      a. Must be performed by a qualified person (such as an occupational physician).
      b. Must be without charge to workers.
      c. Also requires follow-up surveillance for overexposures which may result in chronic disease.
      d. Recommends compliance with other OSHA law (unless specifically superseded by this proposed standard).
   3. Requires an available emergency medical facility which need not be on site. Recommends that personnel trained in first aid be available on site.

H. Requires provisions for record-keeping. (Records are not kept as part of the plan, but the plan must state that records shall be kept.)

1. Document concerns leading to exposure evaluations, the exposure evaluation, and pertinent findings. These must be maintained and kept accessible to employees.
2. Medical consultations must be kept as medical records.

I. Laboratories are required to describe potentially hazardous operations or procedures, which employees may not initiate without specific permission of the employer.

J. The Chemical Hygiene Plan requires a provision for annual evaluation of effectiveness and updating.

REFERENCES

1. American Conference of Governmental Industrial Hygienists. (1986). Threshold Limit Values and Biological Exposure Indices. ACGIH, 6500 Glenway Ave., Building D-7, Cincinnati, Ohio.
2. United States Environmental Protection Agency. (1986). Emergency Planning and Community Right to Know Programs; Interim Final Rule and Proposed Rule, Cross Reference: 40 CFR Part 300; *Federal Register 51*(221):41570-41594.
3. IARC Monographs on the Evaluation of the Carcinogenic Risk of Chemicals to Man. (1985). World Health Organization Publications Center, 49 Sheridan Ave., Albany, New York.
4. U.S. Department of Labor, Occupational Safety and Health Administration. (1983). Chemical Hazard Communication: Final Rule. 29 CFR Part 1910. 120; *Federal Register 48*(228):53280-53348.
5. U.S. Department of Labor, Occupational Safety and Health Administration. (1986). Health and Safety Standards; Occupational Exposure to Toxic Substances in Laboratories; Proposed Rules. 24 CFR Part 1910; *Federal Register 51*(42):22660-22284. U.S. Govt. Printing Office, Washington, D.C.
6. U.S. Department of Labor, Occupational Safety and Health Administration. (1985). Subpart I–Personal Protective Equipment. 29 CFR 1910.134.
7. U.S. Department of Labor, Occupational Safety and Health Administration. (1985). Subpart Z–Toxic and Hazardous Substances. 29 CFR 1910.1000-1047.
8. U.S. Dept. of Health and Human Services, Public Health Service. 1985). Annual Report on Carcinogens, National Toxicology Program, U.S. Government Printing Office, Washington, D.C.

19

Organizing an Effective Training Program to Meet the New OSHA Requirement on HBV/HIV

ROBYN M. GERSHON* *Yale University, New Haven, Connecticut*

DIANE O. FLEMING** and DAVID VLAHOV *Johns Hopkins University, Baltimore, Maryland*

I. INTRODUCTION

The proposed hepatitis B virus (HBV)/human immunodeficiency virus type-1 (HIV-1) OSHA standard was promulgated to minimize the risk of occupational exposure to these bloodborne pathogens. On October 30, 1987, OSHA announced the regulation that will be enforced through the OSHA general duty clause until it becomes a permanent standard (1). The impetus for the development of the standard, which represents OSHA's first involvement with the health care industry, came about as a response to the concerns of health care employees. Many employees are fearful of exposure to HIV-1; one recent survey found that more than 25% of young physicians would not take care of an AIDS patient if they had a choice (2). Recent well-publicized reports of occupationally acquired HIV-1 infections have heightened the fears of health care workers (3-9).

There is evidence that health care employers are not meeting the occupational health needs of their workers. A NIOSH study of hospital employee health conducted in 1976 found that only 8% of the hospitals surveyed had even the basic elements of an effective employee health and safety program (10). Existing programs are usually found within the hospital setting, and these are generally designed to meet the mini-

---

Current affiliations:
*Johns Hopkins University, School of Hygiene and Public Health, Baltimore, Maryland
**Sterling Drug, Great Valley, Pennsylvania

mal standards of the Joint Commission on Accreditation of Hospitals
(JCAH) (11).

The JCHA standards merely require that employees be fit to perform their job duties, be free of communicable diseases, and that
records of employee health be maintained (11).

In contrast, the proposed OSHA standard is much more comprehensive in scope. The standard has a strong employee education and
training component, a medical surveillance component, and an enforcement procedure with targeted inspections and fines of up to $10,000
for each violation. OSHA's directive stipulates that health care employees and employers must both adhere to the current CDC recommended guidelines to minimize the risk of occupationally acquired diseases (12). If the new OSHA standard is fully implemented, then it
may reasonably be expected to have a positive impact on the occupational health needs of the health care industry.

II. ASSESSMENT OF RISK

A. The Risk of Bloodborne Diseases among Health Care Workers

Data on the risk of occupationally acquired infectious disease in the
health care industry have been extremely limited. The Bureau of
Labor Statistics has shown that the health care industry has the
fourth highest injury/illness rate in this country (13). In a 1984
survey, more than 11,000 health care workers were reported to have
had occupationally acquired disease (13). More recent information,
shown independently by Jacobsen and Vesley (14,15), indicates that
the rate of laboratory-acquired infections is approximately 2 per 1,000.

These rates do not differ substantially from the results obtained
through the pioneering work on laboratory acquired infections by Pike
in the 1970s (16).

Needlestick injuries are common among health care workers and
may lead to exposure to bloodborne pathogens (17). Needlestick surveys have shown that nursing personnel have the highest incidence
rates among all health care personnel, accounting in one study for
60-70% of all hospital-based needlestick injuries (18). In one hospital-
based study analyzing 1,339 work-related injuries and illnesses, 249
were caused by needlestick injuries among nurses (19).

Recently Gerberding and co-workers reported on health care
workers with multiple needlestick injuries involving HIV-1-contaminated
needles; in one instance a single health care worker reported 11 accidental needlestick injuries from patients with AIDS (20).

B. The Risk of Occupationally Acquired HBV/HIV-1 Infection

Studies of 1,700 exposed health care workers estimate the risk of
HIV-1 in the health care setting to be approximately 4 in 1,000 (21).

In addition to these studies, using the criteria of Vlahov and Polk, there are an additional 10 case reports of documented seroconversion in the health care setting, indicating that transmission of HIV-1 can and does occur (3-10,21,22). However, these anecdotal case reports generally do not involve a denominator (i.e., a population at risk of seroconver-sion), so that risk per se cannot be estimated.

Table 1 summarizes all of the anecdotally reported occupationally acquired HIV-1 infections in health care workers. As of June 30, 1987, 883 health care workers have been tested for HIV-1 antibody following percutaneous or mucous membrane exposure. One health care worker has been found to be seropositive for HIV-1 antibody (22). An additional 425 health care workers with similar exposures have been tested, and of these three have had documented seroconver-sions. These three cases have been reported by CDC and are also listed in Table 1 (8).

There is well-documented evidence that HBV is transmitted in the health care workplace. Hepatitis B virus has an estimated prevalence of 16% in health care workers (23). Hepatitis B virus infection is the most frequently reported occupationally acquired infection in the health care workplace (23). The prevalence of serologic markers for HBV is highest in emergency room nurses, pathology and surgical staff, and blood bank workers (24). The routes of transmission of HBV in the health care setting are similar to HIV-1, but HBV is found in pa-tients' blood at much higher concentrations, and thus is much more easily transmitted (25,26).

It has been estimated that 200 health care workers die annually from complications arising from HBV infections (27). Fortunately, there is a highly effective and safe vaccine available for immunization against HBV (28).

Other bloodborne diseases may also present risk to the health care worker. Examples of these are typhoid, shigella, herpes type I, cytomegalovirus, syphilis, and many other infectious agents (15). In some cases chemotherapeutic treatments are available; in others, such as HIV-1 infection, there are none.

It is estimated that occupationally acquired infectious diseases in the health care workplace cause at least 250 employee deaths per year and 10 to 20 million dollars in expenses to the industry (29).

III. RISK REDUCTION STRATEGIES

In general, occupational risk may be caused by a variety of factors. These are lack of information, lack of resources (e.g., safety equipment), and lack of sufficient motivation.

Before attempting to train a group of health care workers, the assessment of these issues through informal observations and inter-views should be performed. This educational needs assessment is criti-

TABLE 1 Case Reports of HIV-Infected Health Care Workers

Case #	HCW sex/occup.	Report source	Incident	Seroconversion documented
1.	Female nurse	Anon., 1984 (3)	Needlestick injury	+
2.	Female nurse	Stricof et al., 1986 (4)	Deep intramuscular needlestick	+
3.	Female nurse	Oksenhender et al., 1986 (5)	Superficial needlestick	+
4.	Female student	Neisson-Vernant et al., 1986 (6)	Needlestick injury	+
5.	Female home care provider	CDC, 1986 (7)	Numerous exposures to blood/body fluids	+
6.	Female health care worker	CDC, 1987 (8)	Held bloody gauze on wound for 20 min	+
7.	Female phlebotomist	CDC, 1987 (8)	Blood spill on face (acne present)	+
8.	Female medical technologist	CDC, 1987 (8)	Large blood spill on skin	+
9.	HCW	Weiss, et al., 1988 (9)	Not known	+
10.	HCW	Weiss et al., 1988 (9)	Puncture wound	+

cal to the design of an effective program. If the level of knowledge (e.g., on the transmission of HIV-1) is adequate, then training sessions devoted to basic facts will have only marginal impact. If there is an inadequate commitment to the necessary resources (e.g., sharps containers, disposable items), then a discussion of knowledge and attitudes will be only marginally effective. Thus, while an educational needs assessment may be time-consuming in and of itself, it is nevertheless crucial to maximizing the effectiveness of a training program.

The new OSHA proposal attempts to minimize occupational risk through the following key components of the program:

1. *Risk Assessment*: Assigning work activities to one of three OSHA exposure categories (I, II, III).
2. *Risk Management*: The development of Standard Operating Procedures (SOPs), as required by OSHA; the development of an Educational and Training Program; the development of an employee Medical Surveillance program; the development of a Record Keeping System.

IV. TRAINING AND EDUCATION AS A RISK REDUCTION STRATEGY

A. Introduction

The purpose here is to focus upon the area of education and training, an area with potential to greatly impact the effectiveness of a health and safety program.

To be effective, a training program needs to be well planned and fully supported by high level administration. The responsibility for the program, the maintenance of program records, and the program evaluation must be clearly designated in advance.

The position of training director/coordinator is an important function and should be held by a responsible manager. The director should have a clear set of objectives and goals along with strong administrative support in the form of a realistic budget and adequate staffing. In small health care facilities it may not be feasible to hire a full-time training coordinator. Employees may be rotated into the position of training coordinator for limited periods of time. Even small work sites will need a well-organized record keeping system, which could be computer-based.

Many hospitals and health care institutions must provide training programs, such as those in infection control, to meet the requirements of JCAH and other accrediting agencies. These programs will probably incorporate the training requirements for OSHA. Other institutions that have training programs designed to meet the "right to know" legislation should likewise be capable of adapting their programs to suit the new OSHA requirement. Thus, it may be possible to maximize the usefulness of ongoing programs by extending their scope to cover the OSHA health and safety requirements.

B. Setting up Your Program

There are several aspects to consider when devising an effective training program (30). The four key elements that need to be well developed at the outset are:

1. *The Source*: Who will do the training?
2. *The Message*: What is the message?
3. *The Recipient*: Who is the target audience?

4. *The Response*: What is the desired behavior change or modification, and how will that be evaluated?

The Source

The speaker/trainer must be a credible person to the particular audience, with the authority and subject knowledge to command the interest and respect of the attendees. It is essential for the speaker to
be capable of delivering the message in a highly effective manner.
    The speaker/trainer could be a nurse educator, physician, scien-
tist, health and safety specialist, guest speaker, or training program
director. If there is difficulty in getting skilled speakers, then you
may wish to obtain one of the many excellent audiovisual films or training aids available (see Appendix). If the responsibility of speaking
is delegated to an in-house staff member, it is important to support
attendance at conferences and courses to allow a continual updating
of knowledge on HIV, HBV, and other biohazardous agents.

The Message

Both the content of the message and the manner in which it is delivered
are critical in getting the message across. The message content, that
is, the subject matter that must be covered in training sessions, is
broadly outlined by OSHA (Table 2). CDC has developed a series
of guidelines that detail the recommended practices designed to minimize the risk of occupationally acquired HIV-1 infection (12,31-34).
It should be noted that the specific details for each of the standard
operating procedures (SOPs) and all the built-in safety features pertinent to those SOPs *must* be taught by the first-line supervisor. Only
the immediate supervisor can be sufficiently familiar with the practices
and procedures of each particular job task to provide detailed health
and safety training. For that reason, the initiation of "train-the-trainer"
sessions should be done to assure that the front-line supervisors are
familiar with the necessary health and safety precautions needed for
each activity they supervise.
    There may be an interest in covering more than one health and
safety or infection control topic at each session, but it is best to try
not to deliver too much information at any one time. The message will
be better received if it is clear, straightforward, and limited in content.
    The employees will be more likely to listen to the speaker if you
attend to certain aspects in addition to the actual content of the message.

    Environmental comfort of the audience. The audience will be in a
more receptive frame of mind if room temperature, seating, lighting,
and sound systems are arranged to maximize comfort.

    Message comprehensibility of the audience. It is important to know
what the audiences' beliefs are on the subject matter before you attempt

TABLE 2 Summary of the Required Message Content

1. The modes of transmission of bloodborne infectious agents such as HBV and HIV-1. The risk to the employees and the methods needed to minimize that risk.

2. The differentiation between Category I and Category II tasks.

3. General understanding of the effective use, care, and limitations of personal protective equipment, such as eye goggles and disposable gloves. The effective use of safety equipment, such as Biological Safety Cabinets.

4. Effective emergency response action, including accident reporting

5. Appropriate actions to take and persons to contact if unplanned Category I tasks are encountered.

6. Clear understanding of all Standard Operating Procedures (SOPs) required for performing a particular task.

7. The proper storage, handling, decontamination, and disposal of contaminated/infectious materials.

8. Employee participation in the Employee Health Medical Surveillance Program.

Source: Adapted from Ref. 1.

to train them. By knowing this information, the speaker can better gauge the type of approach to use in order to maximize the effectiveness of the educational session. While not always readily available, one can sometimes extrapolate this information from data on the audiences' educational level, professional standing, and job descriptions. By knowing the extent of the audiences' knowledge, you can tailor the lecture to reinforce their knowledge and strengthen those areas that are weak.

For those employees who speak English as a second language, every effort should be made to present the material in their native language. Similarly, those employees who are functionally illiterate should have the material presented in a manner that they can understand (35). For this particular group, it may be necessary to use films, pictograms, or demonstrations to facilitate the comprehensibility of the material. The needs of such workers should be attended to in a considerate manner.

Message comprehensibility can be enhanced by high quality audio/visual materials. It is critical that slides be legible and informative. Many prepared slides can be purchased at nominal cost.

Receptivity of the audience. The audience will be in a more favor-
able position to accept the training message if they are in a receptive
frame of mind. One way to accomplish this is to conduct the sessions
during times that are not too disruptive to their general work schedule.
Attendance should be during regular (paid) working hours; if that
is not possible, then the session should be held before the start of
the shift and remuneration for that time should be given. Early morning
sessions seem to be the most readily accepted, especially with the added
inducement of a coffee break. The serving of refreshments is highly
recommended.

Frequency and length of the educational program. Educational
programs are very costly in terms of lost work hours, thus every effort should be made to maximize the effectiveness of the training sessions. The new OSHA HBV/HIV standard does not stipulate the number of hours per year of training that are required. In general, basic
health and safety training sessions should be given to category I and
II workers at least once a year. It is best to limit your presentations
to one hour or less to keep the audience from getting restless.

Message reinforcement. It is crucial to reinforce your message.
This can most easily be accomplished by supplementing the training
programs with a take-home flyer. The flyer should contain a brief
summary of the message, the key points to remember, and, most im-
portantly, the procedures to follow in case of an emergency (i.e., a
spill, exposure, accident). It would be useful to print the important
health and safety and emergency contact personnel and their phone
numbers on the flyer.
    In addition to the flyer, it is useful to have posters and signs
displayed throughout the workplace in highly visible areas. These
should be changed and rotated often to keep them effective. Workplace
health and safety campaigns, slogans, contests, and promotions are
all valuable ways of improving the visibility of your training program.
    A scorecard of blood/body fluid exposures and needlestick injuries
might be posted by each department to reflect any decrease brought
about by the "safety awareness" training.

The Recipient

The CDC definition of a health care worker is the following: "Persons
including students and trainees, whose activities involve contact with
patients and patient specimens in a health care setting." The new
OSHA requirement also specifies levels of potential exposure within
this definition. Table 3 lists the three exposure categories as defined
by OSHA. In Table 4, we have attempted to broadly classify job positions within these three categories. It is difficult to clearly charac-

TABLE 3 Exposure Categories

Category I

Tasks that involve exposure to blood, body fluids, or tissues.

All procedures or other job-related tasks that involve an inherent potential for mucous membrane or skin contact with blood, body fluids, or tissues, or a potential for spills or splashes of them, are Category I tasks.

Use of appropriate protective measures should be required of every employee engaged in Category I tasks.

Category II

Tasks that involve no exposure to blood, body fluids, or tissues, but employment may require performing unplanned Category I tasks.

The normal work routine involves no exposure to blood, body fluids, or tissues, but exposure or potential exposure may be required as a condition of employment.

Appropriate protective measures should be readily available to every employee engaged in Category II tasks.

Category III

Tasks that involve no exposure to blood, body fluids, or tissues, and Category I tasks are not a condition of employment.

The normal work routine involves no exposure to blood, body fluids, or tissues (although situations can be imagined or hypothesized under which anyone, anywhere, might encounter potential exposure to body fluids).

Persons who perform these duties are not called upon as part of their employment to perform or assist in emergency medical care or first aid or to be potentially exposed in some other way.

Tasks that involve handling of implements or utensils, use of public or shared bathroom facilities or telephones, and personal contacts such as handshaking are Category III tasks.

---

terize certain job duties as belonging strictly to Category I or II, therefore it is recommended that training for Category I and II be given to all workers in either of those exposure categories.

One question that often comes up is how to identify new employees who will be required to receive this type of health and safety training. New hires must be informed that attendance at these training sessions

TABLE 4 Job Position Classifications Within Each of the Three Categories

---

CATEGORY I: Tasks that involve exposure to blood, body fluids, or tissues.

1. Medical personnel (nurses, physicians, dentists, assistants).
2. Laboratory personnel handling human blood, body fluids, or tissues, such as laboratory technicians, medical technicians, clinical technologists, pharmaceutical and industrial technologists, phlebotomists.
3. Emergency response technicians, fire fighters, law enforcement officers.
4. Morgue, funeral home employees.

CATEGORY II: Tasks that involve no exposure to blood, body fluids, or tissues. Employment may require performing unplanned Category I tasks.

1. Housekeeping services.
2. Transport personnel (transport of specimens).
3. Service and maintenance personnel.
4. Home nursing care.

CATEGORY III: Tasks that involve no exposure to blood, body fluids, or tissues, and Category I tasks are not a condition of employment.

1. Auxiliary staff, such as clerical, administrative, and religious personnel.

---

is mandatory and a condition of employment. Many institutions follow standard stepwise disciplinary measures for employees who do not follow health and safety standard procedures, which would include attendance at training sessions. The personnel department can provide a list of names of the new hires to the training department; usually the job title or description will be sufficient to determine whether these employees need to receive Category I and II training. These new employees should receive their training as soon as possible. If training sessions cannot be scheduled often enough to accommodate frequent hires, then the supervisory staff will be required to provide the training on an interim basis.

Incumbent employees must also be trained. This training should be initiated by the supervisor, who knows which workers are currently performing Category I and Category II tasks.

In addition to the above, it may be helpful to hold informational-educational sessions for all of your employees. HIV-1 and AIDS are subjects of interest to many people, especially those employees that work in a health care setting.

The Result

Documentation of attendance is essential. A responsible person should be delegated the task of assuring that everyone attending the session signs in, including latecomers. In addition, a notice of attendance should include the following information:

date
length of session
speaker
topics covered

The notice should be signed by every attendee. The original notice should be placed in the employee's personnel file, a copy should be sent to the employee's supervisor, and a copy should be kept in the training office. The training coordinator should also keep lists of speakers, subjects covered, pre- and post-test results (if available), evaluations, copies of handouts, and any other pertinent material.

It is essential to evaluate the effectiveness of the training program. By evaluating the program, you can learn how well you have attained your goals and find out what areas need improvement. One simple method of evaluating the effectiveness is to issue pre- and post-tests on the material. There are other more quantitative methods of evaluation, such as measuring the number of work-related injuries and illnesses, the number of sick days, and compensation costs.

Valuable information can be obtained from participant evaluations; these are usually filled out by the attendees at the conclusion of the training.

IV. SUMMARY OF THE PROGRAM DEVELOPMENT

Health care employees who are adequately trained will be able to minimize the potential risks involved with handling infectious materials. Employees who are encouraged to follow the recommended CDC universal precautions can effectively protect themselves from occupationally acquired infectious diseases.

Employers who provide educational and training programs designed to minimize these risks will benefit in several ways. Besides being in compliance with the Federal OSHA regulation, employers may find improved employee-employer relations. Measurable improvements in accident/illness rates will translate into financial benefits.

Careful planning and implementation will result in a successful health and safety training program.

V. APPENDIX: TRAINING MATERIALS

1. National Committee for Clinical Laboratory Standards. 771 E. Lancaster Ave., Villanova, PA 19085. 215-525-2435. Copies of their new guideline "Protection of Laboratory Workers from Infectious Disease Transmitted by Blood and Tissue" (NCCLS Document M29-P) are available for $15.00 at the above address.
2. CDC/NIH 1984. "Biosafety in Microbiological and Biomedical Laboratories," HHS Publication No. (CDC) 84-8395. Washington, D.C. U.S. Government Printing Office. A new update, which includes information on HIV, is now available through NIH)
3. Fleming, D. O. Hazard Control of Infectious Agents. Occupational Medicine State of the Art Reviews. 1987, Vol. 2:499-511.
4. National Audiovisual Center, National Archives and Records Service, Reference section CF, Washington, D.C. 20409. 301-763-1896. This national center has a large number of films/videos for preview and purchase at reasonable rates.
5. National Safety Council, 441 N. Michigan Ave., Chicago, Ill. 60611. This association also has a very large inventory of available audiovisual training aids. A slide tape cassette, "Introduction to Biohazards Control, Stock No. 176.54, is available.

REFERENCES

1. OSHA. (1987). Proposed rules: Occupational exposure to hepatitis B virus and human immunodeficiency virus. *Federal Register 152*:45438-45441.
2. Link, R. N., Feingold, A. R., Charap, M. H., et al. (1988). Concerns of medical and pediatric house officers about acquiring AIDS from their patients. *AJPH. 78*:455-459.
3. Anon. (1984). Needlestick transmission of HTLV-III from a patient infected in Africa. *Lancet 2*:1376-1377.
4. Stricof, R. L., Morse, D. L. (1986). HTLV-III/LAV seroconversion following a deep intramuscular needlestick injury. *N. Engl. J. Med. 314*:1115.
5. Oskenhendler, E., Harzic, M., et al. (1986). HIV infection with seroconversion after a superficial needlestick injury to the finger. *N. Engl. J. Med. 315*:582.
6. Neisson-Vernant, C., Arfi, S., et al. (1986). Needlestick HIV seroconversion in a nurse. *Lancet 2*:814.
7. CDC. (1985). Update: Evaluation of human T lymphotrophic virus type III/lymphadenopathy-associated virus infection in health care personnel–United States. *MMWR 34*:575-578.
8. CDC. (1987). Update: Human immunodeficiency virus infection in health care workers exposed to blood of infected patients. *MMWR 36*:3032-3039.

9. Dr. W. E. Barkley, NIH, personal communication.
10. U.S. Department of Health, Education and Welfare. (1976). Hospital Occupational Health Services Study. VII. Study and Conclusions. NIOSH.
11. Accreditation Manual for Hospitals. (1987). Joint Commission on Accreditation of Hospitals, Chicago, IL.
12. CDC. (1987). Recommendations for prevention of HIV transmission in health care settings. *MMWR*(2S)*36*:3-18.
13. U.S. Department of Labor, Bureau of Labor Statistics. (1986). Occupational injuries and illnesses in the United States by industry, 1984. Bulletin 2259.
14. Jacobson, J. T., Orlob, R. B., and Clayton, J. L. (1985). Infections acquired in clinical laboratories in Utah. *J. Clin. Mic. 21*:486-489.
15. Vesley, D., and Hartmann, H. M. (1988). Laboratory acquired infections and injuries in clinical laboratories: A 1986 survey. *AJPH 78*:1213-1215.
16. Pike, R. M. (1976). Laboratory associated infections: A summary and analysis of 3,921 cases. *Health Lab. Sci. 13*:106-114.
17. Neuberger, J. S., Harris, J. A., Kund, W. D., et al. (1984). Incidence of needlestick injuries in hospital personnel: Implications for prevention. *Am. J. Inf. Control 12*:171-176.
18. McCormick, R., Maki, D. (1981). Epidemiology of needlestick injuries in hospital personnel. *Am. J. Med. 70*:928-932.
19. Braver, E. (1978). Hospital occupational safety and health: A medical record study of housekeepers and a study of workers' compensation claims at a large university hospital. Johns Hopkins University Thesis.
20. Gerberding, J. L., Bryant-LeBlanc, Nelson, K., et al. (1987). Risk of transmitting the human immunodeficiency virus, cytomegalovirus, and hepatitis B virus to health care workers exposed to patients with AIDS and AIDS related conditions. *J. Infect. Dis. 156*:1-8.
21. Vlahov, D., and Polk, B. F. (1987). Transmission of human immunodeficiency virus within the health care setting. *Occupational Medicine State of the Art Reviews 2*:429-450.
22. McCray, E., the Cooperative Needlestick Surveillance Group. (1980). Occupational risk of the acquired immunodeficiency virus syndrome among health care workers. *N. Engl. J. Med. 314*:1127-1132.
23. Nelson, K. (1987). Prevention of hepatitis in health care workers. *Occupational Medicine State of the Art Reviews 2*:451-470.
24. Dienstag, J. L., and Ryan, D. M. (1982). Occupational exposure to hepatitis B virus in hospital personnel: Infection or immunization. *Am. J. Epi. 115*:26-39.

25.   Favero, M. S. (1987). Biological hazards in the laboratory.
      Lab. Med. 18:665-670.
26.   Sande, M. A. (1986). Transmission of AIDS. The case against
      contagion. *N. Engl. J. Med. 314*:380-382.
27.   Palmer, D. L., Barash, M., King, R., and Neir, F. (1983).
      Hepatitis among hospital employees. *West. J. Med. 138*:519-523.
28.   CDC. (1987). Update: Hepatitis B prevention. *MMWR 36*:353-
      360.
29.   Koop, C. E. (1987). Surgeon General's report on acquired im-
      munodeficiency syndrome. US DHHS, Oct., 36pp.
30.   Petty, R. E., and Cacioppo, J. T. (1981). Attitudes and persua-
      sion: Classic and contemporary approaches. W. C. Brown Co.
      Publ., Iowa.

Index